Development of Light Alloys and Their Applications

Development of Light Alloys and Their Applications

Guest Editors

Hua Qian Ang
Gulshan Kumar
Erdem Karakulak

MDPI

Basel • Beijing • Wuhan • Barcelona • Belgrade • Novi Sad • Cluj • Manchester

Guest Editors
Hua Qian Ang
School of Engineering
RMIT University
Melbourne
Australia

Gulshan Kumar
Department of Mechanical Engineering
BITS Pilani Dubai Campus
Dubai
United Arab Emirates

Erdem Karakulak
Brunel Centre for Advanced Solidification Technology
Brunel University London
Uxbridge
UK

Editorial Office
MDPI AG
Grosspeteranlage 5
4052 Basel, Switzerland

This is a reprint of the Special Issue, published open access by the journal *Crystals* (ISSN 2073-4352), freely accessible at: https://www.mdpi.com/journal/crystals/special_issues/08SJE3BXDR.

For citation purposes, cite each article independently as indicated on the article page online and as indicated below:

Lastname, A.A.; Lastname, B.B. Article Title. *Journal Name* **Year**, *Volume Number*, Page Range.

ISBN 978-3-7258-6522-2 (Hbk)
ISBN 978-3-7258-6523-9 (PDF)
https://doi.org/10.3390/books978-3-7258-6523-9

Cover image courtesy of Hua Qian Ang

Contents

About the Editors

Hua Qian Ang

Hua Qian Ang is the founder of EPIC Engineering Comic, a Chartered Professional Engineer with Engineers Australia, and an award-winning engineering educator and researcher. Dr. Ang is deeply passionate about engineering education, with a vision to make it free, accessible, and equitable worldwide. Additionally, Dr. Ang's research focuses on the mechanical properties and deformation behaviour of light alloys. Her interests include additive manufacturing and laser materials processing. She currently holds multiple editorial leadership roles and serves on the editorial boards of several Q1 journals. Dr. Ang also holds multiple volunteer roles focused on supporting community-driven engineering initiatives.

Gulshan Kumar

Gulshan Kumar is a Faculty member in the Department of Mechanical Engineering at BITS Pilani Dubai Campus, United Arab Emirates (UAE). He acquired his doctorate from the Indian Institute of Technology Bombay, India, in the field of Materials and Metallurgy and has experience of more than ten years in academia and research. His research forte includes manufacturing, materials behaviour, residual stress analysis, computational modelling and industrial engineering practices. Dr. Kumar possesses expertise in experimental characterisation techniques such as Scanning Electron Microscopy (SEM), Electron Backscatter Diffraction (EBSD), X-ray Diffraction (XRD), which he used for determining microstructure features, phases and residual stress development in the materials. His competency in FEM simulations using Abaqus, Deform, Ansys, etc., makes him suitable to study deformation mechanisms and damage evolution in hexagonal alloys. In addition to his work in materials and simulations, he also works in lean operations, sustainable manufacturing, Industry 4.0 applications and operations management. He has publications to his credit in reputed International Journals and actively supervises PhD scholars, postgraduate and undergraduate students. He has a strong network of collaboration spanning leading academic and research institutions.

Erdem Karakulak

Erdem Karakulak is a Postdoctoral Researcher in Brunel Centre for Advanced Solidification Technology (BCAST) in Brunel University London. Dr. Karakulak's research is mainly focused on casting and solidification of aluminium and magnesium alloys. With extensive experience spanning over 20 years, he has practical and theoretical expertise in alloy design, casting techniques, heat treatment, materials characterisation and material testing. Throughout his career, he has collaborated with multidisciplinary teams to innovate and optimise processes, leading to enhanced product performance and cost efficiency. He has published several scientific papers in various peer-reviewed journals and conference proceedings.

Preface

This Reprint focuses on light alloys, a class of materials that includes magnesium, aluminium, and titanium. These alloys are commonly classified as light alloys because of their high strength-to-weight and/or stiffness-to-weight ratios, properties that make them especially valuable for modern engineering applications. Owing to these characteristics, light alloys have gained significant commercial and technological importance across a wide range of industries.

Among light alloys, aluminium alloys are the most widely used. Their applications extend from automotive and aerospace structures to everyday products such as cans and foils. In recent years, titanium and magnesium alloys have also attracted increasing attention. Titanium alloys are primarily used in the aerospace industry, where their low density and high strength, particularly at elevated temperatures, offer clear advantages. In addition, their excellent corrosion resistance makes titanium alloys suitable for marine and petrochemical applications.

Magnesium alloys are employed where further reductions in weight are critical. Typical applications include aircraft components, material-handling equipment, and portable power tools. Despite the successful application of light alloys in many fields, challenges and limitations remain. As a result, research and development in this area continue to be active and essential.

This Reprint aims to provide a focused overview of light alloys, their properties, applications, and ongoing research efforts, contributing to a deeper understanding of these important materials and their future potential.

Hua Qian Ang, Gulshan Kumar, and Erdem Karakulak
Guest Editors

crystals

MDPI

Editorial

Development of Light Alloys and Their Applications

Hua Qian Ang [1,*], Gulshan Kumar [2] and Erdem Karakulak [3]

1 School of Engineering, RMIT University, Melbourne, VIC 3000, Australia
2 Department of Mechanical Engineering, Birla Institute of Technology and Science, Pilani, Dubai Campus, Dubai International Academic City, Dubai P.O. Box 345055, United Arab Emirates
3 BCAST, Brunel University London, Uxbridge UB8 3PH, UK
* Correspondence: huaqian.ang@rmit.edu.au

1. Introduction

Magnesium, aluminium and titanium are commonly classified as light alloys because of their high strength-to-weight and/or stiffness-to-weight ratios. Owing to these properties, light alloys have attracted considerable commercial and industrial interest. Among them, aluminium alloys are the most widely used, finding extensive applications not only in the automotive and aerospace sectors but also in everyday products such as packaging cans and foils. In recent years, titanium and magnesium alloys have also seen increasing utilisation. Titanium alloys are predominantly employed in the aerospace industry due to their low density combined with excellent mechanical strength, particularly at elevated temperatures. In addition, their outstanding corrosion resistance makes them suitable for demanding environments, including marine and petrochemical applications. Magnesium alloys, on the other hand, are preferred when further weight reduction is critical, such as in aircraft components, material-handling equipment and portable power tools.

Despite the successful application of light alloys across a broad range of industries, several challenges and limitations remain, including issues related to processing efficiency, performance optimisation, cost effectiveness and environmental sustainability. Addressing these challenges will require continued advances in alloy design, processing technologies, modelling and characterisation methods, as well as closer integration between fundamental research and industrial practice. Against this backdrop, this Special Issue "Development of Light Alloys and Their Applications" comprises ten contributions that collectively reflect the breadth and depth of current research on light alloys and their applications.

The published works cover key topics across aluminium, titanium and magnesium alloys, including advanced joining and solid-state processing techniques, microstructure–property relationships, novel alloy systems, additive manufacturing and hybrid manufacturing routes, and surface engineering and corrosion behaviour (contributions 1 and 2). Several papers focus on aluminium and titanium alloys for structural applications, highlighting the critical roles of welding, friction stir processing, cooling behaviour and fatigue performance in tailoring mechanical properties for demanding service conditions (contributions 3–6). Other contributions report the development and characterisation of emerging aluminium-based alloys designed for specialised functions, such as thermal neutron shielding, and assess the performance of complex dissimilar-metal joints intended for harsh industrial environments (contributions 7 and 8).

In addition, two comprehensive review articles (contributions 9 and 10) provide forward-looking perspectives on biodegradable magnesium alloys for biomedical applications. These reviews not only summarise recent advances in alloy design, surface modification and corrosion control, but also critically examine the remaining scientific, technological

https://doi.org/10.3390/cryst16010011

and regulatory challenges that must be addressed to enable broader clinical adoption. The insights presented highlight the importance of multidisciplinary approaches that integrate materials science, surface engineering, biomechanics and clinical requirements.

Taken together, the contributions in this Special Issue demonstrate the dynamic and evolving nature of light alloy research. They point towards future directions that emphasise intelligent alloy design, process optimisation, sustainability-driven innovation and application-specific performance tailoring. It is anticipated that the findings and perspectives presented here will stimulate further research efforts, foster collaboration between academia and industry, and contribute to the continued advancement of light alloys as key materials for next-generation structural, functional and biomedical applications.

2. Highlights of Novelty and Impact of the Published Papers

The following section highlights the key novelty and significance of each contribution published in this Special Issue. Ugodchikova et al. (contribution 1) explore the effect of particle incorporation on micro-arc oxidation (MAO) coatings on bioinert titanium alloys. Coatings containing β-tricalcium phosphate (CP), wollastonite (CS), and their combination (SP) were fabricated at 350–500 V. Surface morphology, composition, phase structure, and fine structure were characterised using SEM, EDX, XRD, and TEM. Scratch tests showed SP coatings exhibited the highest adhesion (Lc = 22 N), highlighting their potential for durable, load-bearing biomedical implant applications.

Fernández-Calvo and co-workers (contribution 2) demonstrate a hybrid manufacturing approach combining ablation casting and laser metal deposition (LMwD) for Al-5356 alloy, enabling near-net-shape components with complex geometries. Their work validates the process by showing low porosity (<0.04%) and minimal lack-of-union defects, with LMwD regions exhibiting higher strength than ablation-cast areas. These findings highlight the potential of integrating conventional and additive manufacturing to improve mechanical performance, productivity, and reliability in industrial-scale aluminium alloy fabrication.

Elshaghoul and team (contribution 3) investigate Friction Stir Spot Welding (FSSW) of high-strength AA2024-T4 aluminium, focusing on low rotational speeds (700–1300 rpm) and a short 3 s dwell time. Results reveal the optimal 900 rpm achieves maximum load-carrying capacity and stir zone hardness, demonstrating how grain refinement and thermal softening govern joint strength in demanding automotive and aerospace applications.

El-Eraki and collaborators (contribution 4) provide first-time insights into the isolated effect of friction stir processing (FSP) travel speed on hypoeutectic Al-5 wt.% Si alloys. By refining grain structure, FSP significantly enhances hardness, tensile strength, toughness, and wear resistance, with 26 mm/min achieving optimal performance. Their work underscores travel-speed control as key for tailoring cast alloy properties.

Wang et al. (contribution 5) investigate the fatigue behaviour of TC4/Ti60 dissimilar titanium alloy joints fabricated by electron beam welding. Results reveal that the weld zone, strengthened by martensite, resists failure, while fatigue cracks initiate in the TC4 base material. Findings provide key insights into microstructure–fatigue relationships for high-performance structural applications.

Wang and team (contribution 6) investigate gradient cooling effects on Al-2 at.% Nd alloy microstructure and properties in a vacuum. Slower cooling transforms mixed columnar–equiaxed grains into equiaxed grains and forms discontinuous α-Al and $Al_{11}Nd_3$ eutectic phases, reducing phase and grain boundaries. First-principles calculations confirm these changes, correlating microstructural evolution with a 25.5% decrease in hardness.

Amer's study (contribution 7) reports a novel quaternary intermetallic phase, $Al_{21}GdCrTi$, in an Al-Gd-Cr-Ti alloy, alongside detailed analyses of microstructure evolu-

https://doi.org/10.3390/cryst16010011

tion, texture, mechanical strengthening, and neutron absorption. Hot–cold rolling enhances hardness, yield, and tensile strength while maintaining ductility. The alloy exhibits ~100× lower neutron scattering than pure aluminium, providing a foundation for advanced neutron-shielding materials.

Marônek and co-workers (contribution 8) evaluate Cu-Al-AlMg3 structural transition joints fabricated via explosion welding for galvanizer hanger renovation. Microstructural analysis reveals wavy interfaces, localised strain hardening, and Al_2Cu intermetallic formation. Bond strength doubles that of AW1050 alloy, with failure at the AW1050–AW5754 interface, demonstrating the effectiveness of welding parameters and potential for industrial structural applications.

The two reviews on biodegradable magnesium alloys complement each other in scope and focus. Aikin et al. (contribution 9) provide a broad, historical perspective, highlighting alloy design, manufacturing, surface engineering, and clinical translation, while addressing challenges such as unpredictable degradation, safety, and regulatory issues. Liu and team (contribution 10) are more technical and solution-oriented, concentrating on alloying and surface treatments to improve corrosion resistance, mechanical performance, and biocompatibility. Together, they offer both a forward-looking clinical roadmap and targeted strategies for overcoming material limitations, underscoring the continued development of Mg-based biomaterials for medical applications.

3. Conclusions

Collectively, the contributions in this Special Issue underscore the remarkable progress and ongoing challenges in the development of light alloys. From novel aluminium, titanium, and magnesium systems to advanced joining, processing, and surface engineering techniques, these studies highlight both scientific innovation and practical application. The insights provided not only advance fundamental understanding but also pave the way for future high-performance, sustainable, and clinically or industrially relevant light alloy technologies.

Author Contributions: Conceptualization, H.Q.A.; methodology, H.Q.A.; validation, H.Q.A., investigation, H.Q.A.; writing—original draft preparation, H.Q.A.; writing—review and editing, H.Q.A., G.K. and E.K. All authors have read and agreed to the published version of the manuscript.

Acknowledgments: As Guest Editors, we would like to express our sincere appreciation to all authors whose valuable work was published in this Special Issue, contributing to its success, and to all anonymous reviewers who provided their expert evaluations and constructive feedback, supporting the peer review process.

Conflicts of Interest: The authors declare no conflicts of interest.

List of Contributions

1. Ugodchikova, A.V.; Tolkacheva, T.V.; Uvarkin, P.V.; Khimich, M.A.; Sharkeev, Y.P.; Kashin, A.D.; Glukhov, I.A.; Sedelnikova, M.B. Micro-Arc Coatings with Different Types of Microparticles on Titanium Alloy: Formation, Structure, and Properties. *Crystals* **2025**, *15*, 811.
2. Fernández-Calvo, A.I.; Madarieta, M.; Solana, A.; Lizarralde, I.; Rouco, M.; Soriano, C. Study of the Hibridation of Ablation Casting and Laser Wire Metal Deposition for Aluminum Alloy 5356. *Crystals* **2025**, *15*, 134.
3. Elshaghoul, Y.G.; Shalaby, M.F.; El-Sayed Seleman, M.M.; Elkelity, A.; Reyad, H.A.; Ataya, S. The Role of Rotational Tool Speed in the Joint Performance of AA2024-T4 Friction Stir Spot Welds at a Short 3-Second Dwell Time. *Crystals* **2025**, *15*, 1054.
4. El-Eraki, B.; Shalaby, M.F.; El-Sissy, A.; Eisa, A.; Ataya, S.; El-Sayed Seleman, M.M. The Role of Friction Stir Processing Travel Speed on the Microstructure Evolution and Mechanical Performance of As-Cast Hypoeutectic Al-5Si Alloy. *Crystals* **2025**, *15*, 546.
5. Wang, S.; Zhu, X.; Zhai, W.; Gao, Q.; Lu, Y. The Fatigue Behavior of TC4 and Ti60 Dissimilar Titanium Alloy Joints Welded by Electron Beam. *Crystals* **2025**, *15*, 224.

https://doi.org/10.3390/cryst16010011

6. Wang, X.; Zhang, X.; Wu, W.; Sun, S. The Effect of Gradient Cooling Behavior on the Microstructure and Mechanical Properties of Al-2at.% Nd Alloy in a Vacuum Environment. *Crystals* **2025**, *15*, 81.
7. Amer, S.M.; Nikolayev, D.I.; Lychagina, T.A.; El-Khouly, A.; Barkov, R.Y.; Prosviryakov, A.S.; Mikhaylovskaya, A.V.; Glavatskikh, M.V.; Pozdniakov, A.V. Microstructure, Mechanical Properties, Deformation Behavior, and Crystallographic Texture of the Al-Gd-Cr-Ti Quaternary Alloy for Thermal Neutron Absorption. *Crystals* **2025**, *15*, 616.
8. Marônek, M.; Bárta, J.; Bártová, K; Sahul, M.; Sahul, M.; Pašák, M.; Nesvadba, P.; Bezdička, P. Evaluation of Structural Transition Joints Cu-Al-AlMg3 Used in Galvanizer Hangers. *Crystals* **2024**, *14*, 974.
9. Aikin, M.; Shalomeev, V.; Kukhar, V.; Kostryzhev, A.; Kuziev, I.; Kulynych, V.; Dykha, O.; Dytyniuk, V.; Shapoval, O.; Zagorskis, A.; et al. Recent Advances in Biodegradable Magnesium Alloys for Medical Implants: Evolution, Innovations, and Clinical Translation. *Crystals* **2025**, *15*, 671.
10. Liu, Y.; Yin, J.; Zhu, G.Z. Advances in Magnesium-Based Biomaterials: Strategies for Enhanced Corrosion Resistance, Mechanical Performance, and Biocompatibility. *Crystals* **2025**, *15*, 256.

https://doi.org/10.3390/cryst16010011

crystals

MDPI

Article

The Role of Rotational Tool Speed in the Joint Performance of AA2024-T4 Friction Stir Spot Welds at a Short 3-Second Dwell Time

Yousef G. Y. Elshaghoul [1], Mahmoud F. Y. Shalaby [2], Mohamed M. El-Sayed Seleman [3,*], Ahmed Elkelity [1], Hagar A. Reyad [4] and Sabbah Ataya [5,*]

[1] Mechanical Engineering Department, Faculty of Engineering, Suez University, Suez 43221, Egypt; yousef.gamal@eng.suezuni.edu.eg (Y.G.Y.E.); ahmed.elkelity@eng.suezuni.edu.eg (A.E.)

[2] Department of Industrial Engineering, College of Engineering, Imam Mohammad Ibn Saud Islamic University (IMSIU), Riyadh 11432, Saudi Arabia; mfshalaby@imamu.edu.sa

[3] Department of Metallurgical and Materials Engineering, Faculty of Petroleum and Mining Engineering, Suez University, Suez 43221, Egypt

[4] Arabian Steel Industries, Suez 43552, Egypt; h.reyad@arabiansteel.net

[5] Department of Mechanical Engineering, College of Engineering, Imam Mohammad Ibn Saud Islamic University (IMSIU), Riyadh 11432, Saudi Arabia

* Correspondence: mohamed.elnagar@suezuniv.edu.eg (M.M.E.-S.S.); smataya@imamu.edu.sa (S.A.)

Abstract

This study explores Friction Stir Spot Welding (FSSW), a well-established solid-state joining technique, for high-strength aluminum alloys like AA2024-T4, which present significant challenges for conventional welding techniques. This research focuses on the impact of relatively low rotational speeds, specifically within a range of 700 to 1300 rpm, on the mechanical and microstructural properties of the welded joints. By employing a short dwell time of 3 s, this study aims to enhance productivity in the automotive and aerospace industries. The experimental work evaluated the joints' thermal cycles, macrostructure, microstructure, hardness and load-carrying capacity. Results indicated a linear relationship between rotational speed and heat input. Although all welds exhibited a significant grain size reduction in the stir zone (SZ) compared to the base material (29.7 ± 6.1 μm), the SZ grain size increased with rotational speed, ranging from 4.7 ± 1.4 to 8.3 ± 1.3 μm. This study identified 900 rpm as the optimal parameter, achieving the highest load-carrying capacity (7.35 ± 0.4 kN) and a high SZ hardness (99 ± 1.5 HV). These findings confirm that joint strength is a balance between grain refinement and thermal softening. The presence of precipitates and the fractography of the tensile–shear tested specimens were also investigated and discussed.

Keywords: friction stir spot welding (FSSW); AA2024; rotational speed; microstructure; mechanical properties

1. Introduction

The rising global demand to lower greenhouse gas emissions, especially CO_2, and enhance fuel economy in the transportation and aerospace sectors has profoundly influenced material selection in modern industries [1,2]. In this context, lightweight alloys such as magnesium (Mg) and aluminum (Al) have become highly favored for constructing vehicles, shipbuilding and aerospace structural parts. These materials offer compelling advantages, including superior corrosion resistance and an exceptional strength-to-weight ratio [3,4].

Within the Al alloys, the 2xxx series, particularly AA2024, stands out as a material of choice for demanding structural applications [5]. Its excellent strength, lightweight nature, and fatigue resistance make it widely used in critical industries like aerospace, automotive, and defense. However, the inherent characteristics of AA2024, being a heat-treatable alloy, pose significant challenges for conventional fusion welding techniques [1,6]. Welding processes involving melting and solidification often introduce detrimental defects such as hot cracking, porosity, and the formation of brittle secondary phases within the weld zone [7]. The presence of these flaws consistently impairs the mechanical integrity and structural reliability of the fabricated joints. This degradation in performance consequently hinders the wider application of AA2024 in critical and high-performance structures where reliability is paramount [8,9].

Consequently, the scientific and industrial communities have increasingly turned their attention towards solid-state joining alternatives that circumvent these fusion-related issues. Among the various solid-state joining techniques, Friction Stir Welding (FSW) [10–12] and its derivative, Friction Stir Spot Welding (FSSW) [2,13], have emerged as highly effective solutions. FSW operates on the principle of generating localized heat through friction and intense plastic deformation induced by a non-consumable rotating tool. This process facilitates material flow and consolidation below the melting point of the parent material, inherently preventing solidification defects. FSW is renowned for producing joints with refined microstructures and enhanced mechanical properties in both similar [9,14] and dissimilar [15–19] material combinations, thereby improving overall joint quality [20,21]. FSSW, as a specialized variant of FSW, is tailored for localized spot joining of sheet materials. It offers a robust and environmentally friendly alternative to traditional spot joining methods like riveting and resistance spot welding [22–24] providing higher joint efficiency.

The quality and mechanical performance of the FSSW joints is highly dependent on the precise control and optimization of various process parameters, including tool selection (geometry and material) [25–27], plunge depth [28], plunge rate [29], rotational speed [30,31], and dwell time [32]. Building on this, some works have been conducted on the FSSW of Al alloys, including AA2024, to understand the influence of these parameters. For instance, Tiwan et al. [1] studied the effect of using different tool geometries on the dissimilar FSSW joints of AA2024-O and AA6061-T6 Al alloys. The spot-welding process parameters were set at a 5 s fixed dwell time and various rotational speeds of 900, 1400 and 1800 rpm were applied. Their findings highlighted the cylindrical pin as an optimal choice for promoting effective material mixing and upward flow, with 1400 rpm yielding a favorable combination of fine grain size, precipitation hardening, and minimal hook width. Dourandish et al. [2] studied the applicability to use protrusion FSSW to join similar sheet thickness of AA2024-T3 Al alloy. The welding parameters were rotational speeds of 1000, 1600 and 2000 rpm at a constant dwell time of 10 s. They identified 1600 rpm as the optimal speed, achieving the highest maximum load value of 6700 N. Rashkovets et al. [22] investigated the probeless-FSSW of dissimilar 2024-T3 (1.5 mm thick) and 6082-T6 (0.8 mm thick) Al alloy sheets. Their study, conducted at a dwell time of 90 s and rotational speeds ranging from 500 to 2500 rpm, indicated that optimal rotational speeds between 1500 and 2000 rpm maximized the joint's load-bearing capacity. In a limited study, Lewise et al. [33] focused on the FSSW of dissimilar AA2024-T3 and AA7075-T6 sheets (2 mm thick), specifically examining the influence of an 1800 rpm rotational speed, 45 s dwell time, and 3.3 mm plunge depth. They attributed improvements in both hardness and tensile strength to the superior mechanical properties of the starting AA7075-T6 alloy. Zhang et al. [34] also studied the FSSW of dissimilar aluminum alloy sheets (1.6 mm thick), with AA2024-T3 as the upper sheet and AA7075-T6 as the lower sheet. Their work, using a plunge rate of 2 mm/min, a 5 s dwell time, and rotational speeds from 630 to 1000 rpm, demonstrated that increasing rotational speed led to an increase in grain sizes (from 3.3 to 7.8 μm) and an enhancement in the load-bearing capacity of the FSSW joints.

Despite these valuable contributions, a notable gap persists in the published literature concerning the FSSW of similar sheet thicknesses of AA2024 Al alloy, particularly a systematic investigation across a focused relatively low range of rotational speeds at a short dwell time [2]. Previous studies often involve dissimilar alloys, different material thicknesses, or broader parameter ranges that may not fully elucidate the specific behavior of similar AA2024 joints. The precise effect of different rotational speeds on the microstructural development and subsequent mechanical performance of such joints, particularly with a short dwell period, requires more work.

The novelty of the present work lies in its systematic investigation of FSSW on similar 2 mm thick AA2024 Al alloy sheets. This study specifically focuses on the influence of various relatively low rotational speeds of 700, 900, 1100, and 1300 rpm, while maintaining a constant short dwell time of 3 s, which is a key parameter for enhancing productivity. This strategic rotational speed range (700–1300 rpm) is selected for the FSSW of AA2024-T4 to establish a controlled thermal cycle. This control is crucial for balancing two primary objectives: generating sufficient heat for robust metallurgical bonding and dynamic recrystallization in the stir zone (SZ), while simultaneously limiting excessive heat input to minimize detrimental thermal softening from precipitate over-aging. Furthermore, operating within this range enhances process stability by reducing tool wear and mitigating volumetric defects such as excessive flash.

This research aims to comprehensively study the thermal cycle (temperature and heat input energy) during the joining process. Furthermore, it seeks to characterize the microstructural features and to provide a comprehensive evaluation of the resulting mechanical properties, specifically focusing on hardness and tensile–shear strength, of the fabricated joints. A detailed fractographic analysis of the fracture surfaces will also be conducted to elucidate the underlying failure mechanisms. By establishing optimal FSSW parameters for similar AA2024 joints, this study seeks to provide critical data and insights that will contribute to the enhanced reliability and broader industrial application of AA2024 in high-performance structural components.

2. Materials and Methods

The AA2024-T4 sheet materials were supplied by (Dongguan Fujin Metal Materials Co., Ltd., Dongguan, China), with dimensions of 2 (thickness) × 250 (length) × 250 (width) mm. These sheets were then cut into strips of 2 × 40 × 120 mm using a wire cutting machine for the FSSW experiments. The chemical composition and mechanical properties of the as-received material, including tensile and hardness tests, are detailed in Tables 1 and 2, respectively.

Table 1. Chemical Composition of AA2024-T4 (wt.%).

Elements (%)	Cu	Mg	Mn	Fe	Si	Zn	Cr	Ti	Al
Wt.%	3.8	1.2	0.30	0.5	0.50	0.25	0.10	0.15	Bal.

Table 2. Mechanical properties of AA2024-T4 base material.

Yield Stress (MPa)	Ultimate Tensile Strength (MPa)	Hardness (HV)
279 ± 4.5	420 ± 6.4	111 ± 6.0279

FSSW was used to create lap shear joints with a 40 mm overlap. This process used a specially designed fixation system, as shown in Figure 1, to increase process efficiency and welding quality. The fixture was designed to firmly hold multiple lap joint specimens in place, allowing them to be spot-welded sequentially without realignment or repositioning. The multi-sample fixture reduced setup time between welds, enhanced productivity, and reduced human clamping and repositioning errors that affected weld quality.

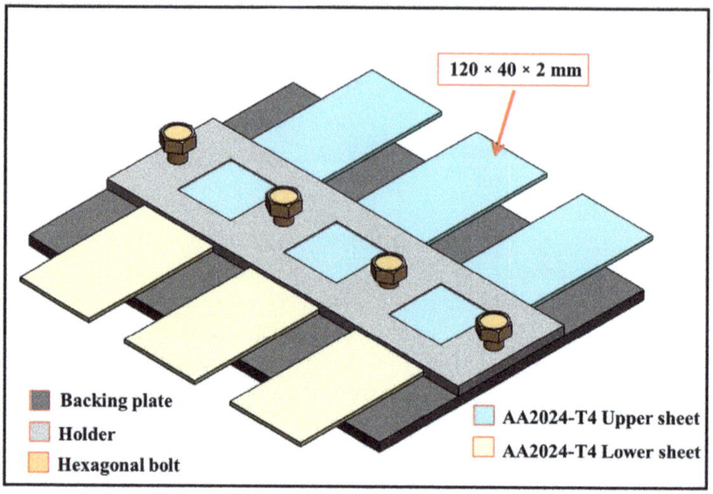

Figure 1. Illustration of the fixation system setup for joining AA2024-T4 in similar FSSW lap joints.

The tool design plays a critical role in heat generation, material flow, and weld quality [35,36]. A welding tool was fabricated from AISI H13 steel, featuring a flat shoulder of 20 mm in diameter. Its tapered pin had a major diameter of 6 mm which decreased to 5 mm over its 3.2 mm length. The specific dimensions of both the pin and the shoulder are illustrated in Figure 2. The FSSW joints were fabricated on an EG-FSW-M1 friction stir welding/processing machine [37,38]. The spot-welding process was achieved in three steps: (1) the plunge and stirring stage with the rotating tool (Figure 3a); (2) the dwell time stage (Figure 3b); and (3) the retreating stage (Figure 3c).

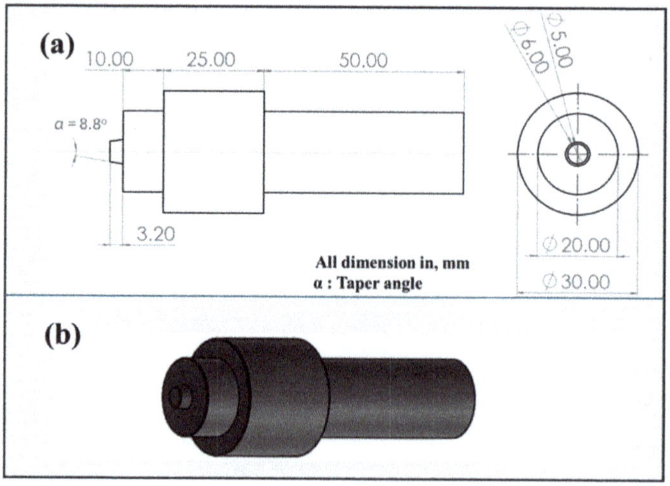

Figure 2. Depicts the design of the welding tool used for FSSW, showing its schematic in (**a**) and its isometric view in (**b**).

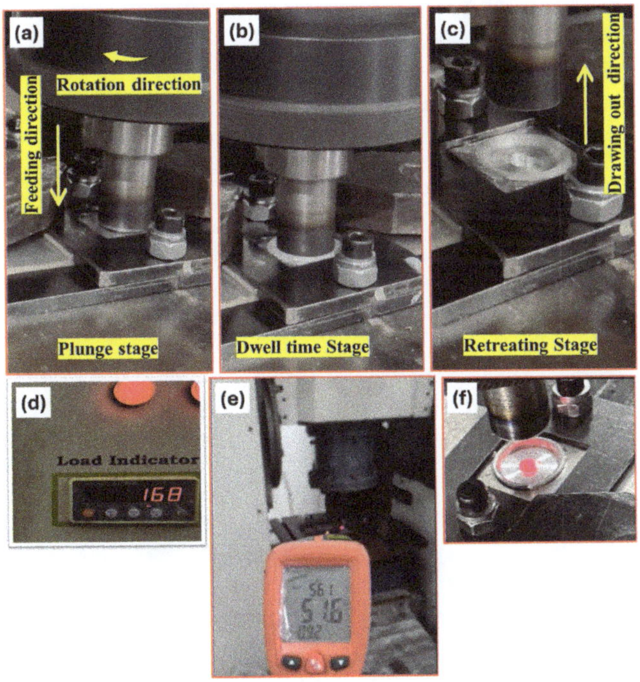

Figure 3. The FSSW process, showing (**a**) the stage of tool's insertion and material stirring, (**b**) the dwell time stage, (**c**) the retreating stage, (**d**) the load indicator, and (**e**,**f**) the temperature measurement setup.

The specimens were joined via Friction Stir Spot Welding (FSSW) in lap joints with a 40 mm overlap. The welding process was conducted with various rotational speeds of 700, 900, 1100, and 1300 rpm and a constant dwell time of 3 s. To ensure consistency, other critical welding parameters, including a plunge depth of 3.4 mm, a plunge rate of 0.1 mm/s, and a tilt angle of 0°, were maintained at fixed values. The axial force exerted by the welding tool throughout the FSSW process was monitored using the integrated load indicator of the EG-FSW-M1 FSW machine, as depicted in Figure 3d. This load indicator offered a visual representation of the instantaneous force exerted during the insertion and holding periods the procedure. The temperature during the FSSW process was also recorded using a dual-laser infrared thermometer (Model 42570, Extech Instruments, FLIR company, Shenzhen, China). The measurement was conducted with a constant emissivity coefficient setting of 0.35, a standard reference value of aluminum alloys. This non-contact instrument, positioned, as seen in Figure 3e,f, functions by sensing the thermal radiation emitted from the weld zone surface. To ensure the statistical reliability and enable direct correlation with the final weld properties, thermal profiles were captured for the same three replicate welds at each rotational speed (700, 900, 1100, and 1300 rpm) that were subsequently used for mechanical and microstructural characterization.

The fabricated FSSW joints were cut for macrostructural assessment, microstructural analysis, and hardness testing. The cross-sectional specimens were ground with SiC sheets up to a final grit of 2400. They were then polished with a cloth and an alumina slurry until a surface smoothness of 0.05 µm was obtained. For microstructural analysis, the polished samples underwent chemical etching using Keller's etchant, a solution consisting of 2.5 mL of HNO_3, 1.5 mL of HCl, 1 mL of HF, and 95 mL of distilled water. All standard laboratory chemicals used in this study, were supplied by El Nasr for Intermediate Chemicals Co.,

Giza, Egypt. The microstructural analysis was performed using an optical microscope (OM) (Model BX41M-LED, Olympus, Tokyo, Japan). The microscope was used for both general microstructural examination and quantitative grain size analysis. For the latter, multiple OM micrographs were taken from distinct areas within the SZ of each weld condition. The reported average grain size and standard deviation for each parameter were derived from measurements of a minimum of 50 grains per condition. Hardness measurements were taken on the transverse cross-sections of the spot-welded joints. A Vickers hardness tester (Model HWDV-75, TTS Unlimited, Osaka, Japan) was employed for the measurements, with a 500 g force applied for a dwell period of 15 s. Indentations were spaced at 1 mm intervals across the weld cross-section to ensure comprehensive data collection. To ensure consistency, the hardness profile for each welding parameter was determined using two separate cross-sectional samples. The average values across the weld zones are plotted for comparison. The tensile–shear test, which assesses load-carrying capacity of the AA2024-T4 FSSW joints, was conducted with a 30-ton universal tensile testing machine (Type-WDW-300D, Dongguan, China) at ambient temperature. The FSSW joints were subjected to tensile–shear testing with a constant loading rate of 0.1 mm/s. Specimen preparation for the tests was conducted in accordance with the ASTM E 8M-04 standard. To ensure statistical reliability, three replicate samples were tested for both each welding parameter condition (700, 900, 1100, and 1300 rpm) and the AA2024-T4 base material (BM), and the average values are reported. Figure 4 provides schematic diagrams of the test specimens, showing the tensile–shear test specimen for the FSSWed AA2024-T4 joint in Figure 4a and the tensile test specimen for the BM in Figure 4b. Following the tensile–shear tests, the fracture surfaces of the separated FSSW specimens were examined using a Quanta FEG 250 scanning electron microscope (SEM) from FEI (Hillsboro, OR, USA). The microscope was equipped with an integrated energy-dispersive spectroscopy (EDS) system, which was utilized for elemental analysis to determine the presence of precipitates.

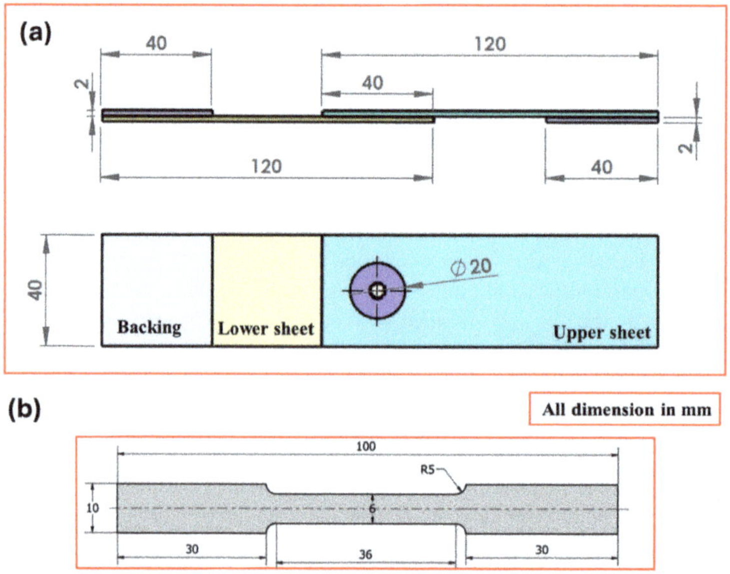

Figure 4. Schematic illustrations of the test specimens. These include: (**a**) the tensile–shear sample for the AA2024-T4 FSSW lap joint and (**b**) the tensile sample for the BM.

3. Results

3.1. Macrostructural Appearance of FSSW Joints

Figure 5 displays the representative top and bottom surface features of the FSSW joints. A visual assessment of the welds, which were fabricated using different tool rotational speeds ranging from 700 to 1300 rpm and a constant dwell time of 3 s, confirmed that the chosen parameters were effective for producing defect-free joints in the AA2024-T4 sheets.

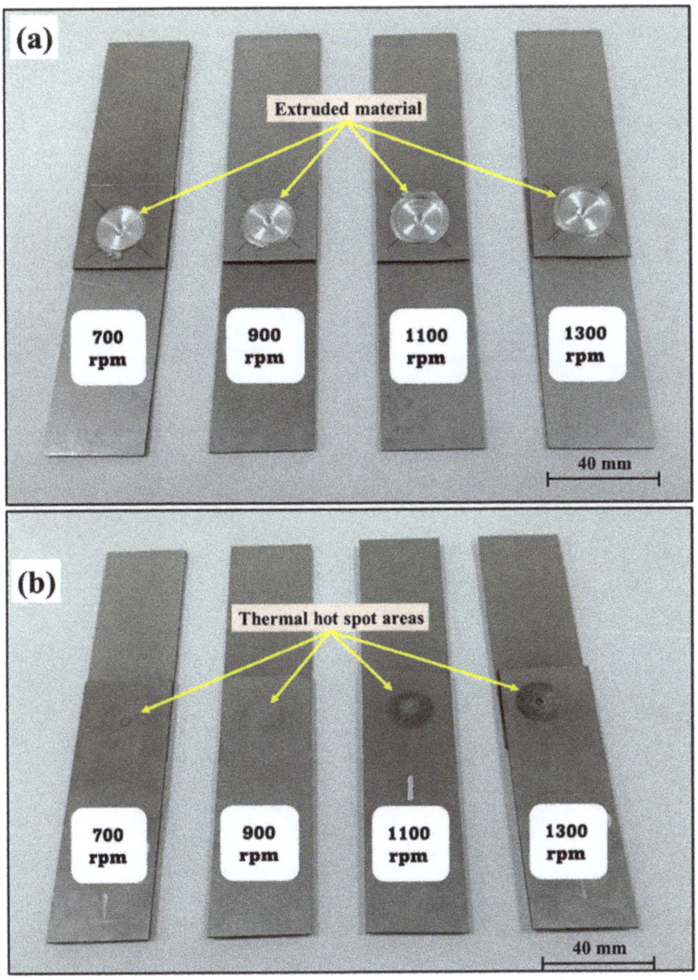

Figure 5. FSSW joints processed at various rotational speeds (700, 900, 1100, and 1300 rpm) and a constant dwell time of 3 s. (**a**) Top view and (**b**) Bottom view.

As seen from the top view in Figure 5a, circular indentations from the shoulder are present at all applied parameters, with extruded material flashing to the sides of the shoulder projection. This extruded material increases with the increase in tool rotational speed [39]. An examination of the FSSW joints' bottom surface revealed visible thermal zones, which correspond to the regions affected by the friction stirring process. These heated areas showed a progressively darker discoloration with increasing rotational speed, which is a direct result of the greater heat input generated during the dwell period, as depicted in Figure 5b.

3.2. Temperature Variation with Rotational Speed in FSSW

This section examines the thermal behavior of the FSSW process on AA2024-T4 sheets. Specifically, it focuses on the influence of tool rotation speed (700–1300 rpm) at a constant dwell time of 3 s on the comparative thermal cycles and relative heat input. A detailed analysis of the measured temperatures provides insight into the crucial thermal-mechanical conditions that govern spot weld formation. As shown in Figure 6, the temperature profiles exhibit a typical three-stage pattern, involving an initial heating stage, the achievement of a peak temperature during the 3 s dwell time, and a final cooling stage. The heating stage is characterized by a slow temperature rise as the tool's shoulder contacts the upper surface of the AA2024-T4 lap joint. As the tool's plunge depth increases, the temperature of the material under processing also rises. Higher rotational speeds accelerate this temperature rise, leading to a quicker achievement of peak temperatures once the tool reaches the target penetration depth of 3.4 mm. These peak temperatures are directly related to the applied rotational speed.

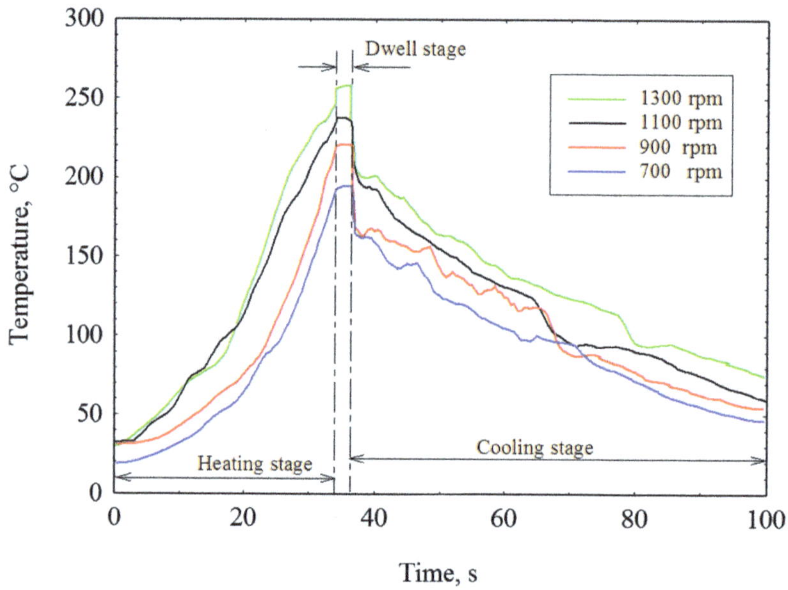

Figure 6. Typical temperature-time profiles during the FSSW process on AA2024-T4 sheets, showing the three distinct stages: heating, peak temperature, and cooling.

The measured peak temperatures were 197 ± 13 °C, 222 ± 10 °C, 238 ± 11 °C, and 259 ± 8 °C for tool rotational speeds of 700 rpm, 900 rpm, 1100 rpm, and 1300 rpm, respectively, as given in Figure 7. The data clearly show that increasing the rotational speed in FSSW results in higher process temperatures due to increased frictional heat generation at higher speeds. While elevated temperatures generally improve material flow and weld quality, it is essential to balance the speed to avoid overheating and associated defects. Following the peak temperature stage, the tool was retracted, and the joint was allowed to cool in the air. The cooling stage is marked by a gradual temperature drop as heat dissipates from the joint. The rate of cooling depends on the amount of heat generated, the surrounding cooling medium, and the welded material's thermal properties [40,41].

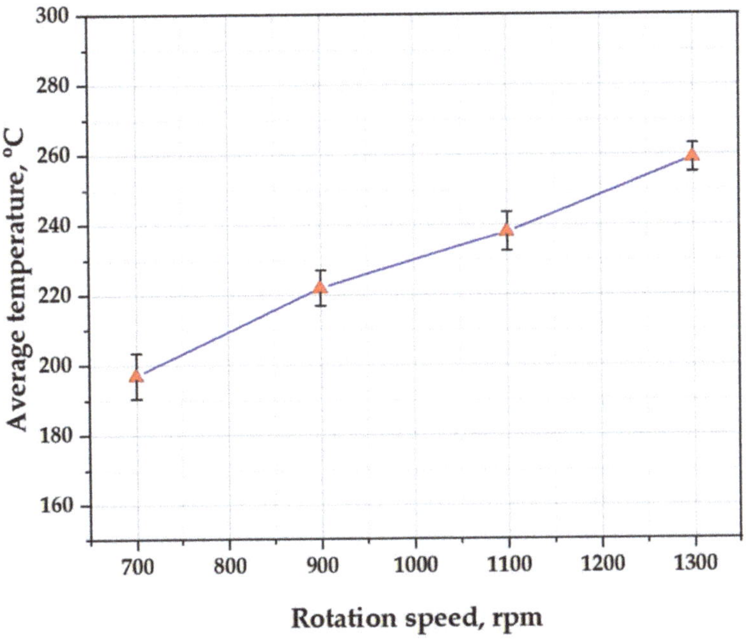

Figure 7. Peak temperatures as a function of tool rotational speed during FSSW of AA2024-T4 joints at various rotational speeds (700–1300 rpm) and a dwell time of 3 s.

3.3. Thermal Energy Generation in FSSW

Based on FSW/FSSW process parameters, the generated heat input energy depends on rotating tool profile and geometry, rotational speed, axial force, dwell time, and friction coefficient. The generated heat input energy (Q_T: watt) was calculated using Equation (1) [42,43]:

$$Q_T = \frac{2}{3} \times \frac{\mu \times \omega \times P}{R_s^2}\left[R_s^3 - R_{PT}^3 + \frac{3}{4} \times \frac{H_P}{\cos\alpha} \times (2 \times R_{PT} - H_P \times \tan\alpha)^2 + (R_{PT} - H_P \times \tan\alpha)^3\right] \quad (1)$$

$$\text{The consumed heat input energy} = Q_T \times t\ (\text{J}) \quad (2)$$

where P (in N) is the axial force, ω (in rad/min) is the angular velocity, calculated as $\omega = (2\pi \times n/60)$, where n is the applied rotational speed in rpm, and R_{PT} (in m) is the pin tip radius, R_S (in m) is the shoulder radius, H_P (in m) is the pin height, μ is the friction coefficient, which describes the interaction between the tool and the aluminum alloy, taken as 0.4 [44,45], α is the design tapered tool angle (in degrees)and t is the dwell time (in s).

By calculating the heat input using the provided equation, the current study quantifies the thermal energy generated at different rotational speeds to provide a crucial understanding of the thermal-mechanical conditions during the AA2024-T4 spot welding process. Figure 8 displays the heat input generated at different rotational speeds for the AA2024-T4 FSSW process. The data reveals a direct correlation, with the thermal energy generated showing a continuous increase as the rotational speed was raised from 700 to 1300 rpm. Specifically, the rotational speed of 1300 rpm exhibited the highest heat input energy of 8580.877 J, while the rotational speed of 700 rpm exhibited the lowest heat input energy of 6796.566 J.

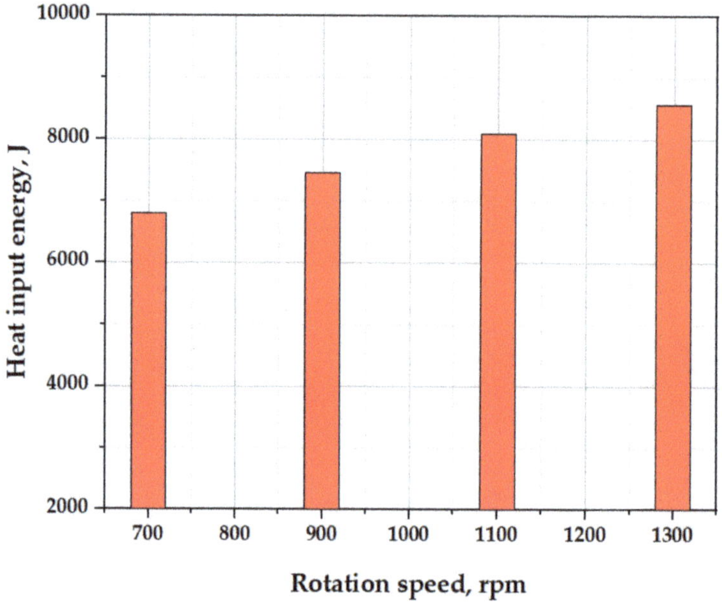

Figure 8. The effect of rotational speed (700–1300 rpm) and a constant dwell time of 3 s on heat input during the FSSW of AA2024-T4 joints.

3.4. Features and Properties of the Produced Spot Joints

3.4.1. Macrographs of the FSSWed Joints

The macrographs of the AA2024-T4 FSSW cross-sections, shown in Figure 9, reveal the disappearance of the interface between the overlapped 4 mm thick AA2024 sheets in the weld zone, which indicates that the spot joining process was successful at the applied FSSW parameters of 700–1300 rpm and 3 s dwell time. The distinct zones characteristic of the FSSW process were found. As a result of the localized heat and severe plastic deformation, four main regions can be identified: the SZ, the thermo-mechanically affected zone (TMAZ), the heat-affected zone (HAZ), and the BM.

The increase in tool rotational speed from 700 to 1300 rpm results in greater heat input, which directly enlarges the size of the SZ, as shown in Figure 9a–d. For instance, the SZ in the joint produced at 900 rpm (Figure 9b) is larger than that at 700 rpm, and this trend continues through 1100 rpm (Figure 9c) up to 1300 rpm. This is a direct result of the greater frictional heat at higher speeds, which softens the material and increases the volume of the stirred zone. A prominent feature in the center of the weld is the keyhole, which formed upon the withdrawal of the tool's pin. This keyhole accurately represents the tool pin's geometry and dimensions. Its size increases with higher rotational speeds (from 700 to 1300 rpm), which is a direct consequence of the greater heat input and resulting material softening around the rotating pin, potentially affecting the joint strength [27,46,47]. The overall macroscopic appearance of the joints indicates that flash formation is directly related to rotational speed. A minimal flash is present at lower rotational speeds (700 and 900 rpm). However, at higher speeds (1100 and 1300 rpm), a more significant flash is observed (Figure 9c,d), which is consistent with an increase in the shoulder projection.

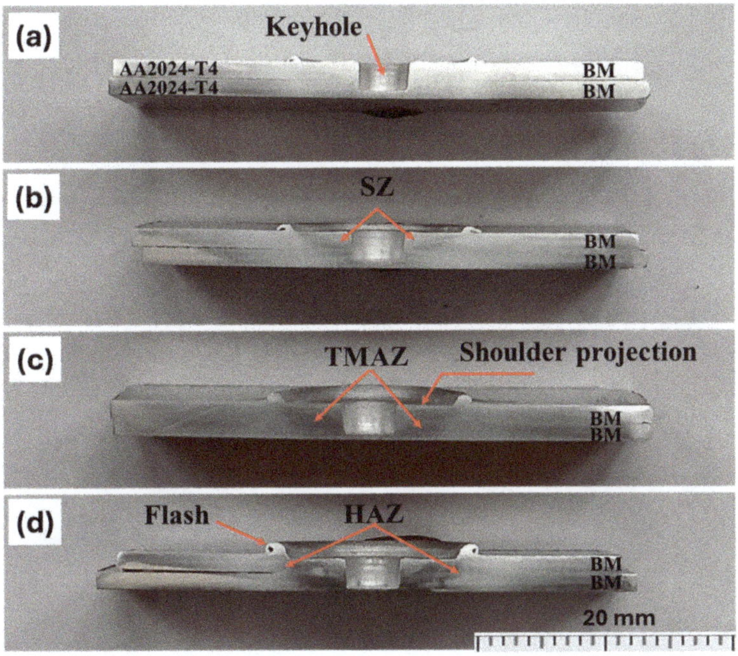

Figure 9. Transverse macrographs of FSSW joints in AA2024-T4 at different rotational speeds, (**a**) 700 rpm, (**b**) 900 rpm, (**c**) 1100 rpm and (**d**) 1300 rpm, all welded at a constant dwell time of 3 s.

3.4.2. Microstructures of the FSSWed Joints

The microstructural evolution during FSSW differs fundamentally from linear FSW, primarily due to the localized, stationary nature of the process. While both utilize a rotating tool to induce plastic deformation and heating, FSSW involves a rapid cycle of plunging, dwelling, and retreating, resulting in a confined, cylindrical weld geometry rather than a continuous seam [48]. This distinction leads to unique material flow patterns (predominantly vertical and rotational mixing around the pin), and a more rapid thermal cycle with less steady-state heat generation compared to FSW. Consequently, the precipitate dissolution, fragmentation, and subsequent dynamic recrystallization (DRX) processes within the weld zone are significantly influenced by this transient heat input profile [39]. Understanding the localized microstructural zones in FSSW is therefore essential for correlating process parameters to final joint properties.

Accordingly, the microstructure of the AA2024-T4 FSSW joints shows significant changes compared to the BM. While the BM is characterized by its elongated grains and the presence of black phase precipitates (Figure 10a), the FSSW process creates distinct zones within the welded joint, as shown in Figure 10b. The microstructural evolution observed in the present FSSW joints, particularly the formation of a fine, dynamically recrystallized equiaxed grain structure in the SZ (Figure 10c–f), aligns with the fundamental mechanisms reported for linear FSW of AA2024 [6,9,16]. This zone is surrounded by the TMAZ, which contains highly deformed and elongated grains. The outermost region, the HAZ, is distinguished by coarsened grains resulting from the thermal cycle of the welding process. This grain refinement in the SZ is a key feature of the FSSW process, as it directly results from the dynamic recrystallization that occurs due to intense plastic deformation and thermal exposure [49]. From Figure 10c–f, the SZ reveals that the grain size is not uniform across all welding conditions; instead, the grain size within the SZ increases as the rotation speed (and thus the heat input) increases. Furthermore, studies on AA2024 FSW often report on the stability and fragmentation precipitates [50,51]. Similarly, a closer examination of the SZ in our work reveals that large, stable precipitates undergo fragmentation and dispersion due to the intense stirring action of the FSSW tool at lower rotational speeds (700 and 900 rpm). Conversely, in alignment with observations in high-heat-input FSW of similar alloys [6], the higher heat inputs in FSSW (1100 and 1300 rpm) appear to promote precipitate coarsening or dissolution within the SZ. This comparative analysis demonstrates that, while the basic metallurgical principles are similar, the specific thermal and strain gradients in FSSW result in distinct microstructural features, which the current study documents in detail for AA2024-T4. Figure 11 presents grain size histograms for the AA2024-T4 BM (Figure 11a) and the stir zones of FSSW joints (Figure 11b–e). The joints were welded at a constant dwell time of 3 s with rotational speeds of 700 rpm, 900 rpm, 1100 rpm, and 1300 rpm. The mean grain size of the BM was 29.7 ± 6.1 μm (Figure 11a), which was significantly reduced in the stir zones to 4.7 ± 1.4 μm (Figure 11b), 5.5 ± 1.5 μm (Figure 11c), 7.5 ± 1.4 μm (Figure 11d), and 8.3 ± 1.3 μm (Figure 11e) for the 700, 900, 1100, and 1300 rpm conditions, respectively.

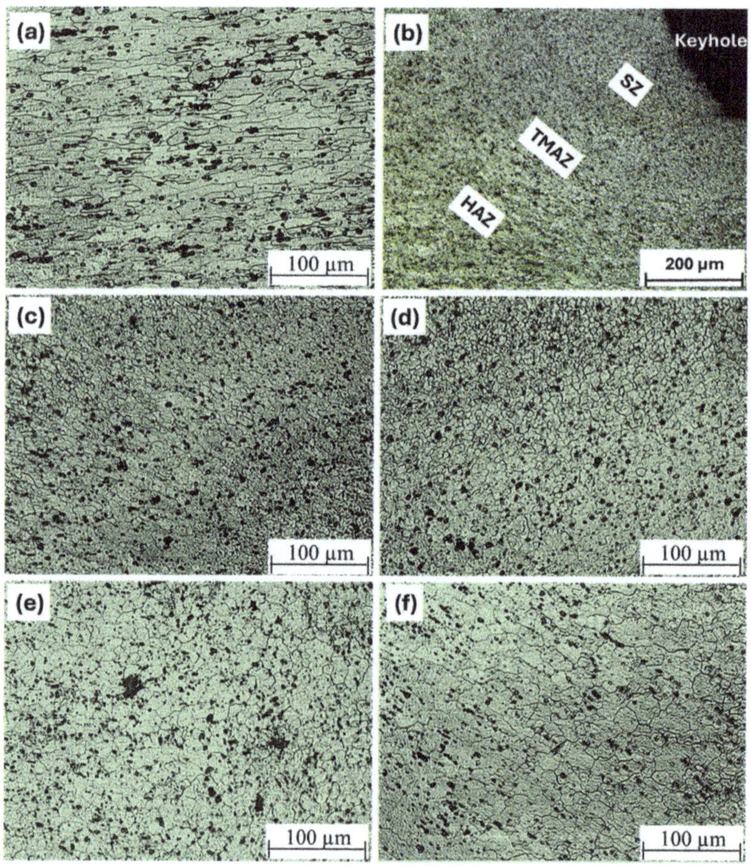

Figure 10. (**a**) Optical micrograph of the AA2024-T4 BM and (**b**) a representative image showing the three distinct zones of the FSSW joint (SZ, TMAZ, and HAZ) for the FSSW joint at 900 rpm and 3 s. (**c**–**f**) Optical micrographs of the stir zones of the FSSW joints welded at (**c**) 700 rpm, (**d**) 900 rpm, (**e**) 1100 rpm, and (**f**) 1300 rpm, at a constant dwell time of 3 s.

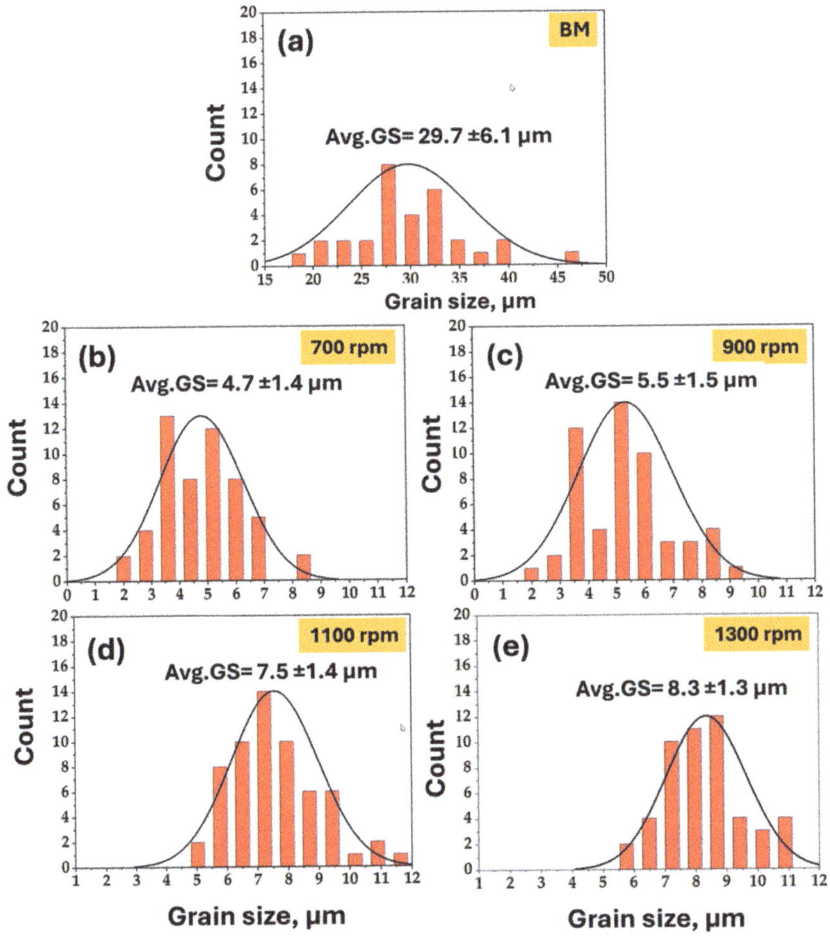

Figure 11. Grain size histograms of (**a**) the AA2024-T4 BM and the stir zones of the FSSW joints welded at a 3 s dwell time and various rotational speeds: (**b**) 700 rpm, (**c**) 900 rpm, (**d**) 1100 rpm, and (**e**) 1300 rpm.

SEM-EDS advanced system was used to identify the microstructural features, specifically the precipitates, in both the AA2024-T4 BM and the FSSW joints. The precipitates appear as bright white phases within the gray aluminum matrix and are irregular in shape. During sample preparation (grinding and polishing processes), some precipitates were observed to be pulled out of the matrix, while others remained attached.

As shown in the micrographs (Figure 12a,b), these hard phases are coarse in the BM. In contrast, the intense stirring action in the SZ of the FSSW joint (Figure 12c) fragments these precipitates into smaller particles and redistributes them in the aluminum matrix. EDS analysis further confirmed the presence of the precipitates, identifying the irregular Al(CuMnFe) phases (Figure 12d), as well as rod- and spherical-shaped Al_2CuMg (Figure 12e) and Al_2Cu (Figure 12f) particles. These findings are consistent with results reported by other researchers for the FSW of AA2024 alloy [50,52,53].

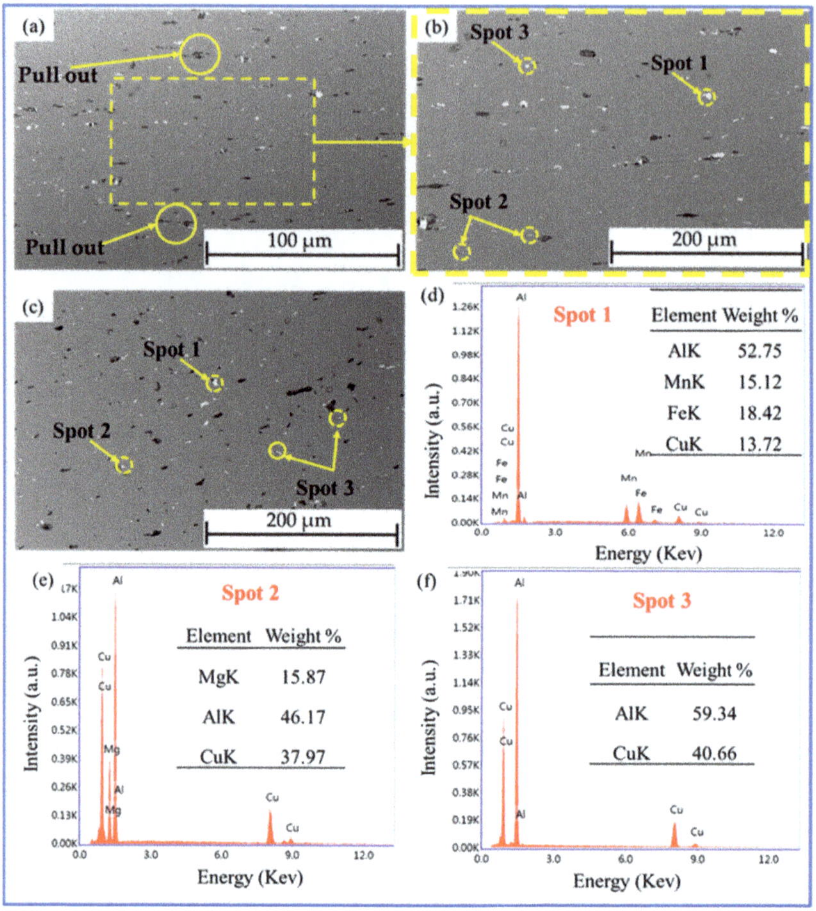

Figure 12. (**a**–**c**) SEM micrographs and (**d**–**f**) EDS analysis of the AA2024-T4 BM and the FSSW joint processed at 900 rpm and 3 s.

3.5. Mechanical Properties

3.5.1. Hardness Test Results

Hardness measurements were conducted on the highly polished cross-sections of the AA2024-T4 spot-welded joints. These specimens were fabricated at a fixed dwell time of 3 s and different rotational speeds of 700, 900, 1100, and 1300 rpm. Figure 13 represents the collected data for all the specimens tested. The distribution of hardness across the weld zones is widely recognized to be a direct consequence of the thermal cycle induced by the FSSW process. The heat generated during welding alters the material's microstructure in specific regions, which is the primary factor controlling the final hardness. The resultant hardness behavior is also significantly affected by the initial state of the starting material. In the current investigation, the BM was in the T4 temper condition, which indicates that it was entirely hardened through age hardening. High thermal exposure during the FSSW process is expected to cause a loss in hardness within the weld zones, primarily as a result of coarsening or re-dissolution of the strengthening precipitates [18,51,54–56]. The AA2024-T4 BM showed a mean hardness value of 111 $\pm$ 6 HV (Table 2). Compared to the BM, the hardness values of the spot weld zones significantly decreased at all the applied rotational speeds (700–1300 rpm). This reduction is attributed to the thermal softening of the AA2024-T4 BM caused by the frictional heat generated during the FSSW processes (Figure 13). Within every spot-welded joint, the lowest hardness values were invariably found in the HAZ. This is a direct consequence of over-aging, where the precipitates that contribute to hardening coarsen and dissolve, leading to maximum softening in this specific region. In contrast to the HAZ, an increased hardness was seen in the SZ for all the welds. This is mainly attributable to a dynamically recrystallized, equiaxed fine-grain structure and the possibility of reprecipitation of hardening phases throughout the cooling cycle [57]. The hardness values of the stir zones were measured as 105 $\pm$ 2.4 HV, 99 $\pm$ 1.5 HV, 94 $\pm$ 2.5 HV, and 88 $\pm$ 3.4 HV for the specimens welded at 700, 900, 1100, and 1300 rpm, respectively. This clear decrease in hardness with increasing rotational speed (and thus increasing heat input) is consistent with the microstructural observations shown in Figure 11. The 700-rpm specimen achieved the highest SZ hardness because it possesses the smallest grain size. This relationship between grain size and hardness directly correlates with the Hall-Petch relation [58,59]. As shown in Figure 13, the TMAZ displayed a hardness value that was intermediate to the SZ and HAZ. The notable increase in hardness within the TMAZ relative to the adjacent HAZ can be directly linked to the considerable plastic deformation that occurs in this region, which greatly increases the material's dislocation density [49].

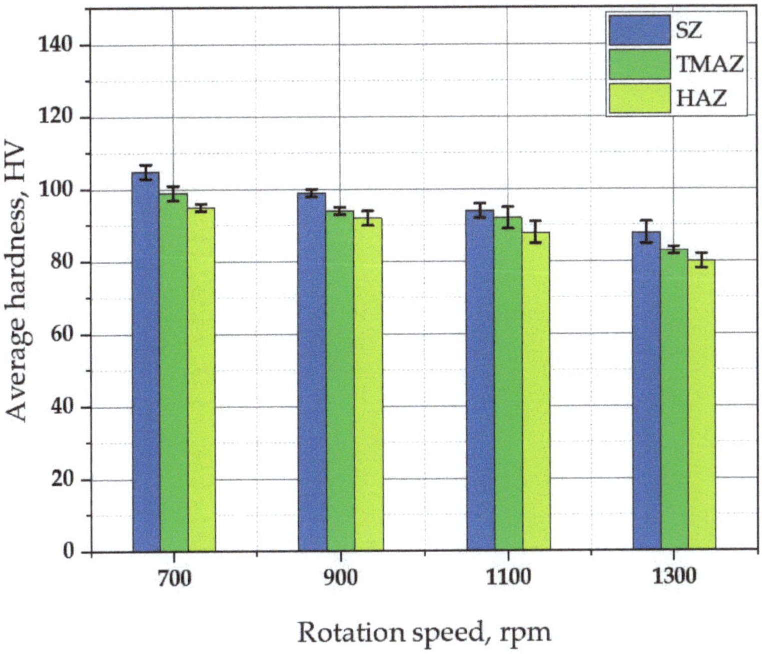

Figure 13. Shows the effect of rotational speed on the average hardness of the SZ, TMAZ, and HAZ for the AA2024-T4 FSSW joints welded at various rotational speeds (700–1300 rpm) and a 3 s dwell time.

3.5.2. Load-Carrying Capacity and Fracture Behavior

The tensile–shear test results, as depicted in Figure 14, show the distinct load versus displacement curves for FSSW AA2024-T4 lap joints produced at various rotational speeds (700 to 1300 rpm) and a constant dwell time of 3 s. These curves provide a comprehensive view of the joint's mechanical response of each joint under tensile shear loading, from initial deformation to ultimate failure. The data summarized in Figure 15, which represents the maximum load-carrying capacity of AA2024-T4 FSSW joints as a function of tool rotational speed. Based on these Figures 13 and 14, the optimal strength was achieved at a rotational speed of 900 rpm, which resulted in the highest load-carrying capacity of 7.3 ± 0.4 kN. In comparison, the joints produced at 700 rpm showed a slightly lower capacity of 6.1 ± 0.6 kN, while those at 1100 rpm and 1300 rpm showed significantly reduced strengths of 5.8 ± 0.5 kN and 5.1 ± 0.3 kN, respectively. This parabolic relationship suggests that there is an ideal range of heat input and material flow for achieving the maximum joint strength. The tensile test results can be directly correlated with the previously observed hardness (Figure 13) and microstructural changes (Figures 10 and 11). The significant drop in joint strength at higher rotational speeds (1100 and 1300 rpm) is a direct consequence of the excessive heat input. As discussed in the hardness analysis, this higher temperature leads to more pronounced over-aging and coarsening of the strengthening precipitates in the HAZ. This over-softened HAZ acts as the weakest link in the joint, dictating its overall load-carrying capacity. Conversely, the joint produced at the lowest speed of 700 rpm, while possessing the highest hardness in the SZ due to its very fine grain structure, did not achieve the maximum tensile strength. This suggests that the heat input at 700 rpm was likely insufficient to produce a fully bonded, defect-free interface, leading to premature failure despite a strong SZ. The peak strength at 900 rpm represents an ideal balance: the heat input was high enough to promote adequate material flow and grain refinement, while not being so high as to excessively soften the critical HAZ. This balance resulted in a joint with an optimized balance of strength and ductility, leading to the highest load-carrying capacity.

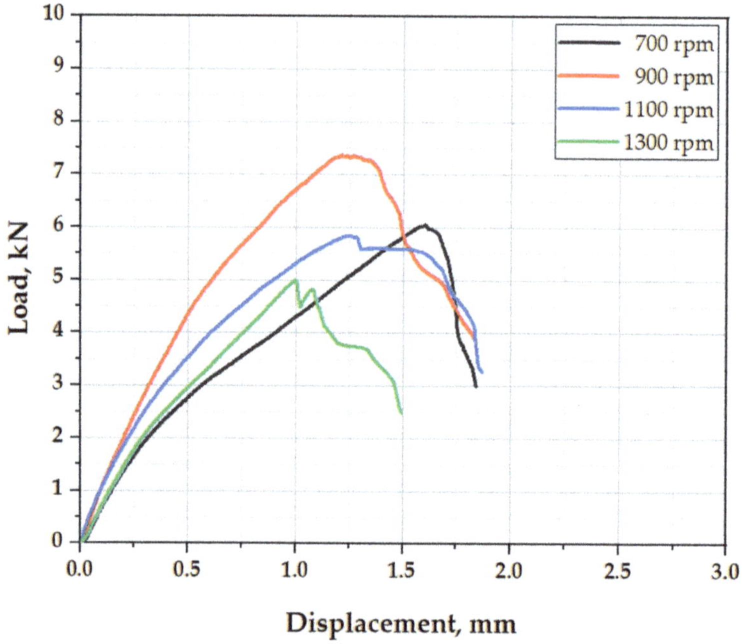

Figure 14. Load–displacement curves for the AA2024-T4 FSSW lap joints produced at various rotational speeds (700–1300 rpm) and a constant dwell time of 3 s.

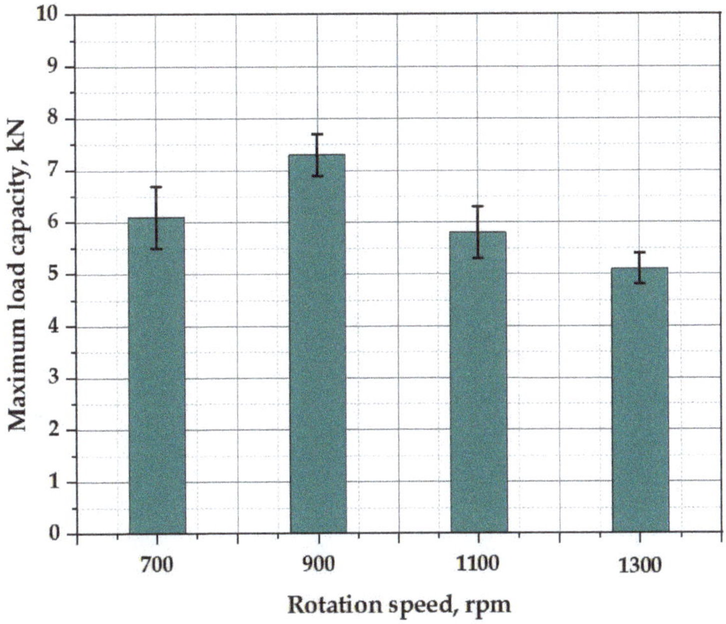

Figure 15. Maximum load-carrying capacity of the AA2024-T4 FSSW joints as a function of tool rotational speed at a fixed dwell time of 3 s.

The fracture location of a FSSW joint is a critical indicator of its mechanical performance and is highly dependent on the welding parameters [48,60,61]. Analysis of the fracture surfaces, as shown in Figure 16, provides a clear visual correlation between rotational speed, joint strength, and failure mechanism. The joint produced at 700 rpm (Figure 16a) exhibits a fracture path indicative of insufficient bonding, likely due to inadequate heat input and material flow, which explains its lower strength of 6.1 ± 0.6 kN. In contrast, the optimal joint strength of 7.3 ± 0.4 kN was achieved at 900 rpm (Figure 16b), where the fracture surface reveals a plug-type failure through the SZ or TMAZ, signifying a strong metallurgical bond that forced failure into the parent material. As rotational speed increased further, excessive heat input led to a severe decline in strength. The fracture for the joint at 1100 rpm (Figure 16c) shows a failure mode transitioning into the excessively softened HAZ, corresponding to its reduced load capacity of 5.8 ± 0.5 kN. This trend culminated in the joint at 1300 rpm (Figure 16d), which displayed the lowest strength (5.1 ± 0.3 kN) and a fracture path completely confined within the over-aged and weakest region of the HAZ.

Figure 17 presents the fracture surfaces of both the base metal (BM) and the FSSW joints. The base metal (Figure 17a,b) exhibits a classic ductile fracture mode, characterized by an aluminum matrix with dimples of varying sizes. The presence of micro-voids, formed by the pull-out of strengthening precipitates, is also evident. These fracture surface features are directly correlated with the known microstructural characteristics of the AA2024-T4 base metal [62]. Conversely, the fracture surfaces of the tensile–shear tested FSSW joints (Figure 17c–f) reveal a more complex failure mechanism. Failure initiated at the stress concentration site of the hook's tip, with microcracks propagating horizontally along the joint interface. The subsequent shearing action ultimately resulted in final failure through the stir zone (SZ). Examination of the lower sheet's fracture surface across all joints showed consistent features, including numerous micro-voids and a high density of small, deep dimples. This morphology is a direct consequence of grain refinement within the SZ during the FSSW process and confirms that the failure, while influenced by the joint geometry, remained fundamentally ductile in nature.

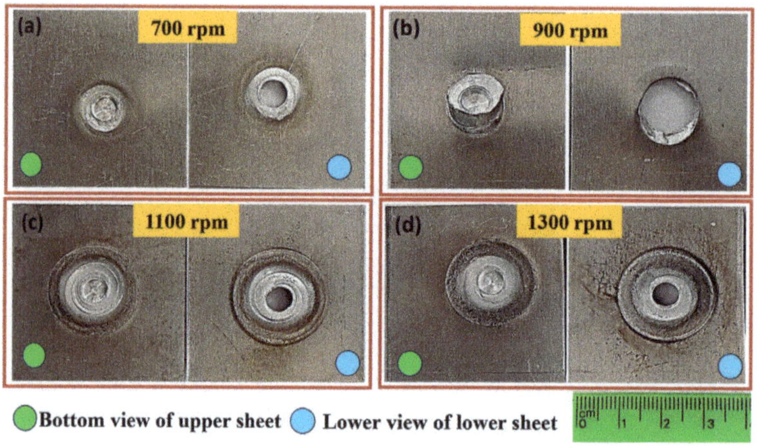

Figure 16. Images of the inner fracture surfaces of the AA2024-T4 FSSW joints after tensile–shear testing for welds fabricated at (**a**) 700 rpm, (**b**) 900 rpm, (**c**) 1100 rpm, and (**d**) 1300 rpm and 3 s welding time.

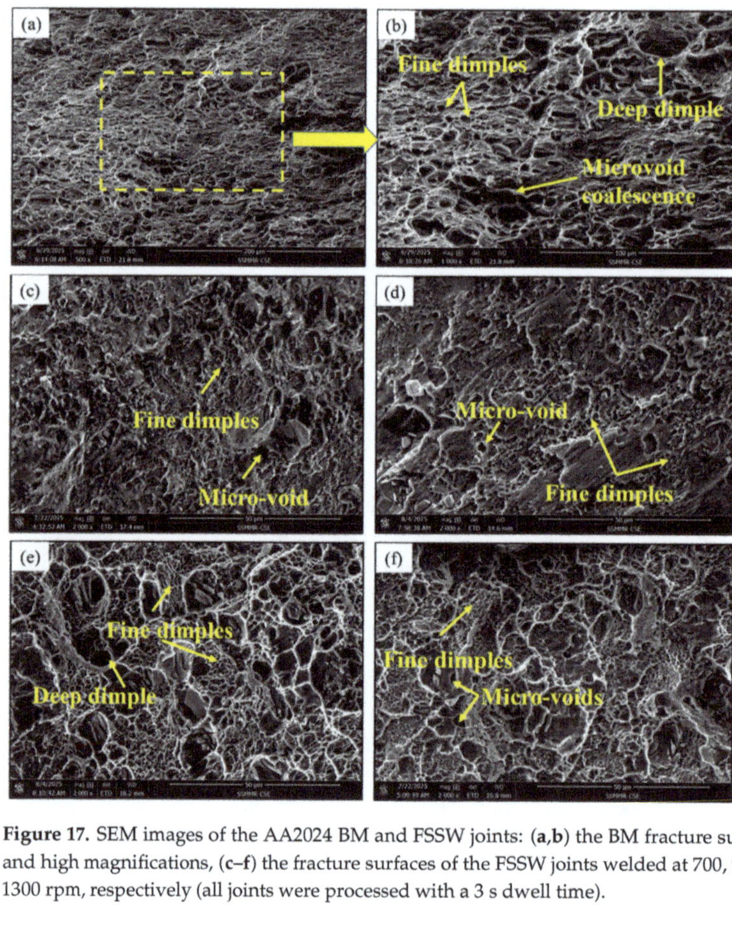

Figure 17. SEM images of the AA2024 BM and FSSW joints: (**a**,**b**) the BM fracture surfaces at low and high magnifications, (**c**–**f**) the fracture surfaces of the FSSW joints welded at 700, 900, 1100 and 1300 rpm, respectively (all joints were processed with a 3 s dwell time).

Finally, the optimal parameter set identified in this study (900 rpm, 3 s) underscores the process efficiency of FSSW for AA2024-T4. The 3 s dwell time is a key driver of productivity, enabling rapid joining cycles. Crucially, achieving the peak load-carrying capacity at this intermediate rotational speed (rather than at the maximum energy input of 1300 rpm) defines an energy-efficient process window. This indicates an efficient conversion of process energy into joint integrity while avoiding the strength degradation associated with excessive thermal softening. Compared to a previous study on a similar alloy that employed longer dwell times and higher rotational speeds [2], the parameters identified here offer an advantageous synergy of reduced process time and enhanced mechanical performance.

4. Conclusions

This study comprehensively investigated the influence of relatively low rotational speeds (700–1300 rpm) and a constant short dwell time of 3 s on the mechanical performance and microstructural evolution of 2 mm thick AA2024-T4 spot-lap welds. The main findings from the analysis of heat input, grain structure, hardness, tensile shear strength, and fracture morphologies are summarized below.

1. By using a short, fixed dwell time of 3 s and a range of tool rotational speeds (700–1300 rpm), this study shows that FSSW is an effective method for joining 2 mm thick AA2024 sheets in similar lap spot welds. The rapid cycle time achieved with these parameters is a key finding, as it indicates a significant advantage for productivity in industrial settings.
2. An optimal tool rotational speed of 900 rpm was identified, producing joints with the highest tensile–shear strength of 7.3 $\pm$ 0.4 kN, as it achieved an ideal balance between sufficient material flow and controlled heat input.
3. This study concluded that increasing the rotational speed from 700 to 1300 rpm resulted in a corresponding increase in the SZ grain size. The base material's (BM) mean grain size of 29.7 $\pm$ 6.1 μm was significantly refined in all joints, but the refinement was most pronounced at 700 rpm, yielding the smallest grain size of 4.7 $\pm$ 1.4 μm. Conversely, the largest SZ grain size of 8.3 $\pm$ 1.3 μm was observed at 1300 rpm, directly correlating with the higher heat input at this speed.
4. The fracture location and mode are directly governed by the rotational speed: joints at 700 rpm failed via interfacial fracture due to insufficient bonding, while joints at the optimal 900 rpm failed by a plug-type fracture through the nugget, indicating superior bond strength.
5. Microstructural analysis confirms that failure in all cases remained ductile, characterized by dimpled fracture surfaces; however, the significantly finer and deeper dimples in the stir zone fracture surfaces are a direct consequence of grain refinement during the FSSW process.
6. The findings of this research offer valuable insights for the practical implementation of FSSW, particularly for manufacturing applications in the aerospace and automotive industries where high-strength, lightweight structures are essential.

Author Contributions: Conceptualization, Y.G.Y.E., M.M.E.-S.S. and S.A.; methodology, Y.G.Y.E., H.A.R. and M.M.E.-S.S.; software, H.A.R. and A.E.; validation, Y.G.Y.E., M.M.E.-S.S. and H.A.R.; formal analysis, A.E. and S.A.; investigation, Y.G.Y.E., H.A.R. and M.M.E.-S.S.; resources, A.E.; data curation, H.A.R. and S.A.; writing—original draft preparation, Y.G.Y.E., S.A. and M.M.E.-S.S.; writing—review and editing, Y.G.Y.E., M.M.E.-S.S., M.F.Y.S. and S.A.; visualization, M.F.Y.S. and A.E.; supervision, M.M.E.-S.S.; project administration, M.F.Y.S. and S.A.; funding acquisition, M.F.Y.S. and S.A. All authors have read and agreed to the published version of the manuscript.

Funding: This work was supported and funded by the Deanship of Scientific Research at Imam Mohammad Ibn Saud Islamic University (IMSIU) (grant number: IMSIU-DDRSP2503).

Data Availability Statement: The original contributions presented in this study are included in the article. Further inquiries can be directed to the corresponding author.

Conflicts of Interest: The author Hagar A. Reyad was employed by the company Arabian Steel Industries. The remaining authors declare that the research was conducted in the absence of any commercial or financial relationships that could be construed as potential conflict of interest.

References

1. Tiwan; Ilman, M.N.; Kusmono; Sehono. Microstructure and Mechanical Performance of Dissimilar Friction Stir Spot Welded AA2024-O/AA6061-T6 Sheets: Effects of Tool Rotation Speed and Pin Geometry. *Int. J. Lightweight Mater. Manuf.* **2023**, *6*, 1–14. [CrossRef]
2. Dourandish, S.; Mousavizade, S.M.; Ezatpour, H.R.; Ebrahimi, G.R. Microstructure, Mechanical Properties and Failure Behaviour of Protrusion Friction Stir Spot Welded 2024 Aluminium Alloy Sheets. *Sci. Technol. Weld. Join.* **2018**, *23*, 295–307. [CrossRef]
3. Anton Savio Lewise, K.; Edwin Raja Dhas, J. FSSW Process Parameter Optimization for AA2024 and AA7075 Alloy. *Mater. Manuf. Process.* **2022**, *37*, 34–42. [CrossRef]
4. Siddiquee, A.N.; Pandey, S.P.; Khan, N.Z. Friction Stir Welding of Austenitic Stainless Steel: A Study on Microstructure and Effect of Parameters on Tensile Strength. *Mater. Today Proc.* **2015**, *2*, 1388–1397. [CrossRef]
5. Sabry, I. Exploring the Effect of Friction Stir Welding Parameters on the Strength of AA2024 and A356-T6 Aluminum Alloys. *J. Alloys Metall. Syst.* **2024**, *8*, 100124. [CrossRef]
6. Zhang, C.; Huang, G.; Zhang, D.; Sun, Z.; Liu, Q. Microstructure and Mechanical Properties in Dissimilar Friction Stir Welded AA2024/7075 Joints at High Heat Input: Effect of Post-Weld Heat Treatment. *J. Mater. Res. Technol.* **2020**, *9*, 14771–14782. [CrossRef]
7. Ahmed, M.M.Z.; Abdelazem, K.A.; El-Sayed Seleman, M.M.; Alzahrani, B.; Touileb, K.; Jouini, N.; El-Batanony, I.G.; Abd El-Aziz, H.M. Friction Stir Welding of 2205 Duplex Stainless Steel: Feasibility of Butt Joint Groove Filling in Comparison to Gas Tungsten Arc Welding. *Materials* **2021**, *14*, 4597. [CrossRef]
8. Zhang, C.; Huang, G.; Cao, Y.; Zhu, Y.; Li, W.; Wang, X.; Liu, Q. Microstructure and Mechanical Properties of Dissimilar Friction Stir Welded AA2024-7075 Joints: Influence of Joining Material Direction. *Mater. Sci. Eng. A* **2019**, *766*, 138368. [CrossRef]
9. Pang, Q.; Zhang, J.H.; Huq, M.J.; Hu, Z.L. Characterization of Microstructure, Mechanical Properties and Formability for Thermomechanical Treatment of Friction Stir Welded 2024-O Alloys. *Mater. Sci. Eng. A* **2019**, *765*, 138303. [CrossRef]
10. Kashaev, N.; Ventzke, V.; Çam, G. Prospects of Laser Beam Welding and Friction Stir Welding Processes for Aluminum Airframe Structural Applications. *J. Manuf. Process.* **2018**, *36*, 571–600. [CrossRef]
11. Albaijan, I.; Ahmed, M.M.Z.; El-Sayed Seleman, M.M.; Touileb, K.; Habba, M.I.A.; Fouad, R.A. Optimization of Bobbin Tool Friction Stir Processing Parameters of AA1050 Using Response Surface Methodology. *Materials* **2022**, *15*, 6886. [CrossRef] [PubMed]
12. Gain, S.; Acharyya, S.K.; Sanyal, D.; Das, S.K. Influence of Welding Parameters on Austenitic Stainless Steel Pipe Weldments Produced by Friction Stir Welding. *Trans. Indian Inst. Met.* **2024**, *77*, 3927–3939. [CrossRef]
13. Deng, L.; Li, S.; Ke, L.; Liu, J.; Kang, J. Microstructure and Fracture Behavior of Refill Friction Stir Spot Welded Joints of AA2024 Using a Novel Refill Technique. *Metals* **2019**, *9*, 286. [CrossRef]
14. Yadav, S.K.; Yusufzai, M.Z.K.; Mishra, S.; Singh, D.K. Influence of Multiple Friction Stir Welding Passes on the Microstructure and Mechanical Properties of AA2024 Age-Hardened Aluminum Alloy. *Proc. Inst. Mech. Eng. Part L J. Mater. Des. Appl.* **2024**, *239*, 1292–1305. [CrossRef]
15. Moradi, M.M.; Jamshidi Aval, H.; Jamaati, R.; Amirkhanlou, S.; Ji, S. Microstructure and Texture Evolution of Friction Stir Welded Dissimilar Aluminum Alloys: AA2024 and AA6061. *J. Manuf. Process.* **2018**, *32*, 1–10. [CrossRef]
16. Tang, J.; Shen, Y. Numerical Simulation and Experimental Investigation of Friction Stir Lap Welding between Aluminum Alloys AA2024 and AA7075. *J. Alloys Compd.* **2016**, *666*, 493–500. [CrossRef]
17. Sadeesh, P.; Venkatesh, K.M.; Rajkumar, V.; Avinash, P.; Arivazhagan, N.; Devendranath, R.K.; Narayanan, S. Studies on Friction Stir Welding of Aa 2024 and Aa 6061 Dissimilar Metals. *Procedia Eng.* **2014**, *75*, 145–149. [CrossRef]
18. Ahmed, M.M.Z.; Jouini, N.; Alzahrani, B.; Seleman, M.M.E.S.; Jhaheen, M. Dissimilar Friction Stir Welding of AA2024 and AISI 1018: Microstructure and Mechanical Properties. *Metals* **2021**, *11*, 330. [CrossRef]
19. Ghazaei Najafabadi, M.A.; Abdollah-Zadeh, A.; Mofid, M.A. Post-Weld Heat Treatment Effects on Microstructure and Mechanical Properties of Underwater Friction Stir-Welded AA 2024/6061 Lap Joints. *Mater. Chem. Phys.* **2025**, *339*, 130789. [CrossRef]

20. Jabraeili, R.; Jafarian, H.R.; Khajeh, R.; Park, N.; Kim, Y.; Heidarzadeh, A.; Eivani, A.R. Effect of FSW Process Parameters on Microstructure and Mechanical Properties of the Dissimilar AA2024 Al Alloy and 304 Stainless Steel Joints. *Mater. Sci. Eng. A* **2021**, *814*, 140981. [CrossRef]
21. Rathinasuriyan, C.; Puviyarasan, M.; Sankar, R.; Selvakumar, V. Effect of Process Parameters on Weld Geometry and Mechanical Properties in Friction Stir Welding of AA2024 and AA7075 Alloys. *J. Alloys Metall. Syst.* **2024**, *7*, 100091. [CrossRef]
22. Rashkovets, M.; Dell'Avvocato, G.; Contuzzi, N.; Palumbo, D.; Galietti, U.; Casalino, G. On the Role of Rotational Speed in P-FSSW Dissimilar Aluminum Alloys Lap Weld. *Weld. World* **2025**, *69*, 2095–2107. [CrossRef]
23. Yuan, W.; Mishra, R.S.; Webb, S.; Chen, Y.L.; Carlson, B.; Herling, D.R.; Grant, G.J. Effect of Tool Design and Process Parameters on Properties of Al Alloy 6016 Friction Stir Spot Welds. *J. Mater. Process Technol.* **2011**, *211*, 972–977. [CrossRef]
24. Gerlich, A.; Avramovic-Cingara, G.; North, T.H. Stir Zone Microstructure and Strain Rate during Al 7075-T6 Friction Stir Spot Welding. *Metall. Mater. Trans. A* **2006**, *37*, 2773–2786. [CrossRef]
25. Tozaki, Y.; Uematsu, Y.; Tokaji, K. Effect of Tool Geometry on Microstructure and Static Strength in Friction Stir Spot Welded Aluminium Alloys. *Int. J. Mach. Tools Manuf.* **2007**, *47*, 2230–2236. [CrossRef]
26. Palanivel, R.; Koshy Mathews, P.; Murugan, N.; Dinaharan, I. Effect of Tool Rotational Speed and Pin Profile on Microstructure and Tensile Strength of Dissimilar Friction Stir Welded AA5083-H111 and AA6351-T6 Aluminum Alloys. *Mater. Des.* **2012**, *40*, 7–16. [CrossRef]
27. Ataya, S.; Ahmed, M.M.Z.; El-Sayed Seleman, M.M.; Hajlaoui, K.; Latief, F.H.; Soliman, A.M.; Elshaghoul, Y.G.Y.; Habba, M.I.A. Effective Range of FSSW Parameters for High Load-Carrying Capacity of Dissimilar Steel A283M-C/Brass CuZn40 Joints. *Materials* **2022**, *15*, 1394. [CrossRef]
28. Piccini, J.M.; Svoboda, H.G. Effect of Pin Length on Friction Stir Spot Welding (FSSW) of Dissimilar Aluminum-Steel Joints. *Procedia Mater. Sci.* **2015**, *9*, 504–513. [CrossRef]
29. Awang, M.; Mucino, V.H. Energy Generation during Friction Stir Spot Welding (FSSW) of Al 6061-T6 Plates. *Mater. Manuf. Process.* **2010**, *25*, 167–174. [CrossRef]
30. Jo, D.S.; Kahhal, P.; Kim, J.H. Optimization of Friction Stir Spot Welding Process Using Bonding Criterion and Artificial Neural Network. *Materials* **2023**, *16*, 3757. [CrossRef] [PubMed]
31. Sehono, S.; Putra, I.R. Microstructure and Mechanical Performance of Friction Stir Spot Welding to Riveting Process Aluminum AA2024-T3 Plate: Effect of Tool Rotation Speed. *AIP Conf. Proc.* **2025**, *3250*, 070003. [CrossRef]
32. Fujimoto, M.; Koga, S.; Abe, N.; Sato, S.Y.; Kokawa, H. Analysis of Plastic Flow of the Al Alloy Joint Produced by Friction Stir Spot Welding. *Weld. Int.* **2009**, *23*, 589–596. [CrossRef]
33. Anton Savio Lewise, K.; Edwin Raja Dhas, J.; Pandiyarajan, R.; Sabarish, S. Metallurgical and Mechanical Investigation on FSSWed Dissimilar Aluminum Alloy. *J. Alloys Metall. Syst.* **2023**, *2*, 100010. [CrossRef]
34. Zhang, G.; Xiao, C.; Ojo, O.O. Dissimilar Friction Stir Spot Welding of AA2024-T3/AA7075-T6 Aluminum Alloys under Different Welding Parameters and Media. *Def. Technol.* **2021**, *17*, 531–544. [CrossRef]
35. Sadoun, A.M.; Wagih, A.; Fathy, A.; Essa, A.R.S. Effect of Tool Pin Side Area Ratio on Temperature Distribution in Friction Stir Welding. *Results Phys.* **2019**, *15*, 102814. [CrossRef]
36. Ratković, N.; Jovanović Pešić, Ž.; Arsić, D.; Pešić, M.; Džunić, D. Tool Geometry Effect on Material Flow and Mixture in FSW. *Adv. Technol. Mater.* **2022**, *47*, 33–36. [CrossRef]
37. Seleman, M.M.E.S.; Ahmed, M.M.Z.; Ramadan, R.M.; Zaki, B.A. Effect of FSW Parameters on the Microstructure and Mechanical Properties of T-Joints between Dissimilar Al-Alloys. *Int. J. Integr. Eng.* **2022**, *14*, 1–12. [CrossRef]
38. Elfishawy, E.; Ahmed, M.M.Z.; El-Sayed Seleman, M.M. Additive Manufacturing of Aluminum Using Friction Stir Deposition. In *TMS 2020 149th Annual Meeting & Exhibition Supplemental Proceedings*; Springer International Publishing: Cham, Switzerland, 2020; pp. 227–238.
39. Mubiayi, M.P.; Akinlabi, E.T. An Overview on Friction Stir Spot Welding of Dissimilar Materials. In *Transactions on Engineering Technologies: World Congress on Engineering and Computer Science 2014*; Springer: Dordrecht, The Netherlands, 2015; pp. 537–549, ISBN 9789401772365.
40. Lu, X.; Luan, Y.; Meng, X.; Zhou, Y.; Zhao, N.; Liang, S.Y. Temperature Distribution and Mechanical Properties of FSW Medium Thickness Aluminum Alloy 2219. *Int. J. Adv. Manuf. Technol.* **2022**, *119*, 7229–7241. [CrossRef]
41. Chao, Y.J.; Qi, X.; Tang, W. Heat Transfer in Friction Stir Welding–Experimental and Numerical Studies. *J. Manuf. Sci. Eng.* **2003**, *125*, 138–145. [CrossRef]
42. Schmidt, H.; Hattel, J.; Wert, J. An Analytical Model for the Heat Generation in Friction Stir Welding. *Model Simul. Mat. Sci. Eng* **2004**, *12*, 143–157. [CrossRef]
43. Ranjan Kumar, R.; Kumar, S.; Kumar, A. Study on Effect of Tool Geometry on Energy and Temperature of Friction Stir Welding. *Int. J. Civ. Eng. Technol.* **2017**, *8*, 742–754.
ilton, C.; Dymek, S.; Sommers, A. A Thermal Model of Friction Stir Welding in Aluminum Alloys. *Int. J. Mach. Tools Manuf.* [CrossRef]

45. Javadi, M.; Tajdari, M. Experimental Investigation of the Friction Coefficient between Aluminium and Steel. *Mater. Sci.-Pol.* **2006**, *24*, 305–310.
46. Correia, A.N.; Santos, P.A.M.; Braga, D.F.O.; Baptista, R.; Infante, V. Effects of Friction Stir Welding Process Control and Tool Penetration on Mechanical Strength and Morphology of Dissimilar Aluminum-to-Polymer Joints. *J. Manuf. Mater. Process.* **2023**, *7*, 106. [CrossRef]
47. Di Bella, G.; Borsellino, C.; Chairi, M.; Campanella, D.; Buffa, G. Effect of Rotational Speed on Mechanical Properties of AA5083/AA6082 Friction Stir Welded T-Joints for Naval Applications. *Metals* **2024**, *14*, 1410. [CrossRef]
48. Bagheri, B.; Mahdian Rizi, A.A.; Abbasi, M.; Givi, M. Friction Stir Spot Vibration Welding: Improving the Microstructure and Mechanical Properties of Al5083 Joint. *Metallogr. Microstruct. Anal.* **2019**, *8*, 713–725. [CrossRef]
49. Pang, Q.; Zhao, M.; Zhang, Z. The Effect of High-Temperature Deformation on the Mechanical Properties and Corrosion Resistance of the 2024 Aluminum Alloy Joint after Friction Stir Welding. *Materials* **2024**, *17*, 2969. [CrossRef]
50. Staszczyk, A.; Sawicki, J. Comparison of Mechanical Behaviour of Microstructures of 2024 Aluminium Alloy Containing Precipitates of Different Morphologies. *IOP Conf. Ser. Mater. Sci. Eng.* **2020**, *743*, 012052. [CrossRef]
51. Ehrich, J.; Staron, P.; Karkar, A.; Roos, A.; Hanke, S. Precipitation Evolution in the Heat-Affected Zone and Coating Material of AA2024 Processed by Friction Surfacing. *Adv. Eng. Mater.* **2022**, *24*, 2201019. [CrossRef]
52. Irmer, D.; Yildirim, C.; Sennour, M.; Esin, V.A.; Moussa, C. Effect of Second-Phase Precipitates on Deformation Microstructure in AA2024 (Al–Cu–Mg): Dislocation Substructures and Stored Energy. *J. Mater. Sci.* **2024**, *59*, 18978–19002. [CrossRef]
53. Morozova, I.; Królicka, A.; Obrosov, A.; Yang, Y.; Doynov, N.; Weiß, S.; Michailov, V. Precipitation Phenomena in Impulse Friction Stir Welded 2024 Aluminium Alloy. *Mater. Sci. Eng. A* **2022**, *852*, 143617. [CrossRef]
54. Mohammadtaheri, M.; Haddad-Sabzevar, M.; Mazinani, M. The Effects of Heat Treatment and Cold Working on the Microstructure of Aluminum Alloys Welded by Friction Stir Welding (FSW) Technique. *Adv. Mater. Res.* **2012**, *409*, 287–292. [CrossRef]
55. Staszczyk, A.; Sawicki, J.; Adamczyk-Cieslak, B. A Study of Second-Phase Precipitates and Dispersoid Particles in 2024 Aluminum Alloy after Different Aging Treatments. *Materials* **2019**, *12*, 4168. [CrossRef]
56. Habba, M.I.A.; Alsaleh, N.A.; Badran, T.E.; El-Sayed Seleman, M.M.; Ataya, S.; El-Nikhaily, A.E.; Abdul-Latif, A.; Ahmed, M.M.Z. Comparative Study of FSW, MIG, and TIG Welding of AA5083-H111 Based on the Evaluation of Welded Joints and Economic Aspect. *Materials* **2023**, *16*, 5124. [CrossRef]
57. El Rayes, M.M.; Soliman, M.S.; Abbas, A.T.; Pimenov, D.Y.; Erdakov, I.N.; Abdel-Mawla, M.M. Effect of Feed Rate in FSW on the Mechanical and Microstructural Properties of AA5754 Joints. *Adv. Mater. Sci. Eng.* **2019**, *2019*, 156176. [CrossRef]
58. Baudin, T.; Bozzi, S.; Brisset, F.; Azzeddine, H. Local Microstructure and Texture Development during Friction Stir Spot of 5182 Aluminum Alloy. *Crystals* **2023**, *13*, 540. [CrossRef]
59. Khalafe, W.H.; Sheng, E.L.; Bin Isa, M.R.; Omran, A.B.; Shamsudin, S.B. The Effect of Friction Stir Welding Parameters on the Weldability of Aluminum Alloys with Similar and Dissimilar Metals: Review. *Metals* **2022**, *12*, 2099. [CrossRef]
60. Padhy, G.K.; Wu, C.S.; Gao, S. Friction Stir Based Welding and Processing Technologies–Processes, Parameters, Microstructures and Applications: A Review. *J. Mater. Sci. Technol.* **2018**, *34*, 1–38. [CrossRef]
61. Kosturek, R.; Torzewski, J.; Wachowski, M.; Śnieżek, L. Effect of Welding Parameters on Mechanical Properties and Microstructure of Friction Stir Welded AA7075-T651 Aluminum Alloy Butt Joints. *Materials* **2022**, *15*, 5950. [CrossRef] [PubMed]
62. Urrego, L.F.; García-Beltrán, O.; Arzola, N.; Araque, O. Mechanical Fracture of Aluminium Alloy (AA 2024-T4), Used in the Manufacture of a Bioproducts Plant. *Metals* **2023**, *13*, 1134. [CrossRef]

crystals MDPI

Article

Micro-Arc Coatings with Different Types of Microparticles on Titanium Alloy: Formation, Structure, and Properties

Anna V. Ugodchikova [1], Tatiana V. Tolkacheva [2], Pavel V. Uvarkin [2], Margarita A. Khimich [3], Yurii P. Sharkeev [2], Alexander D. Kashin [2], Ivan A. Glukhov [2] and Mariya B. Sedelnikova [2,*]

[1] Research Institute of Technical Physics and Automation, 115230 Moscow, Russia; ugodch99@gmail.com
[2] Laboratory of Physics of Nanostructured Biocomposites, Institute of Strength Physics and Materials Science of SB RAS, 634055 Tomsk, Russia; tolkacheva@ispms.ru (T.V.T.); uvarkin@ispms.ru (P.V.U.); sharkeev@ispms.ru (Y.P.S.); kash@ispms.ru (A.D.K.); gia@ispms.ru (I.A.G.)
[3] Laboratory of Nanobioengineering, Institute of Strength Physics and Materials Science of SB RAS, 634055 Tomsk, Russia; khimich@ispms.ru
* Correspondence: mariyased@ispms.ru

Abstract

This study examines the effects of electrolyte composition, specifically the incorporation of dispersed particles, on the properties and formation kinetics of micro-arc oxidation (MAO) coatings on a bioinert titanium alloy. Coatings with particles of β-tricalcium phosphate (CP), wollastonite (CS), and combined coatings containing both types of particles (SP) were obtained. The MAO process was carried out using a Micro-Arc 3.0 unit in pulsed potentiostatic anode mode, with the process voltage ranging from 350 to 500 volts. The surface morphology and internal structure of the coatings were examined using scanning electron microscopy. The elemental composition of the coatings was determined by the EDX method, while the phase composition and fine structure of the coatings were investigated by XRD and TEM methods, respectively. The adhesion properties of the coatings were determined by means of scratch testing. When the MAO process voltage was increased to 500 V, the thickness of CP, CS, and SP coatings increased to 80, 50, and 50 μm, respectively. Notably, SP coatings demonstrated the highest adhesion strength (critical load L_c = 22 N), indicating their potential for use in load-bearing medical implants, where preventing delamination under mechanical stress is critical.

Keywords: micro-arc oxidation; titanium alloy; tricalcium phosphate and wollastonite microparticles; electrolyte conductivity; biocoating; adhesion strength

1. Introduction

Titanium and its alloys are gradually replacing stainless steels and Co-Cr-Mo alloys, becoming the most common materials for orthopedic and dental implants. These materials offer a number of advantages, including high mechanical strength, corrosion resistance, and biocompatibility with body tissues [1]. Nevertheless, the TiO_2 oxide film formed on the surface of a titanium implant does not possess sufficient electrostatic potential to adsorb protein molecules and thereby initiate the osteointegration processes [2]. In some cases, this results in fibrous tissue formation at the bone–implant interface, subsequent implant loosening, and eventual loss of the implant [3]. In an effort to enhance the biological activity of titanium implants, numerous researchers have put forward a variety of surface modification methods [4–6]. Such modification not only enhances the structure of the

 https://doi.org/10.3390/cryst15090811

implant surface, thereby increasing its hydrophilic and bioactive properties [7–11], but also endows the implant surface with specific osteogenic or antibacterial properties [12,13].

Among the various methods of surface modification of titanium implants, electrochemical methods such as electropolymerization, electrophoretic deposition, and electrothermal polarization are particularly prominent [2]. However, the most sophisticated and technologically advanced method of coating the surface of titanium and its alloys is micro-arc oxidation, otherwise referred to as plasma electrolytic oxidation [14–16]. The MAO process is carried out at voltages exceeding the breakdown voltage of the dielectric oxide layer. Thus, plasma discharge channels with local temperatures ranging from 2000 to 10,000 K are formed [17]. During the coating growth process, short-term microdischarges are steadily formed across the surface [18]. Unlike conventional anodizing, the peculiarity of the MAO process is that these high-temperature and high-pressure conditions within the discharge channels trigger plasma-chemical reactions, resulting in the synthesis of new phases [19].

Studies have shown that modifying the chemical composition of the electrolyte, substrate material, application time, and electrophysical parameters (e.g., voltage, current density, pulse duration, and pulse repetition rate) makes it possible to obtain coatings with different structures, compositions, and properties [20–23].

The electrolyte composition plays a pivotal role in MAO coating formation [24–28], since it directly influences the breakdown voltage and final voltage/current density during the MAO process, which determine the coating's resulting morphology and physicochemical properties [29]. Electrolytes used in MAO can be either true solutions, where all components dissolve completely, or suspensions containing both dissolved and dispersed insoluble particles [30,31]. The most common insoluble components of electrolytes are calcium phosphates, as the compounds closest in mineralogical and chemical composition to bone [32–34]. However, recent studies have demonstrated that calcium and magnesium silicates also exhibit high biological activity and activate osseointegration processes [35]. Wollastonite is a calcium silicate $CaSiO_3$ with a chain structure in which silicon–oxygen tetrahedrons are interconnected by their vertices into chains, and the chains are cross-linked by calcium ions [36]. Wollastonite exists in two mineral forms: a low-temperature form, β-wollastonite, and a high-temperature form, α-wollastonite (or pseudowollastonite) [37]. The presence of calcium (Ca^{2+}) and silicon (Si^{4+}) ions in wollastonite ($CaSiO_3$) has been shown to be instrumental in its high activity in the process of biomineralization, specifically the formation of the hydroxyapatite (HA) layer [19,38]. Prior studies were directed towards identifying the formation patterns, examining the structure and properties of coatings with β-TCP and wollastonite particles on the surface of bioresorbable magnesium alloys [19,30,38,39]. The data regarding the MAO process kinetics, specifically the interaction between the magnesium substrate and an electrolyte containing dispersed particles, have been obtained.

The objective of the present study was to examine how the electrolyte composition containing β-TCP and wollastonite particles affects the MAO process on the surface of a bioinert titanium alloy, as well as the formation of morphology, structure, and physicomechanical properties of coatings.

2. Materials and Methods

2.1. Sample Preparation

The substrates utilized for coating were 10 × 10 × 1 mm Grade 2 titanium plates, which were obtained through electrospark cutting. Titanium samples were subjected to preliminary mechanical treatment, namely grinding with 400 and 600 grit abrasive sandpaper. Subsequently, the samples were subjected to an ultrasonic cleaning process (Elmasonic S, Elma, Germany) in ethanol for a duration of 60 min. This was followed by drying in a desiccator for 30 min at a temperature of 150 °C. Three electrolyte compositions were prepared for experiments on coating deposition by micro-arc oxidation: I—with the addition of β-tricalcium phosphate (β-TCP) particles to obtain calcium phosphate coatings (CP coatings); II—with the addition of wollastonite particles to obtain calcium silicate coatings (CS coatings); and III—with the addition of both types of particles (β-TCP and wollastonite) to obtain hybrid silicate–phosphate coatings (SP coatings) (Table 1). Wollastonite MIVOLL® 05-96 (CJSC GEOKOM, Kaluga, Russia) and tricalcium phosphate CAFOS DB chemical plant "Budenheim" Dr. Oetker (Budenheim, Germany) were used in the work. In addition, water-soluble components such as $Na_2HPO_4{\cdot}12H_2O$ (LLC "Sigma Track", Khimki, Russia), NaOH (JSC "Kaustik" Volgograd, Russia), Na_2SiO_3 (LLC "Sigma Track", Khimki, Russia), and NaF (LLC "Sigma Track", Khimki, Russia) were introduced into the electrolyte composition.

Table 1. Electrolyte compositions.

Compound	Content in the Electrolyte Composition, g/L		
	I	II	III
$CaSiO_3$ (wollastonite)	–	90	45
β-$Ca_3(PO_4)_2$ (β-TCP)	90	–	45
$Na_2HPO_4{\cdot}12H_2O$	20	–	10
NaOH	10	10	10
Na_2SiO_3	–	5	2.5
NaF	5	5	5
pH	11	12	11
Coating designator	CP	CS	SP

The micro-arc oxidation process was conducted on the Micro-Arc 3.0 unit (ISPMS SB RAS, Tomsk, Russia) [39], operating within the anodic potentiostatic mode. The process parameters utilized are outlined below: pulse duration of 100 μs, pulse repetition rate of 50 Hz, and process duration of 10 min. The process voltage was varied in increments of 50 V, ranging from 350 V to 500 V.

2.2. Structural, Morphological, and Mechanical Properties of Coatings

The coatings were examined using scanning electron microscopy (SEM) (LEO EVO 50, Zeiss, Oberkochen, Germany) with an energy dispersive analysis (EDX) attachment (INCA, Oxford Instruments, Abingdon, UK) and transmission electron microscopy (TEM) (JEM-2100, Jeol Ltd., Tokyo, Japan) in "Nanotech" center at ISPMS SB RAS (Tomsk, Russia). Surface roughness was quantified using a Model-296 profilometer (Moscow, Russia) via the *Ra* parameter, which represents the average roughness value across multiple measurement sections. The phase composition of the coatings was investigated by X-ray phase analysis on a DRON-7 diffractometer (IC Burevestnik, Nizhniy Novgorod, Russia) ("Nanotech" center). Imaging was carried out in the range of angles 2θ = 5–90° with a scanning step of 0.02° in CoKα radiation (λ = 0.17902 nm). The Joint Committee on Powder Diffraction Standards (JCPDS) database was used for phase identification and interpretation of the obtained data. The adhesion strength of the coatings was evaluated using a Revetest RST scratch tester (CSM Instruments SA, Peseux, Switzerland). The device was outfitted with a Rockwell diamond indenter with a radius of 200 μm. The indenter was moved over the specimen surface at a speed of 1 mm/min while the load was linearly increased from 0.5 to 30 N. The length of the scratch was 5 mm, and the loading rate was 5.9 N/min. The laboratory conductometric liquid analyzer "MULTITEST KSL" (JSC SPE "SEMIKO", Novosibirsk, Russia) was used to measure the specific electrical conductivity of electrolytes. Instrument error was 0.001 S/m. Each measurement was performed at least five times. The temperature of electrolytes was measured and controlled by a thermometric measuring channel included in the analyzer kit. The measurements were carried out at room temperature. The duration of the measurement of specific conductivity was no more than 10 s. The coating thickness was measured using a micrometer LEGIONER EDM-25-0.001 (Legioner, Löningen, Germany) and by analyzing SEM images of the coatings cross-sections.

3. Results and Discussions

3.1. Coating Formation

Wollastonite ($CaSiO_3$), a calcium silicate mineral, is characterized by its chain-like crystal structure [36] (Figure 1a), which gives it its distinctive elongated shape. During the grinding of wollastonite powder and the splitting of larger particles, the formation of small needle-shaped crystals occurs. Natural wollastonite of MIVOLL® 05-96 grade, with an average particle length of 20 μm, was utilized in this study. The β-TCP powder particles, measuring between 2 and 5 μm, exhibit an isometric morphology (Figure 1b).

(**a**) (**b**)

Figure 1. SEM images of wollastonite (**a**) and β-TCP powders (**b**).

In previous studies [19], the results of measuring the zeta potential (ζ, mV) of β-TCP and wollastonite particles as a function of the medium's pH were obtained, and it was found that the absolute value of the zeta potential |ζ| for both types of particles increases with increasing pH. As shown earlier, within the pH range of 8 to 11, β-TCP particles were characterized by a higher absolute zeta potential value than wollastonite particles. In the presented work, alkaline electrolytes with a pH of 11–12 were used (Table 1). At such pH, the ζ values of β-TCP and wollastonite particles reached −47 mV and −43 mV, respectively [19]. High |ζ| implies higher electrophoretic mobility of β-TCP particles compared to wollastonite particles in the electrolyte dispersion medium under the influence of an electric field.

The results of measuring the specific conductivity of the electrolytes (Figure 2a) indicate that electrolyte I, containing β-TCP particles, exhibits a higher value of 2.87 S/m, while electrolyte II, comprising wollastonite particles, demonstrates the lowest specific conductivity of 2.6 S/m. Therefore, it was determined that the incorporation of dispersed particles of β-TCP with elevated modulus of zeta potential into the composition of electrolytes enhances the specific conductivity of electrolytes. The specific conductivity of the electrolyte, in turn, affects the magnitude of the decrease in the potential at which the dielectric coating breakdown occurs [25], which affects the morphological features of the surface, structure, and properties of the coatings. The higher absolute value of zeta potential and, accordingly, higher electrophoretic mobility of β-TCP particles result in their faster diffusion to the coating/electrolyte interface in response to the electric field when compared with wollastonite particles. Furthermore, the isometric shape of β-TCP crystals favors their higher mobility compared to wollastonite particles having an elongated shape.

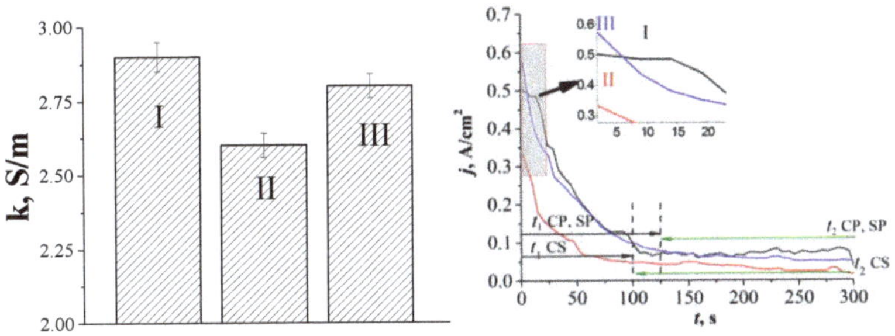

(a) (b)

Figure 2. Specific electrical conductivity of electrolytes with dispersed particles (**a**), plots of current density variation during the MAO process at a voltage of 400 V (**b**).

Thus, β-TCP particles are more actively involved in the formation of a quasicathode (a layer of negatively charged counterions formed at the coating/electrolyte interface). This results in a more substantial accumulation of charge, which in turn impacts the initiation of micro-arc discharges.

The dependences of current density (j, A/cm^2) on the duration during the formation of all types of coatings are described by exponential functions (Figure 2b). To investigate how the electrolyte composition affects the current density of the MAO process, the voltage was kept constant at 400 V. Two periods can be conventionally distinguished on the graphs: t_1—the period of rapid decrease in current density and t_2—the period during which the current density practically does not change. The duration of the t_1 period for the CP and SP coating process in electrolytes I and III lasts approximately 125 s, while the duration of this period for the CS coating formation process in electrolyte II is 100 s. At the same time, higher initial values of current density correspond to the MAO processes in electrolytes I and III (0.50 and 0.58 A/cm^2, respectively).

Figure 3 presents plots of the current density and coating thickness versus time at a constant process voltage value of 400 V. The growth rate of the coatings is also shown in Figure 3.

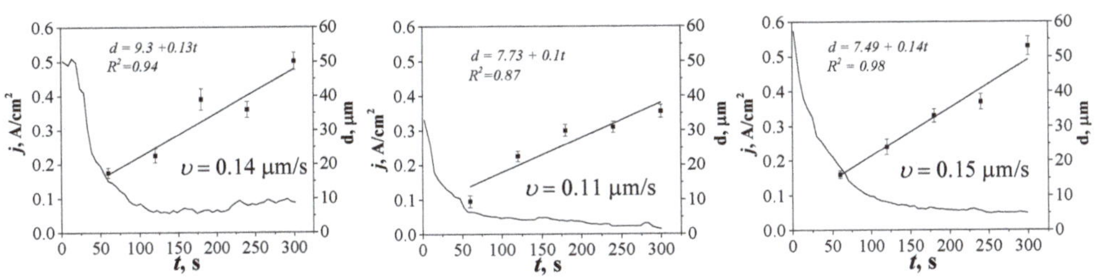

(a) (b) (c)

Figure 3. Graphs of change in current density of the MAO process and in thickness of CP (**a**), CS (**b**), and SP (**c**) coatings versus time.

The graphs show that higher current densities during the MAO process in electrolytes I and III contribute to a higher coating thickness growth rate υ, μm/s (0.14 and 0.15 μm/s, respectively) (Figure 3a,c). Furthermore, the morphology of the dispersed particles involved in the MAO process is also likely to influence the rate of coating growth. Wollastonite particles having a specific elongated shape are less intensively involved in the combustion regions of micro-arc discharges than β-TCP particles having an isometric shape.

3.2. Morphology, Structure, and Elemental Composition of Coatings

The analysis of the SEM images of the coating surface showed that all coatings have a developed surface morphology with a large number of pores, which is explained by the mechanism of the MAO process (Figures 4 and 5). A close examination of CP coatings formed at voltages of 400 V and 500 V reveals the presence of numerous isometric particles measuring between 3 and 5 μm. These particles exhibit a morphology consistent with that of β-TCP particles.

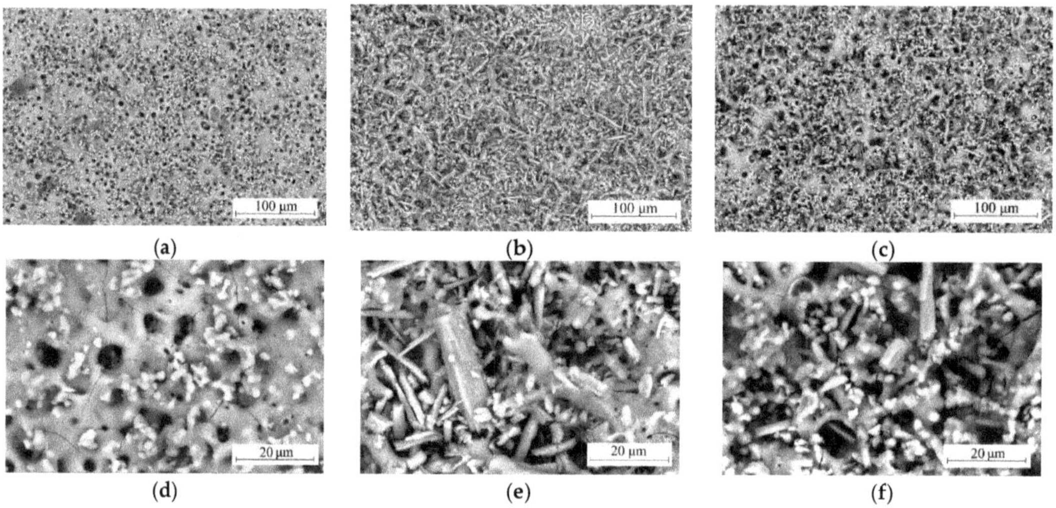

Figure 4. SEM images of CP (**a**,**d**), CS (**b**,**e**), and SP (**c**,**f**) coatings synthesized at 400 V.

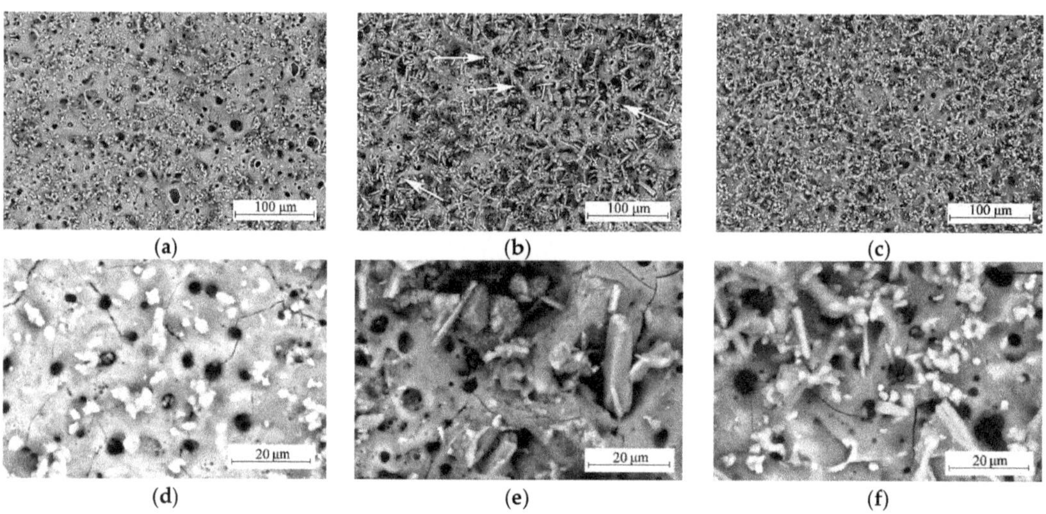

Figure 5. SEM images of CP (**a**,**d**), CS (**b**,**e**), and SP (**c**,**f**) coatings synthesized at 500 V.

The surface of CS coatings synthesized at 400 V is represented by a pronounced crystalline layer consisting of needle-shaped microparticles up to 20 μm long, characteristic of wollastonite (Figure 4b,e). At 500 V, more powerful micro-arc discharges promote intensive melting of particles on the surface of CS coatings, as well as their fusion with each other (Figure 5b—indicated by arrows).

Isometric particles similar to β-TCP and elongated particles similar to wollastonite are observed on the surface of SP coatings formed at both 400 V and 500 V (Figure 4c,f and Figure 5c,f). Increasing the voltage from 400 to 500 V leads to a noticeable decrease in the concentration of both types of particles on the surface of all coatings (Figure 5d–f), which is related to the temperature impact of intense micro-arc discharges on the particles.

Figure 6a–c presents the SEM images of the cross-sections of the coatings. It can be noted that the internal structure of the coatings is porous, with a uniform distribution of pores throughout the volume. The evaluation of the elemental composition of the coatings revealed that the elements P and Si were distributed uniformly across the entire thickness of the coatings. Furthermore, the analysis revealed an absence of a gradient in concentrations for these elements, as illustrated in Figure 6d–f.

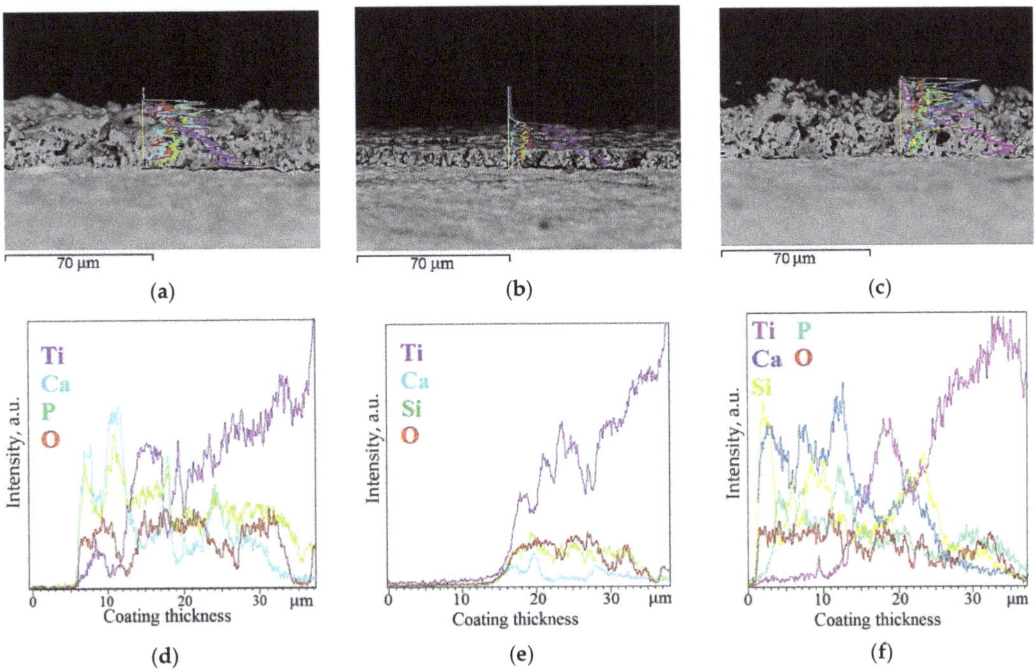

Figure 6. SEM images of coating cross-sections (**a–c**) and distribution of elements over coating cross-sections (**d–f**): (**a**,**d**)—CP coating, (**b**,**e**)—CS coating, (**c**,**f**)—SP coating; voltage during MAO process −400 V.

The increase in Ca content is observed in the direction from the substrate to the surface, which is associated with the growth of coatings and the more intense interaction of dispersed particles from the electrolyte with the formed coating. The more intense the discharges become, the more pronounced this effect is. Conversely, the Ti content exhibits an increase in the opposite direction, from the surface to the substrate.

The SEM images of coatings (Figure 7a–c) show regions 1—with isometric particles, 2—without particles (Figure 7a), 3—with needle particles, 4—without particles (Figure 7b), 5—with isometric particles, 6—with needle particles, and 7—without particles (Figure 7c). The results of the study of the elemental composition of the particles on the surface of the coatings (Figure 7d–f) indicate that they correspond to β-TCP and wollastonite, as evidenced by the high content of the elements Ca and P in the case of CP coatings and Ca and Si in the case of SP coatings.

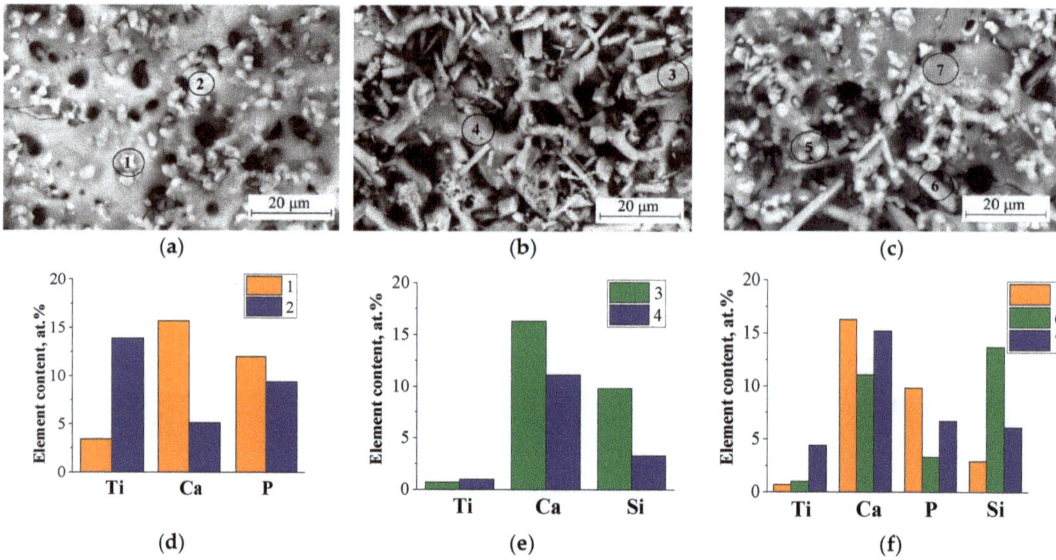

Figure 7. SEM images with highlighted investigated areas of CP (**a**), CS (**b**), and SP (**c**) coatings; histograms of chemical elements distribution on the surface of CP (**d**), CS (**e**), and SP (**f**) coatings formed at 400 V.

Table 2 presents the chemical element content of each highlighted area in Figure 7a–c.

Table 2. Content of chemical elements in highlighted areas of coatings, at. %.

Elements	1	2	3	4	5	6	7
O	67.8 ± 1.2	68.4 ± 1.1	68.1 ± 0.9	63.7 ± 1.3	69.2 ± 0.8	69.0 ± 1.2	65.4 ± 1.1
Na	1.2 ± 0.1	3.3 ± 0.9	0.7 ± 0.03	4.6 ± 0.7	4.1 ± 0.4	1.9 ± 0.6	5.4 ± 0.9
P	12 ± 0.9	9.4 ± 0.5	–	–	9.3 ± 0.7	3.3 ± 0.4	6.0 ± 0.5
Si	–	–	15.7 ± 0.5	14.6 ± 0.8	2.0 ± 0.2	13.7 ± 0.9	5.9 ± 0.3
Ca	15.7 ± 0.9	5.1 ± 0.2	14.7 ± 0.7	15.7 ± 0.6	13.5 ± 0.4	11.1 ± 0.3	10.1 ± 0.7
Ti	3.4 ± 0.2	13.9 ± 0.8	0.7 ± 0.09	2.3 ± 0.1	1.9 ± 0.08	1.0 ± 0.05	7.3 ± 0.6
Coating designator	CP		CS		SP		

It is important to note that in the case of SP coatings, P and Si are simultaneously present in both needle-like particles similar to wollastonite and isometric particles similar to β-TCP. Under the influence of high temperatures near the channels of micro-arc discharges, surface diffusion of element P into wollastonite particles and Si into β-TCP particles occurs. Thus, the structure and chemical composition of particles can undergo changes associated with the formation of new phases and compounds.

It was found that the Ca element content on the surface of CS coatings (area 4 in Figure 7b) is almost three times higher than on the surface of CP coatings (area 2 in Figure 7a). Ca is initially present in the electrolyte only as β-TCP ($Ca_3(PO_4)_2$) and wollastonite ($CaSiO_3$). The high calcium content in CS coatings can be explained by the lower melting point of wollastonite, 1540 °C [40]. In the process of MAO, more intensive melting of wollastonite particles may occur in the nearby regions of the micro-arc discharge channel.

3.3. *Phase Composition and Fine Structure of Coatings*

The phase composition of the coatings was analyzed by the X-ray diffraction method. To study the effect of the coating deposition duration on their phase composition, the voltage of the MAO process was maintained constant at 400 V. As indicated by the X-ray phase analysis (Figure 8), it was found that all the synthesized coatings have an amorphous crystalline structure, since the X-ray diffraction patterns of the coatings show both the presence of diffraction maxima indicating the presence of crystalline phases and the presence of diffuse scattering regions in the angle range of 10–45°.

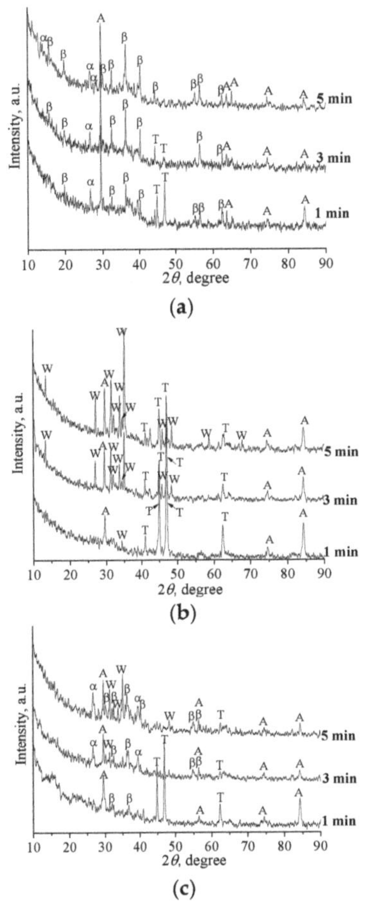

Figure 8. X-ray diffraction patterns of CP coatings (**a**), CS coatings (**b**), and SP coatings (**c**) formed at 400 V and application times of 1, 3, and 5 min: β—β-TCP, α—α-TCP, W—wollastonite, T—α-Ti, A—anatase (TiO_2).

The following were identified as crystalline phases in the coatings: β-TCP (PDF #09-0169), α-TCP (PDF #09-0348), wollastonite (PDF #43-1460), and titanium oxide (TiO_2) in anatase modification (PDF #21-1272). In addition, reflections from the α-Ti titanium substrate (PDF #44-1294) were observed in the X-ray diffraction images. The low-temperature β- TCP phase at 1125 °C transforms into α-TCP phase, which is stable in the temperature range of 1150–1470 °C. At temperatures above 1470 °C, the α-TCP phase reversibly transforms into α′-TCP [41]. There are reports that the α′-TCP compound is stable up to the temperature of 1765 °C [41].

In CP coatings, the presence of β-TCP, α-TCP, and TiO_2 (anatase) is identified. In CS coatings, the presence of wollastonite and anatase is observed. In SP coatings, β-TCP, α-TCP, wollastonite and anatase are present. More intense and narrower peaks corresponding to α-TCP are observed in the X-ray diffraction patterns of SP coatings compared to CP coatings, indicating a more intense β→α-TCP phase transition. Such a transition can only occur in regions close to the micro-arc discharge channels, where the temperature reaches a value corresponding to the transition, but no melting of the particles has yet occurred. In addition, in the X-ray diffraction patterns of CS coatings, the wollastonite peaks have a higher intensity than for SP coatings. It is important to note that TiO_2 is present in the coatings only in the anatase modification, which is metastable [42,43], indicating a short-term temperature impact during the micro-arc discharge process.

In order to understand the mechanism of coating structure formation in the dynamics of the MAO process, X-ray diffraction patterns of the coatings deposited at the MAO process voltage of 400 V for 1, 3, and 5 min were obtained (Figure 8).

The analysis of the X-ray diffraction patterns of CP coatings showed that most of the crystalline phase is formed during the initial minute of the MAO process (Figure 8a). With increasing deposition time, only the intensity of the peaks corresponding to α-, β-TCPs increases, and the intensity of the α-Ti substrate peaks decreases, which is associated with the increase in the thickness of the coatings.

Anatase and substrate material peaks are observed in the CS X-ray diffraction patterns of the coatings obtained at 1 min of the MAO process (Figure 8b). Only one small wollastonite reflex can be seen in the X-ray diffraction pattern. The other wollastonite peaks appear only after 3 min of the deposition process. This is explained by the fact that due to the specific morphology of particles and lower zeta potential value, wollastonite particles diffuse slower to the substrate/oxide layer/electrolyte interface, and in the first period of the MAO process, the coating formation proceeds with the participation of only soluble compounds, which leads to the formation of the oxide layer.

The analysis of X-ray diffraction patterns of SP coatings showed that at 1 min of the process, the first two peaks corresponding to β-TCP appear (Figure 8c). More intensive growth of crystalline compounds is observed after 3 min of the MAO process, which is confirmed by the increase in the number and intensity of the peaks corresponding to β-TCP and wollastonite.

The fine structure of the coatings obtained in all three electrolytes was investigated using transmission electron microscopy. Figure 9 shows light-field and dark-field TEM images of CP coating particle fragments as well as their microdiffraction patterns. The β-TCP and anatase phases are predominantly observed in CP coating fragments (Figure 9b). In dark-field images, their nanocrystallites of 5–10 nm in size can be observed in the reflections of (101) anatase and (214) β-TCP (Figure 9b).

3.4. *Physico-Mechanical Properties of Coatings*

In this work, the thickness (d, μm) and roughness (Ra, μm) dependences of CP, CS, and SP coatings on the voltage (U, V) of the MAO process were investigated. As illustrated in Figure 10, an increase in voltage is observed to correspond with an increase in thickness and roughness of the coatings, irrespective of the electrolyte type. However, the most prominent thickness increase (from 27 to 80 μm) is observed for CP coatings with the addition of β-TCP microparticles (Figure 10a). The higher value of current density, characteristic for the MAO process with β-TCP particles, is a contributing factor. (Figure 2b). The thicknesses of CP and SP coatings vary from 20 to 50 μm and 40 to 50 μm, respectively (Figure 10b,c). Moreover, it should be noted that the thickness value of CS coatings in the voltage range of 350–450 V varies within 20–40 μm, and only at 500 V increases to 50 μm. However, the highest roughness Ra values correspond to CS coatings (Figure 10b). As the voltage increases from 350 to 500 V in CS coatings, the roughness increases from 3.0 to 5.5 μm, which may be due to the elongated needle shape of the particles contributing to the formation of higher roughness. The wollastonite particles are deposited on the surface and form a rough relief coating, which is also observed in the SEM images (Figures 4 and 5). In the case of CP and SP coatings, the roughness values vary from 2.5 to 4.8 μm and 3.5 to 5 μm, respectively, when varying the MAO process voltage in the range of 350–500 V (Figure 10a,c).

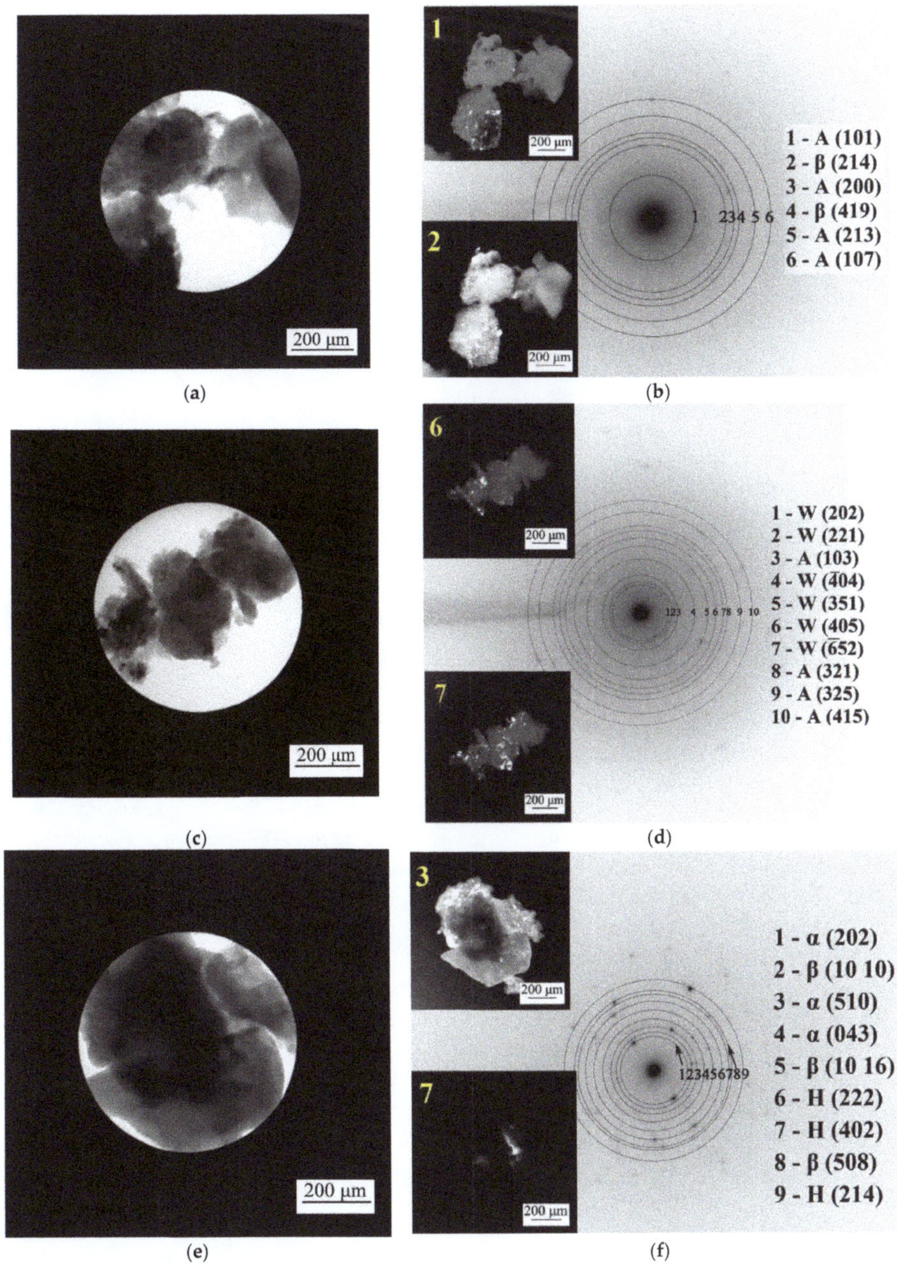

Figure 9. Bright-field TEM images of particle fragments of CP (**a**), CS (**c**), and SP (**e**) coatings, microdiffraction patterns and dark-field TEM images of particle fragments of CP (**b**), CS (**d**), and SP (**f**) coatings: A—anatase, β—β-TCP, α—α-TCP, W—wollastonite, H—hydroxyapatite.

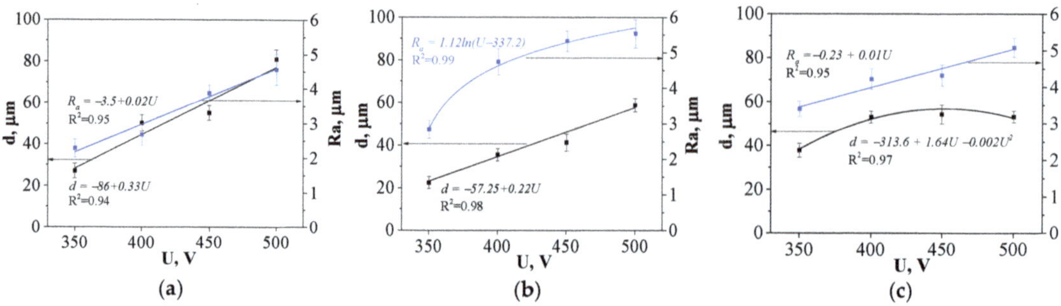

Figure 10. Graphs of thickness and roughness of CP (**a**), CS (**b**), and SP (**c**) coatings on titanium alloy versus the MAO process voltage.

The adhesion strength of the coatings was analyzed by means of scratch testing. Figure 11 shows optical images of scratches formed during scratch testing to evaluate the adhesion strength of all types of coatings formed at 500 V.

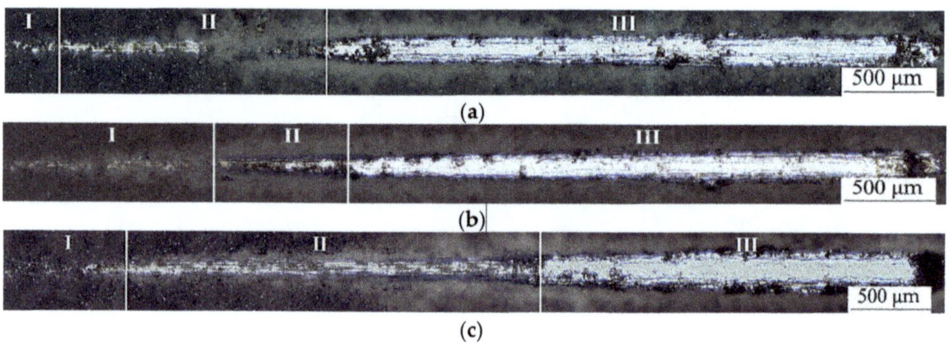

Figure 11. Optical images of scratches: (**a**) on CP coating, (**b**) on CS coating, (**c**) on SP coating, on titanium alloy at MAO voltage of 500 V.

In the context of assessing the adhesion strength of a coating, three primary categories of failure can be identified: cohesive, adhesive, and mixed [27]. Section I (Figure 11) characterizes the initial fracture (cracking) of the coating, at which cohesive deformation (fracture of the coating itself) occurs. The boundary between plots I and II corresponds to the coating spalling, which is defined as cohesive–adhesive deformation. The boundary between sections II and III corresponds to the complete detachment of the coating, which is characterized by adhesive-type deformation. The minimum load at which the coating peels from the substrate is denoted as the critical load, Lc.

As a result of analyzing the optical images of the surface of coatings with indenter scratches, it can be noted that CS coatings have the longest section I (Figure 11b). Earlier, it was revealed that CS coatings are characterized by the highest roughness value, which reaches 5.5 µm at the MAO process voltage of 500 V. Presumably, the high roughness of the coatings determines the long-term cohesive deformation. The shortest section III (complete detachment) is observed for the SP coating (Figure 11c). The data of observation of optical images of tracks after scratch testing correlate with the results of determining the critical load withstood by the coating before delamination. It was found that the highest value of critical load, equal to 22 N, is characterized by SP coatings (Figure 12). For CP and CS coatings formed at 500 V MAO process voltage, the critical load values reach 10.0 N and 10.3 N, respectively. At the same time, in the region of low values of MAO process voltage (350–400 V), the critical load for all types of coatings varies in the range of 9–12 N. Increasing the voltage up to 450–500 V does not affect the adhesion strength of CP and CS coatings, while for SP coatings an increase in the critical load up to 16–22 N is observed. This fact can be explained by the presence of two types of particles in the structure and the formation of nanocrystalline phases—α-TCP and HA, which contribute to the cohesive strength of coatings.

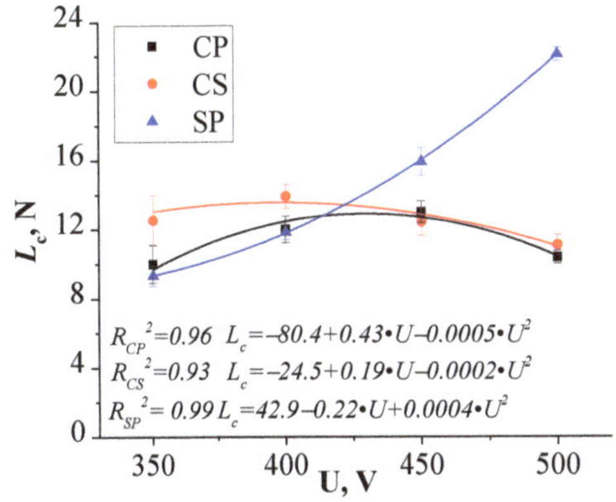

Figure 12. Dependence of the critical load measured for CP, CS, and SP coatings on the MAO process voltage.

4. Conclusions

Coatings were formed on the surface of Grade 2 titanium alloy in electrolytes of different compositions: I (with β-TCP particles)—calcium phosphate coatings (CP), II (with wollastonite particles)—calcium silicate coatings (CS), and III (with β-TCP and wollastonite particles simultaneously)—hybrid silicate–phosphate coatings (SP). β-TCP particles were present on the surface of CP coatings, wollastonite particles were present on the surface of CS coatings, and in SP coatings both particles were present simultaneously.

(1) The specific conductivity of electrolytes I and III was found to be 2.87 and 2.52 Sm/m, respectively, while the specific conductivity of electrolyte II (with wollastonite particles) took an intermediate value of 2.6 Sm/m. This phenomenon can be explained by the fact that β-TCP particles included in the electrolytes I and III are characterized by a more negative zeta potential value, which enhances their electrophoretic properties.

(2) It was found that in the medium of electrolytes with higher specific conductivity, the MAO process occurred at higher values of initial current density (0.50 and 0.58 A/cm^2, for CP and SP coatings, respectively), and a higher growth rate of coatings was observed (0.14 and 0.15 μm/s for CP and SP coatings, respectively). It is evident that with a higher specific conductivity of the electrolyte, more intense micro-arc discharges take place, and therefore, the rate of coating growth increases.

(3) The coatings were found to have an amorphous crystalline structure. The following were identified as the main crystalline phases: β-TCP, α-TCP, wollastonite, and TiO_2 (anatase).

(4) Crystallization of calcium phosphate compounds in CP and SP coatings started during the first minute of the MAO process. In CS coatings, only oxide coating was formed during the first minutes of the MAO process, which is explained by the lower mobility of wollastonite particles due to their needle shape and their lower zeta potential value.

(5) When the voltage of the MAO process was increased to 500 V, the thickness of CP, CS, and SP coatings increased to 80, 50, and 50 μm, respectively, and the roughness of these coatings by *Ra* parameter increased to 4.8, 5.5 and 5.0 μm, respectively.
(6) The values of critical load during scratch testing for CP, CS, and SP coatings synthesized at 500 V were 10.0, 10.3, and 22.0 N, respectively. Thus, the combined coatings formed in the electrolyte with the addition of β-TCP and wollastonite particles were characterized by the highest adhesion strength. Therefore, these coatings are the most promising for use in load-bearing medical implants where it is critical to prevent delamination under mechanical loads.

Author Contributions: Conceptualization, M.B.S. and A.V.U.; methodology, M.B.S. and T.V.T.; software, M.A.K., P.V.U., I.A.G. and A.D.K.; validation, M.B.S. and A.V.U.; formal analysis, P.V.U., A.V.U. and I.A.G.; investigation, A.D.K., T.V.T. and I.A.G.; resources, Y.P.S.; data curation, M.A.K., A.D.K. and T.V.T.; writing—original draft preparation, M.B.S. and A.V.U.; writing—review and editing, A.V.U. and A.D.K.; visualization, A.D.K., M.A.K. and P.V.U.; supervision, Y.P.S.; project administration, M.B.S.; funding acquisition, Y.P.S. All authors have read and agreed to the published version of the manuscript.

Funding: The work was performed in accordance with the government research assignment for ISPMS SB RAS, project FWRW-2021-0007.

Data Availability Statement: The original contributions presented in the study are included in the article; further inquiries can be directed to the corresponding author.

Acknowledgments: The authors express their appreciation for the valuable contributions of A.I. Tolmachev from the Institute of Strength Physics and Materials Science SB RAS (Tomsk, Russia) for his assistance in the preparation of experimental materials.

Conflicts of Interest: The authors declare no conflicts of interest.

Abbreviations

The following abbreviations are used in this manuscript:

MAO	Micro-arc oxidation
CP	Calcium phosphate coating
CS	Calcium silicate coating
SP	Silicate phosphate coating

References

1. Zhang, L.C.; Chen, L.Y.; Wang, L.Q. Surface modification of titanium and titanium alloys: Technologies, developments, and future interests. *Adv. Eng. Mater.* **2020**, *22*, 1901258. [CrossRef]
2. Xu, J.; Zhang, J.; Shi, Y.; Tang, J.; Huang, D.; Yan, M.; Dargusch, M.S. Surface modification of biomedical Ti and Ti alloys: A Review on current advances. *Materials* **2022**, *15*, 1749. [CrossRef] [PubMed]
3. Xiao, M.; Chen, Y.M.; Biao, M.N.; Zhang, X.D.; Yang, B.C. Bio-functionalization of biomedical metals. Mater. *Sci. Eng. C Mater. Biol. Appl.* **2017**, *70*, 1057–1070. [CrossRef]
4. Li, Y.Y.; Zou, L.M.; Yang, C.; Li, Y.H.; Li, L.J. Ultrafine-grained Ti-based composites with high strength and low modulus fabricated by spark plasma sintering. *Mater. Sci. Eng. A* **2013**, *560*, 857–861. [CrossRef]
5. Liu, C.F.; Lee, T.H.; Liu, J.F.; Hou, W.T.; Li, S.J.; Hao, Y.L.; Pan, H.B.; Huang, H.H. A unique hybrid-structured surface produced by rapid electrochemical anodization enhances bio-corrosion resistance and bone cell responses of beta-type Ti-24Nb-4Zr-8Sn alloy. *Sci. Rep.* **2018**, *8*, 6623. [CrossRef]
6. Liu, G.C.; Li, R.Y.; Liang, H.J.; Qin, Y.G. Research advance in the osteointegration of surface bioactive coating on titanium alloys. *Chin. J. Tissue Eng. Res.* **2017**, *21*, 969–974. [CrossRef]
7. Mosas, K.K.A.; Chandrasekar, A.R.; Dasan, A.; Pakseresht, A.; Galusek, D. Recent advancements in materials and coatings for biomedical implants. *Gels* **2022**, *8*, 323. [CrossRef]

8. Jugowiec, D.; Kot, M.; Moskalewicz, T. Electrophoretic deposition and characterisation of chitosan coatings on near-β titanium alloy. *Arch. Metall. Mater.* **2016**, *61*, 657–664. [CrossRef]
9. Wang, W.; Wang, Z.; Fu, Y.; Dunne, N.; Liang, C.; Luo, X.; Liu, K.; Li, X.; Pang, X.; Lu, K. Improved osteogenic differentiation of human amniotic mesenchymal stem cells on gradient nanostructured Ti surface. *J. Biomed. Mater. Res. Part A* **2020**, *108*, 1824–1833. [CrossRef]
10. Lin, Q.; Zhou, Y.; Yin, M.; Cheng, Y.; Wei, Y.; Hu, Y.; Lian, X.; Chen, W.; Huang, D. Hydroxyapatite/tannic acid composite coating formation based on Ti modified by TiO_2 nanotubes. *Colloids Surf. B Biointerfaces* **2020**, *196*, 111304. [CrossRef]
11. Bajda, S.; Liu, Y.; Tosi, R.; Cholewa-Kowalska, K.; Krzyzanowski, M.; Dziadek, M.; Kopyscianski, M.; Dymek, S.; Polyakov, A.V.; Semenova, I.P.; et al. Laser cladding of bioactive glass coating on pure titanium substrate with highly refined grain structure. *J. Mech. Behav. Biomed. Mater.* **2021**, *119*, 104519. [CrossRef] [PubMed]
12. Filipović, U.; Dahmane, R.G.; Ghannouchi, S.; Zore, A.; Bohinc, K. Bacterial adhesion on orthopedic implants. *Adv. Colloid Interface Sci.* **2020**, *283*, 102228. [CrossRef]
13. Fosca, M.; Streza, A.; Antoniac, I.V.; Vadalà, G.; Rau, J.V. Ion-doped calcium phosphate-based coatings with antibacterial properties. *J. Funct. Biomater.* **2023**, *14*, 250. [CrossRef] [PubMed]
14. Abbas, A.; Kung, H.-P.; Lin, H.-C. Effects of electrical parameters on micro-arc oxidation coatings on pure titanium. *Micromachines* **2023**, *14*, 1950. [CrossRef]
15. Yasui, T.; Hayashi, K.; Fukumoto, M. Behaviors of micro-arcs, bubbles, and coating growth during plasma electrolytic oxidation of titanium alloy. *Materials* **2023**, *16*, 360. [CrossRef]
16. Nahum, E.Z.; Lugovskoy, S.; Lugovskoy, A.; Kazanski, B.; Sobolev, A. The study of hydroxyapatite growth kinetics on CP–Ti and Ti65Zr treated by plasma electrolytic oxidation process. *J. Mater. Res. Technol.* **2023**, *24*, 2169–2186. [CrossRef]
17. Hussein, R.O.; Nie, X.; Northwood, D.O.; Yerokhin, A.; Matthews, A. Spectroscopic study of electrolytic plasma and discharging behaviour during the plasma electrolytic oxidation (PEO) process. *J. Phys. D Appl. Phys.* **2010**, *43*, 105203. [CrossRef]
18. Matykina, E.; Berkani, A.; Skeldon, P.; Thompson, G.E. Real-time imaging of coating growth during plasma electrolytic oxidation of titanium. *J. Electrochim. Acta* **2007**, *53*, 1987–1994. [CrossRef]
19. Sedelnikova, M.; Bakina, O.; Ugodchikova, A.; Tolkacheva, T.; Khimich, M.; Uvarkin, P.; Kashin, A.; Miller, A.; Egorkin, V.; Schmidt, J.; et al. The role of microparticles of β-TCP and wollastonite in the creation of biocoatings on Mg0.8Ca Alloy. *Metals* **2022**, *12*, 1647. [CrossRef]
20. Galvis, O.A.; Quintero, D.; Castaño, J.G.; Liu, H.; Thompson, G.E.; Skeldon, P.; Echeverría, F. Formation of grooved and porous coatings on titanium by plasma electrolytic oxidation in H_2SO_4/H_3PO_4 electrolytes and effects of coating morphology on adhesive bonding. *Surf. Coat. Technol.* **2015**, *269*, 238–249. [CrossRef]
21. Li, Q.; Yang, W.; Liu, C.; Wang, D.; Liang, J. Correlations between the growth mechanism and properties of micro-arc oxidation coatings on titanium alloy: Effects of electrolytes. *Surf. Coat. Technol.* **2017**, *316*, 162–170. [CrossRef]
22. Abbas, A.; Wang, T.-Y.; Lin, H.-C. Effects of electrolyte compositions and electrical parameters on micro-arc oxidation coatings on 7075 aluminum alloy. *J. Compos. Sci.* **2023**, *7*, 472. [CrossRef]
23. Bykova, A.D.; Markov, M.A.; Kuznetsov, Y.A.; Kravchenko, I.N.; Belyakov, A.N.; Makarov, A.M. Effect of electrolyte composition on the structure and tribological properties of ceramic coatings obtained by micro-arc oxidation. *Refract. Ind. Ceram.* **2024**, *64*, 541–549. [CrossRef]
24. Simchen, F.; Sieber, M.; Kopp, A.; Lampke, T. Introduction to plasma electrolytic oxidation—An overview of the process and applications. *Coatings* **2020**, *10*, 628. [CrossRef]
25. Fattah-Alhosseini, A.; Keshavarz, M.K.; Molaei, M.; Gashti, S.O. Plasma electrolytic oxidation (PEO) process on cmmercially pure Ti surface: Effects of electrolyte on the microstructure and corrosion behavior of coatings. *J. Metall. Mater. Trans. A* **2018**, *49*, 4966–4979. [CrossRef]
26. Sikdar, S.; Menezes, P.V.; Maccione, R.; Jacob, T.; Menezes, P.L. Plasma electrolytic oxidation (PEO) process—Processing, properties, and applications. *Nanomaterials* **2021**, *11*, 1375. [CrossRef] [PubMed]
27. Wang, C.; Chen, J.; He, J.; Jiang, J.; Zhang, Q. Effect of electrolyte concentration on the tribological performance of MAO coatings on aluminum alloys. *J. Front. Chem. Eng.* **2020**, *14*, 1065–1071. [CrossRef]
28. Fattah-Alhosseini, A.; Molaei, M.; Babaei, K. Influence of electrolyte composition and voltage on the microstructure and growth mechanism of plasma electrolytic oxidation (PEO) coatings on tantalum: A Review. *J. Anal. Bioanal. Electrochem.* **2020**, *12*, 517–535.
29. Sandhyarani, M.; Prasadrao, T.; Rameshbabu, N. Role of electrolyte composition on structural, morphological and in-vitro biological properties of plasma electrolytic oxidation films formed on zirconium. *J. Appl. Surf. Sci.* **2014**, *317*, 198–209. [CrossRef]
30. Sedelnikova, M.B.; Ugodchikova, A.V.; Uvarkin, P.V.; Chebodaeva, V.V.; Tolkacheva, T.V.; Schmidt, J.; Sharkeev, Y.P. Structural-morphological and adhesive properties of calcium-phosphate coatings formed on a magnesium alloy by micro-arc oxidation in an electrolyte containing disperse particles. *Russ. Phys. J.* **2021**, *64*, 830–837. [CrossRef]

31. Kashin, A.D.; Sedelnikova, M.B.; Chebodaeva, V.V.; Uvarkin, P.V.; Luginin, N.A.; Dvilis, E.S.; Kazmina, O.V.; Sharkeev, Y.P.; Khlusov, I.A.; Miller, A.A.; et al. Diatomite-based ceramic biocoating for magnesium implants. *Ceram. Int.* **2022**, *48*, 28059–28071. [CrossRef]
32. Awasthi, S.; Pandey, S.K.; Arunan, E.; Srivastava, C. A Review on hydroxyapatite coatings for biomedical application: Experimental and theoretical perspectives. *J. Mater. Chem. B* **2021**, *9*, 228–249. [CrossRef]
33. Mamaeva, A.; Kenzhegulov, A.; Panichkin, A.; Abdulvaliyev, R.; Fischer, D.; Bakhytuly, N.; Toiynbaeva, N. Influence of current duty cycle and voltage of micro-arc oxidation on the microstructure and composition of calcium phosphate coating. *Coatings* **2024**, *14*, 766. [CrossRef]
34. Molaei, M.; Fattah-alhosseini, A.; Nouri, M.; Nourian, A. Systematic optimization of corrosion, bioactivity, and biocompat-ibility behaviors of calcium-phosphate plasma electrolytic oxidation (PEO) coatings on titanium substrates. *Ceram. Int.* **2022**, *48*, 6322–6337. [CrossRef]
35. Sayed, M.; Mahmoud, E.M.; Bondioli, F.; Naga, S.M. Developing porous diopside/hydroxyapatite bio-composite scaffolds via a combination of freeze-drying and coating process. *Ceram. Int.* **2019**, *45*, 9025–9031. [CrossRef]
36. Bragg, W.L.; Claringbull, G.F. *Crystal Structure of Minerals*; Cornell University Press: Ithaca, NY, USA, 1965; Volume 4.
37. Palakurthy, S.; Reddy, K.V.G.; Samudrala, R.K.; Azeem, P.A. In vitro bioactivity and degradation behaviour of β-wollastonite derived from natural waste. *Mater. Sci. Eng. C* **2019**, *98*, 109–117. [CrossRef]
38. Sedelnikova, M.B.; Ugodchikova, A.V.; Tolkacheva, T.V.; Chebodaeva, V.V.; Cluklhov, I.A.; Khimich, M.A.; Bakina, O.V.; Lerner, M.I.; Egorkin, V.S.; Schmidt, J.; et al. Surface modification of $Mg_{0.8}Ca$ alloy via wollastonite micro-arc coatings: Significant improvement in corrosion resistance. *Metals* **2021**, *11*, 754. [CrossRef]
39. Sedelnikova, M.B.; Mayer, V.V.; Bakina, O.V.; Kashin, A.D.; Uvarkin, P.V.; Khimich, M.A.; Luginin, N.A.; Glukhov, I.A.; Tolkacheva, T.V.; Ugodchikova, A.V.; et al. Antibacterial amorphous-crystalline coatings based on wollastonite and ZnO particles. *Crystals* **2024**, *14*, 886. [CrossRef]
40. Osborn, E.F.; Muan, A.; Levin, E.M.; Robbins, C.R.; McMurdie, H.F. *Phase Diagrams for Ceramists*; American Ceramic Society: Columbus, OH, USA, 1964; p. 598.
41. Brazete, D.; Torres, P.M.C.; Abrantes, J.C.C.; Ferreira, J.M.F. Influence of the Ca/P ratio and cooling rate on the allotropic $\alpha \leftrightarrow \beta$-tricalcium phosphate phase transformations. *J. Ceram. Int.* **2018**, *44*, 8249–8256. [CrossRef]
42. Kim, S.A.; Hussain, S.K.; Abbas, M.A.; Bang, J.H. High-temperature solid-state rutile-to-anatase phase transformation in TiO_2. *J. Solid State Chem.* **2022**, *315*, 123510. [CrossRef]
43. Karunadasa, K.S.P.; Manoratne, C.H. Microstructural view of anatase to rutile phase transformation examined by in-situ high-temperature X-ray powder diffraction. *J. Solid State Chem.* **2022**, *314*, 123377. [CrossRef]

crystals

Review

Recent Advances in Biodegradable Magnesium Alloys for Medical Implants: Evolution, Innovations, and Clinical Translation

Mykyta Aikin [1], Vadim Shalomeev [1], Volodymyr Kukhar [2], Andrii Kostryzhev [3,*], Ihor Kuziev [4], Viktoriia Kulynych [5], Oleksandr Dykha [6], Volodymyr Dytyniuk [6], Oleksandr Shapoval [5], Alvydas Zagorskis [7], Vadym Burko [8], Olha Khliestova [8], Viacheslav Titov [9] and Oleksandr Hrushko [10]

1 Department of Physical Materials Science, National University Zaporizhzhia Polytechnic, 69011 Zaporizhzhia, Ukraine; fitone14@gmail.com (M.A.); shalomeev@zp.edu.ua (V.S.)
2 Department of Metallurgy and Production Organization, Technical University "Metinvest Polytechnic" LLC, 69008 Zaporizhzhia, Ukraine; volodymyr.kukhar@mipolytech.education
3 Centre for Microscopy and Microanalysis, University of Queensland, St. Lucia, Brisbane, QLD 4072, Australia
4 Department of Transport Technologies, Kremenchuk Mykhailo Ostrohradskyi National University, 39600 Kremenchuk, Ukraine; ikuzev@kdu.edu.ua
5 Department of Mechanical Engineering, Kremenchuk Mykhailo Ostrohradskyi National University, 39600 Kremenchuk, Ukraine; vkulinich@kdu.edu.ua (V.K.); shapoval_a@kdu.edu.ua (O.S.)
6 Department of Tribology, Automobiles and Materials Science, Khmelnytskyi National University, 29000 Khmelnytskyi, Ukraine; duhaov@khmnu.edu.ua (O.D.); vodytyniuk@khmnu.edu.ua (V.D.)
7 Department of Environmental Protection and Water Engineering, Vilnius Gediminas Technical University, 10223 Vilnius, Lithuania; alvydas.zagorskis@vilniustech.lt
8 Department of Heat Power Engineering and Environmental Protection, Pryazovskyi State Technical University, 49044 Mariupol, Ukraine; burko_v_a@pstu.edu (V.B.); hlestova_o_a@pstu.edu (O.K.)
9 Department of Aircraft Manufacturing Technology, National Technical University of Ukraine "Igor Sikorsky Kyiv Polytechnic Institute", 03056 Kyiv, Ukraine; vat.kpi@gmail.com
10 Department for Integrated Sensor Systems, University for Continuing Education Krems (Danube University Krems), 2700 Wiener Neustadt, Austria; oleksandr.hrushko@donau-uni.ac.at
* Correspondence: a.kostryzhev@uq.edu.au

Abstract

Biodegradable magnesium alloys have emerged as promising alternatives to permanent metallic implants due to their unique combination of mechanical compatibility with bone and complete resorption, addressing the persistent issues of stress shielding and secondary removal surgeries. This review critically examines the historical development of magnesium-based biomaterials, highlighting advances in alloy design, manufacturing processes, and surface engineering that now enable tailored degradation and improved clinical performance. Drawing on recent clinical and preclinical studies, we summarize improvements in corrosion resistance, mechanical properties, and biocompatibility that have supported the clinical translation of magnesium alloys across a variety of orthopedic and emerging medical applications. However, challenges remain, including unpredictable in vivo degradation kinetics, limited long-term safety data, lack of standardized testing protocols, and ongoing regulatory uncertainties. We conclude that while magnesium-based biomaterials have advanced from experimental concepts to clinically validated solutions, further progress in personalized degradation control, real-time monitoring, and harmonized regulatory frameworks is needed to fully realize their transformative clinical potential.

Crystals **2025**, *15*, 671 https://doi.org/10.3390/cryst15080671

Keywords: biodegradable magnesium alloys; orthopedic implants; corrosion resistance; surface engineering; clinical translation; rare-earth-free alloys; additive manufacturing; regulatory challenges; biocompatibility; smart medical devices

1. Introduction

Medical implants made from titanium and stainless steel have saved countless lives, but they create a fundamental problem: these permanent metals are much stiffer than human bone. This mismatch causes "stress shielding"—where the implant bears most mechanical loads, causing the surrounding bone to weaken and deteriorate over time [1,2]. Additionally, these foreign materials trigger chronic inflammation, infection risks, and require surgical removal in 30% of patients, particularly in growing children [3,4]. These limitations have motivated researchers to develop implants that temporarily support healing tissue and then safely dissolve away—eliminating the need for removal surgery [5,6].

Magnesium emerges as an ideal candidate for such "biodegradable" implants. As the human body's fourth most abundant mineral, magnesium naturally comprises 50–60% of bone content and participates in over 300 biological processes [7–9]. Its mechanical stiffness closely matches that of bone—unlike titanium, which is 4–29 times stiffer—reducing stress shielding while maintaining adequate strength [4,10]. Most importantly, when magnesium implants degrade, the body can safely process and excrete the byproducts through normal kidney function, avoiding permanent foreign material accumulation [11].

The concept is not new. In 1878, surgeon Edward Huse used magnesium wire for blood vessel surgery [12,13], and in 1906, Albin Lambotte implanted magnesium plates to fix broken bones [4,14]. However, these early attempts failed spectacularly—the implants dissolved completely within 8 days, producing hydrogen gas bubbles that interfered with healing [15,16]. For perspective, these early materials corroded at rates exceeding 400 mm per year, while successful bone healing requires implants to maintain strength for at least 3–6 months [7,17]. These failures discouraged clinical use for nearly a century.

Recent advances in alloy development and clinical practice have contributed to the increasing clinical application of magnesium implants, such as MAGNEZIX® screws, which have now been used in over 25,000 patients with degradation times that approximate bone healing timelines [18,19]. Advanced alloy compositions like Mg-2Zn-1Mn achieve degradation rates of just 0.36 mm per year, representing a 100–1000 fold improvement over historical materials [20]. Manufacturing innovations using laser-based 3D printing can now create implants with microscopic grain structures (1–3 μm), which may contribute to improved strength and corrosion resistance [21,22].

Despite these advances, significant scientific debates persist. Researchers disagree about using rare-earth elements (like neodymium or yttrium) in alloy compositions—some argue these elements enhance performance and remain safe at low concentrations [23,24], while others advocate complete avoidance due to potential accumulation in organs over time [25]. Another major challenge involves testing reliability: the same material can show degradation rates varying by 10-fold depending on whether it is tested in laboratory dishes or living tissue [26,27]. Furthermore, the lack of standardized testing methods and inconsistent regulatory requirements across countries hampers global adoption—only eight magnesium implant designs have received regulatory approval worldwide as of 2023 [22].

This review critically examines how biodegradable magnesium implants evolved from repeated failures to emerging clinical success while identifying remaining barriers to widespread adoption. We systematically analyze (1) historical lessons from early failures that informed modern solutions; (2) recent innovations including new alloys achieving bone-like strength (>315 MPa) [28], "smart" protective coatings that self-repair damage [29], and manufacturing precision with less than 2% variation between batches [30]; (3) unresolved challenges including the absence of patient safety data beyond 5 years, unpredictable patient-to-patient variations in degradation, and lack of international standards; and (4) future opportunities for "intelligent" implants containing sensors to monitor healing or release therapeutic drugs on demand. Our analysis indicates that biodegradable magnesium is progressing from experimental research toward broader clinical application. Remaining challenges are largely engineering- and regulation-related, rather than purely scientific. Continued collaborative efforts in personalized medicine, real-time monitoring, and regulatory harmonization may further expand the potential applications of magnesium-based implants in orthopedics and other medical fields.

2. Historical Context and Challenges of Biodegradable Magnesium Alloys in Medical Implants

The vision of magnesium as the ideal biodegradable implant material emerged over a century ago when pioneering surgeons recognized its unique potential: a metal that could provide structural support during healing and then gracefully disappear as natural tissue took over. Yet, this compelling concept remained frustratingly elusive for generations, with early attempts resulting in spectacular failures—implants that dissolved within days and generated alarming gas bubbles. Only after decades of persistent scientific advancement has magnesium evolved from an experimental material to one with increasing clinical acceptance and improved control over its properties. This chapter discusses the historical evolution from Lambotte's pioneering work to modern commercial success, the fundamental advantages of magnesium's biological compatibility, and the critical lessons learned from early failures that enabled today's breakthrough achievements.

2.1. Clinical Need for Biodegradable Implants

The development of biodegradable implants is driven by the need to overcome the mechanical incompatibility between traditional metallic implants and bone tissue, a mismatch that often leads to stress shielding and subsequent bone loss [1,2]. Magnesium alloys achieve a more balanced clinical performance than titanium alloys, stainless steel, polymers, or biocomposites, particularly excelling in modulus match, density compatibility, and biodegradability.

Magnesium alloys rate highest for matching the mechanical properties and density of bone [3,5–8,10,31,32] while also offering good strength, biocompatibility, and full biodegradability (see Table 1). In contrast, titanium and stainless steel exhibit poor compatibility in these areas and lack biodegradability, often requiring secondary removal surgery [3,7,8].

Table 1. Comparison of common orthopedic implant materials.

Property	Magnesium Alloys	Titanium Alloys	Stainless Steel
Elastic Modulus	Good	Fair	Poor
Density Match to Bone	Excellent	Poor	Poor
Strength	Moderate (sufficient for many indications)	High	High
Biocompatibility	Good (no major toxicity, favorable)	Good (no major toxicity)	Good (no major toxicity)
Biodegradability	Complete resorption	Non-biodegradable	Non-biodegradable
No Removal Surgery	Yes	No	No

Note: Mechanical property values are given in Section 3.2. References: [3–6,8–14,33–41].

Beyond mechanical problems such as stress shielding, permanent implants are associated with chronic complications, including inflammation, infection, allergic reactions, and imaging artifacts [3,4,6]. The capacity of magnesium alloys for complete biodegradation eliminates the need for removal surgery and addresses a major limitation of conventional implant materials [3,4].

2.2. Magnesium's Biological Foundation

Figure 1 illustrates magnesium's unique position as the body's fourth most abundant cation, providing fundamental advantages over inert metals detailed in Table 2. The body distribution analysis shows that 55% of total magnesium resides in bone tissue (11.5–16.5 g), with additional stores in soft tissues (27%, 5.7–8.1 g) and muscle (18%, 3.8–5.4 g). This natural abundance, combined with magnesium's role as cofactor for 300–325 enzymatic reactions, enables safe renal excretion of degradation products unlike permanent foreign materials [7–9,11].

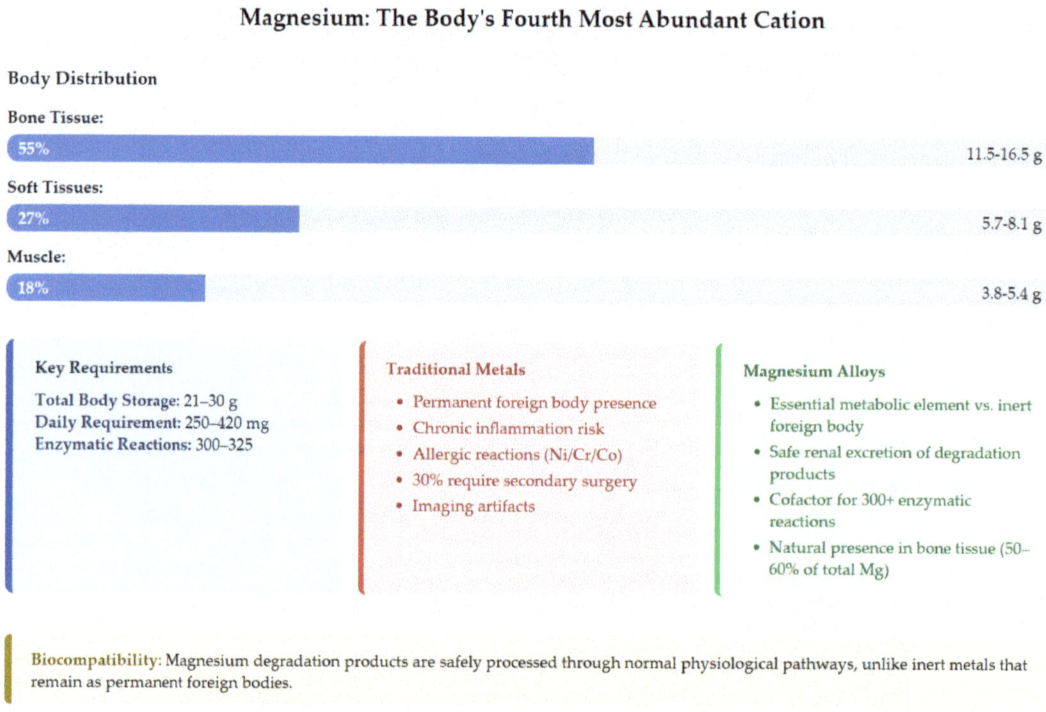

Figure 1. Magnesium distribution in the human body and clinical advantages comparison, highlighting its essential metabolic role versus traditional metals' permanent foreign body presence.

Table 2. Clinical advantages comparison.

Feature	Traditional Metals	Magnesium Alloys	Clinical Impact
Post-healing presence	Permanent	Complete resorption	Eliminates removal surgery
Secondary surgery rate	30% (general), 10–20% (pediatric)	Rare/Never	Reduced morbidity/cost
Biological role	Inert foreign body	Essential metabolic element	Enhanced integration
Stress shielding	Severe (4–29× stiffness)	Minimal (1.3–6× stiffness)	Preserved bone density

References: [3,4,7,10,11,33–36].

The clinical advantages comparison in Figure 1 emphasizes key benefits outlined in Table 2: magnesium serves as an essential metabolic element versus inert foreign body presence, enables safe renal excretion of degradation products, acts as a cofactor for 300+ enzymatic reactions, maintains natural presence in bone tissue (50–60% of total body magnesium), and eliminates the 30% secondary surgery rate associated with traditional metals [3–5,8,34,37–40].

2.3. Historical Development: From Early Pioneers to Modern Success

Magnesium's clinical history began with Edward C. Huse's 1878 use of magnesium wire ligatures for vascular control [12,13,41]. Lambotte's landmark 1906 application of magnesium plates for tibial fractures marked the first orthopedic use, though complete degradation within 8 days revealed the fundamental challenge of degradation control that would persist for over a century [4,11,14,15].

Historical applications consistently demonstrated excessive degradation rates that undermined clinical utility. Pure magnesium exhibited complete dissolution in 10–14 days in muscle tissue, with some cases showing total resorption in 3–5 weeks under acidic conditions [4,41]. The fundamental degradation mechanism produces both ionic and gaseous byproducts through the reaction: $Mg + 2H_2O \rightarrow Mg(OH)_2 + H_2\uparrow$, generating approximately 1 mL of hydrogen gas per 1 mg of degraded magnesium [5,7,16,17].

The historical progress timeline in Section 2.3 illustrates a substantial extension in degradation times, from 8 days in early 20th-century applications to controlled degradation periods of 3–24 months in more recent studies. Despite frequent gas cavity formation and inflammatory responses in early applications, infection rates remained surprisingly low even under pre-antibiotic conditions, suggesting inherent "biotolerance" properties [4,42].

2.4. Early Clinical Failures and Critical Lessons

Early magnesium alloys failed due to excessively rapid degradation that preceded tissue healing. Pure magnesium demonstrated degradation rates of 407 mm/year and corrosion currents reaching 86.06 $\mu A/cm^2$ in simulated body fluid—far exceeding the target <0.5 mm/year for orthopedic applications [4,7,43]. Clinical studies showed 50% strength loss within 4 weeks, compromising structural integrity before sufficient healing [2,3,43].

Hydrogen gas production created significant complications, with acceptable evolution rates below 0.01 $mL/cm^2/day$ to prevent subcutaneous gas pockets [38,44]. X-ray imaging revealed gas accumulation at bone–implant interfaces persisting up to 8 weeks post-implantation [45]. Concurrent alkalization raised local pH above 11.5, while pH values reached 8.70 in culture media under high-Mg conditions—substantially above physiological levels [46,47].

High early release of Mg^{2+} ions reduced cell viability by >30% according to ISO 10993-5:2009 [48] thresholds, with human mesenchymal stem cells tolerating concentrations only up to 8.0×10^{-4} M before ATP production declined [47,49]. Some experimental conditions reached 4.1×10^{-2} M—well above physiological tolerance [47]. These unpredictable responses, combined with lack of standardized evaluation protocols, created substantial regulatory hurdles that delayed market entry for decades [2,4,8].

2.5. Modern Commercial Achievements

Contemporary success demonstrates the validity of historical lessons learned, with dramatic improvements shown in Table 3. MAGNEZIX® procedures now exceed 25,000 cases with 36-month follow-up data showing successful bone integration. The bone–implant contact measurements of 0.42–0.61 versus 0.24–0.35 for titanium represent quantifiable improvements in osseointegration [38,50].

Table 3. Historical vs. modern performance comparison.

Parameter	Historical (Pre-1950)	Modern (2013-Present)	Improvement Factor
Degradation Rate	Complete in 8 days–5 weeks	0.02–0.31 mm/year	100–1000× control
Tensile Strength	Insufficient data	Up to 365 MPa	Bone-compatible
Clinical Cases	Anecdotal reports	25,000+ documented	Systematic validation
Gas Evolution	Uncontrolled cavities	<0.01 mL/cm^2/day	Managed evolution
Regulatory Status	No approvals	Multiple CE marks, FDA IDEs	Commercial viability

References: [4,7,14,33,38,44].

MAGNEZIX® screws (Syntellix AG, Hannover, Germany) became the first CE-certified resorbable metallic orthopedic implants in 2013, demonstrating controlled resorption and bone replacement [3–5,33,38,51]. K-MET screws (K-MET Co., Ltd., Seoul, Republic of Korea) achieved complete healing and resorption at 6–12 months in 53 patients, while cardiovascular applications like AMS stents demonstrated safe 4-month degradation [3,37].

MAGNEZIX® is highlighted here as an illustrative example of a clinically successful magnesium-based implant. Direct comparison with alternative biomaterials such as titanium alloys or polymers is not the primary focus of this review as magnesium-based materials are fundamentally distinct in their combination of biodegradability, osteoconductivity, and mechanical compatibility with bone. A comparative evaluation among different magnesium alloys and products would provide valuable insights but would require comprehensive data collection across manufacturers and clinical studies—a substantial undertaking that could be the subject of future dedicated reports.

2.6. Key Takeaways: Foundation for Modern Development

The historical journey from 1878 to the present, illustrated through the performance evolution shown in Section 2, demonstrates that transformative biomaterial innovation requires persistent effort across multiple decades. Magnesium alloys are approaching a favorable balance between degradation rate (3–24 months), mechanical compatibility (1.3–6× better modulus match), and reduced need for secondary surgery (see Figure 2).

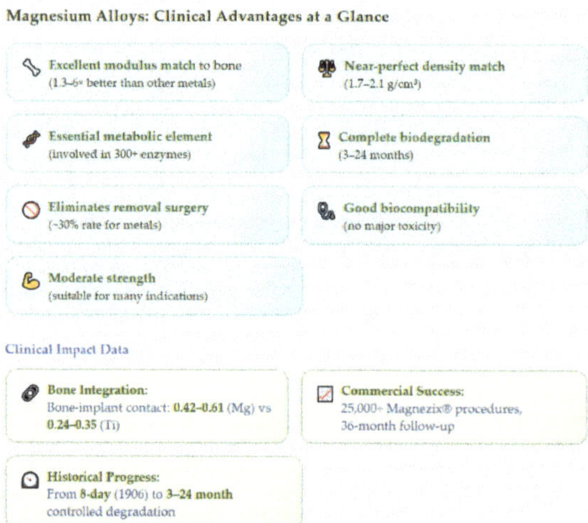

Figure 2. Clinical advantages and impact of magnesium alloys in orthopedic implants [3,4,7–9,14,18,19,38,50].

Modern magnesium implants achieve controlled degradation rates (0.02–0.31 mm/year), predictable mechanical properties (up to 365 MPa tensile strength), and excellent clinical outcomes (>25,000 successful procedures), as documented in the clinical impact data. However, the transition from experimental promise to widespread clinical adoption continues to require addressing challenges in degradation predictability, manufacturing consistency, regulatory standardization, and application expansion.

These historical foundations, quantified through the comprehensive performance comparisons, establish the scientific and clinical basis for examining persistent challenges (Section 3), major recent advances (Section 4), and future directions that will guide the field toward widespread clinical adoption (Section 5).

3. Persistent and Emerging Challenges

The development of biodegradable magnesium implants has advanced beyond proof-of-concept to reveal complex interconnected challenges spanning materials science, manufacturing, and regulatory domains that must be addressed for clinical translation.

This chapter examines representative obstacles limiting magnesium-based biodegradable implant development, though these examples do not encompass the full scope of challenges in this evolving field. Key barriers are organized into five domains: material degradation control and environmental sensitivity; mechanical property limitations; safety and biocompatibility concerns; manufacturing constraints; and regulatory complexities. Understanding these interconnected technical barriers is essential for identifying viable pathways toward clinical implementation.

3.1. Material-Related Challenges

As established in Section 2, magnesium alloys offer superior compatibility with bone and key advantages over traditional metallic implants. However, their clinical adoption is limited by significant material-related challenges—chief among them, unpredictable degradation kinetics.

Precisely controlling the degradation rates of magnesium implants remains difficult, with aggressive corrosion observed especially in additively manufactured magnesium scaffolds. These scaffolds can reach Mg^{2+} ion concentrations incompatible with typical bone healing timelines [52].

Degradation rates vary dramatically across testing methodologies and alloy systems, as shown in Figure 3. For instance, rates can differ by nearly an order of magnitude between immersion and electrochemical tests [26,27]. Clinical studies show extreme variability: AZ31 (1.17 mm/year), AZ91D (1.38 mm/year), LAE442 (1.01 mm/year), and Mg-6Zn (2.32 mm/year) in animal models, often exceeding optimal healing windows [8].

Moreover, scaffold geometry plays a decisive role in early degradation behavior. Certain architectures, such as diamond-structured scaffolds, exhibit more rapid initial corrosion compared to biomimetic or gyroid designs, although degradation rates tend to converge over time. This geometry-dependent variability complicates the development of patient-specific implants, where design optimization is critical.

Environmental factors also play a significant role in magnesium degradation. Physiological temperatures (37 °C) double degradation rates compared to room temperature (20 °C), and increased chloride concentrations—common in the body—accelerate corrosion by converting protective $Mg(OH)_2$ layers to $MgCl_2$ [50,53]. Additionally, hydrogen gas generated during magnesium corrosion can persist for weeks and may interfere with tissue integration [5,8,54].

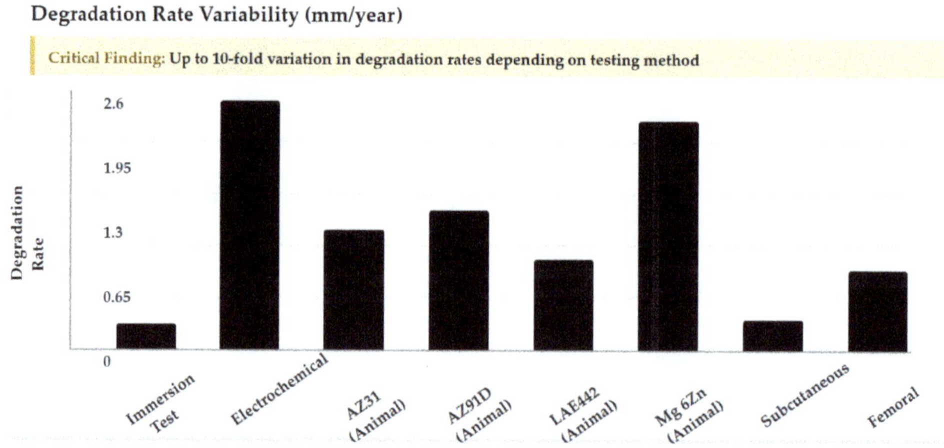

Figure 3. Degradation rate variability showing up to 10-fold variation in degradation rates depending on testing method, from immersion tests to electrochemical evaluations across different alloy systems.

3.2. Mechanical Limitations and Structural Integrity

Although Section 2 established that magnesium alloys have a modulus closer to bone than traditional metals, their absolute mechanical strength remains a limiting factor for many applications.

Pure magnesium has an elastic modulus of ~45 GPa and a yield strength of about 130 MPa. Even improved magnesium alloys display tensile strengths ranging from 150 to 400 MPa and elastic modulus values of 41–45 GPa, with elongation capacities of 2–20%. In contrast, stainless steel and titanium alloys have much higher tensile strengths (480–834 MPa and 550–1400 MPa, respectively) and moduli (200–230 GPa for steel and 110–114 GPa for titanium) [4] (see Table 4).

Table 4. Mechanical properties comparison of magnesium alloys vs. traditional materials [3,4,7,10,11,33–36].

Material	Density (g/cm^3)	Elastic Modulus (GPa)	Tensile Strength (MPa)	Elongation (%)
Pure Mg	1.74	45	130–220	6.4
Mg Alloys	1.74–2.0	41–45	150–365	2–20
Stainless Steel	7.9–8.1	200–230	480–834	12–50
Ti Alloys	4.4–4.5	110–114	550–1400	8–15
Cortical Bone	1.8–2.1	15–30	130–180	1–3

The mechanical performance of AM magnesium scaffolds is highly dependent on their architecture. For example, with 75% porosity and 800 μm pore size, gyroid designs achieve a plateau stress of 32.34 ± 1.36 MPa and a modulus of 0.760 ± 0.020 GPa, while biomimetic designs achieve only 7.47 ± 0.56 MPa and 0.207 ± 0.018 GPa (see Table 5). These values, while promising for certain applications (porous, bone-mimicking structures), remain well below the strength of dense metals and only approach the lower range of cancellous bone, highlighting the ongoing challenge for load-bearing orthopedic sites.

Table 5. Mechanical properties of AM magnesium scaffolds by architecture [4].

Architecture	Plateau Stress (MPa)	Young's Modulus (GPa)	Porosity (%)
Gyroid	32.34 ± 1.36	0.760 ± 0.020	75
Diamond	16.20 ± 0.66	0.466 ± 0.035	75
Biomimetic	7.47 ± 0.56	0.207 ± 0.018	75

3.3. Safety and Biocompatibility Concerns

Material Availability and Toxicity Constraints: Only pure magnesium and WE43 powders are widely available for additive manufacturing, reflecting regulatory restrictions due to toxicity. AZ91D is discouraged for clinical use because its 9 wt% aluminum content is linked to neurotoxicity and Alzheimer's disease concerns [4,50,55].

Rare-Earth Element Toxicity: Alloys like WE43 contain rare-earth elements (REEs) such as neodymium and yttrium, raising biocompatibility questions due to limited long-term safety data [23,25]. Natural background levels of REEs are below 47 μg in humans, but full implant degradation can lead to accumulation in bone and liver [23]. To stay within safety guidelines (daily intake below 4.2 mg), neodymium and yttrium in a 6.9 g implant should not exceed 5 wt% if fully degraded in three months. Short-term studies show low cytotoxicity at these levels, but long-term monitoring remains essential (see Table 6) [5,56].

Table 6. Toxicity Thresholds for common alloying elements [23–25,53,57,58].

Element	TD_{50} (mol/L)	Classification	Daily Limit (mg/day)	Impurity Limit (ppm)	Clinical Concern
Aluminum	-	Moderate–Severe	1.0	-	Neurotoxicity, Alzheimer's
Yttrium	2.54×10^{-4}	Moderate	0.016	-	Bone/liver accumulation
Zinc	9.28×10^{-5}	Severe	15–40	-	Essential but toxic at high levels
Nickel	4.15×10^{-5}	Severe	-	20–50	Cytotoxicity, allergenic
Copper	4.59×10^{-5}	Severe	2–3	100–300	Corrosion acceleration
Iron	-	Moderate	-	35–50	Cathodic acceleration
Manganese	-	Moderate	2–5	-	Neurological effects at high doses
Lithium	-	Moderate	0.5–2.0	-	Therapeutic but narrow window
Total REE	-	Moderate	4.2	-	Accumulation in bone/liver

Impurity Control Requirements: Iron (35–50 ppm), nickel (20–50 ppm), and copper (100–300 ppm) must be tightly controlled as these impurities can increase cytotoxicity and accelerate corrosion [59] (see Table 6). Advanced refining methods, like carbon-based filters, help reduce impurities [60], but consistently achieving these low levels is technically demanding and energy-intensive [5,53,61].

3.4. Manufacturing and Processing Challenges

3.4.1. Production and Reproducibility Limitations

Gas Atomization Hazards: Producing magnesium powder is hazardous due to its high reactivity with oxygen and moisture, creating explosion risks. This limits supplier availability and results in low yields. From the wide particle size range produced, only the 20–70 μm fraction is suitable for additive manufacturing, increasing costs and limiting scalability [5]

Process Instability: Magnesium's low vapor temperature and high vapor pressure cause powder splash during processing, leading to defects such as porosity, inclusions, and microstructural irregularities. These defects reduce mechanical strength and corrosion resistance [5]. As an alternative, advanced forging techniques—such as upsetting of large ingots with ledges—have been investigated to enhance internal uniformity and mitigate typical casting-related defects [62].

Coating Process Variability: Micro-arc oxidation coatings show uneven thickness (18–22 μm) due to sensitivity to substrate microstructure, with corrosion current reductions varying by 1–6 orders of magnitude based on processing conditions [33]. PLA coatings also exhibit significant thickness variation (1.6–41.8 μm) depending on application technique and cycle number [63].

Since many of these inconsistencies are thermally driven, the implementation of dynamic temperature control systems—like those applied in precision drawing technologies—could significantly improve process stability and coating uniformity [64].

3.4.2. Customization and Complex Geometry Constraints

Traditional Manufacturing Limitations: Conventional casting and machining cannot achieve precision required for complex, interconnected porous architectures essential for bone-mimetic scaffolds. While selective laser melting enables the creation of these complex structures, reproducibility and quality control for such intricate geometries remain unresolved [4].

Surface Treatment Limitations: Most coating development focuses on simple geometries, with effectiveness on complex, patient-specific shapes largely untested [55]. Surface treatments lack validated protocols for irregular surfaces common in personalized implants. Insights from superplastic forming simulations and surface deformation analyses, particularly in shell-type geometries, contribute to addressing this limitation and optimizing treatment uniformity [65–67].

3.5. Clinical Translation and Regulatory Challenges

3.5.1. Inconsistent Experimental and Clinical Outcomes

Environmental Testing Variability: Pure magnesium shows degradation rates from 0.26 mm/year (immersion) to 2.52 mm/year (electrochemical)—a ten-fold difference based solely on methodology [26,27]. In vivo variability spans from 0.33 mm/year subcutaneously to 0.86 mm/year in femoral implantation [50].

Cell Viability Inconsistencies: Mg-1Zn-Mn extruded alloy achieved 100% viability, while Mg-1Sr as-rolled showed 50–84% viability [43]. Corrosion current density reductions span three orders of magnitude between studies (10^{-4} to 10^{-7} A/cm^2) [68–71].

3.5.2. Insufficient Long-Term Biocompatibility Data

Limited Study Duration: Most biocompatibility assessments of biodegradable magnesium alloys are based on short- to mid-term animal studies, typically ranging from 4 to 24 weeks [72–75], with only a few studies extending up to one year or beyond [76]. Human clinical data remain scarce. As a result, there is a lack of comprehensive data on the complete long-term degradation profile and the full spectrum of tissue responses, including potential systemic effects and accumulation of alloying elements, especially for periods exceeding one year post-implantation [76].

Coating System Longevity: The protective effect of common surface coatings on magnesium alloys remains limited in duration. MgF_2 coatings offer short-term corrosion resistance but are reported to degrade completely within 4 weeks, after which long-term biological responses remain unclear [33]. Hydroxyapatite (HA) coatings typically provide in vivo protection for about 8 weeks before rapid substrate degradation resumes [55]. Similarly, calcium phosphate (CaP) coatings on ZK60 alloy maintain corrosion resistance for approximately 29 days, after which their protective effect is lost [6].

Rare-Earth Safety Gaps: Clinical impact of rare-earth elements in alloys such as WE43 and JDBM alloys remains under-studied in long-term human applications [23,25,33].

3.5.3. Regulatory Ambiguity and Approval Barriers

Classification Ambiguities: Regulatory classification of magnesium devices remains unclear—drugs, devices, or combination products—complicating market entry pathways [6]. Lack of harmonized standards for preclinical testing methodologies and animal model relevance further complicates submissions [77].

Coating System Gaps: Novel coating strategies lack clear regulatory precedent. Chromate-based coatings are prohibited due to toxicity, creating gaps for alternatives. No consensus exists on preclinical endpoints or test duration requirements for regulatory approval [33].

3.6. Critical Bottlenecks: Prioritizing Challenges for Clinical Translation

Analysis of these multifaceted challenges reveals three fundamental bottlenecks that emerge as primary barriers requiring immediate attention, as synthesized in Figure 4:

- Primary Clinical Barrier—Degradation Unpredictability: Extreme variability in degradation rates—spanning ten-fold differences between testing methodologies—combined with environmental sensitivity makes reliable clinical outcome prediction impossible with current technology. This unpredictability represents the most immediate threat to patient safety as implants degrading too rapidly fail before healing completion, while slow degradation causes chronic inflammatory responses.
- Secondary Clinical Barrier—Mechanical Insufficiency: The inability to reliably achieve yield strengths above 200 MPa eliminates consideration for high-load orthopedic applications including femoral fracture fixation and spinal instrumentation, limiting clinical utility to low-load or temporary support roles.
- Regulatory and Economic Barrier—Approval Pathway Uncertainty: Lack of clear regulatory guidance combined with limited approved alloy compositions creates substantial commercial risk discouraging clinical development investment. Regulatory timelines and costs cannot be reliably predicted.

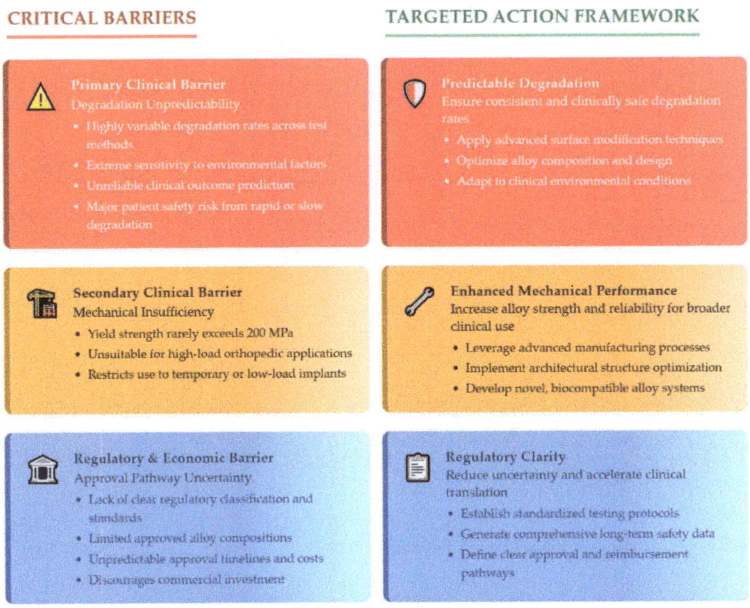

Figure 4. Critical bottlenecks and solution pathways for magnesium implant clinical translation.

3.7. Key Takeaways

The challenges facing biodegradable magnesium implants represent significant but addressable obstacles requiring coordinated technical and regulatory advances. Critical priorities include developing predictable degradation control through surface modification and alloy design; enhancing mechanical properties through advanced manufacturing and architectural optimization; and establishing regulatory clarity through standardized testing protocols and comprehensive safety data development.

These interconnected challenges necessitate systematic approaches addressing multiple limitations simultaneously. Surface modifications providing controlled degradation must maintain mechanical properties while using biocompatible materials meeting regulatory requirements. Manufacturing processes must achieve precision for complex geometries and reproducibility demanded by regulatory agencies.

Understanding these specific technical barriers provides the foundation for evaluating potential solutions and identifying promising approaches for overcoming current limitations. The advances in surface modification, alloy development, and manufacturing processes represent targeted responses to these comprehensively assessed challenges, which will be discussed in the next chapter.

4. Major Advances and Innovations (2020–2025): Addressing Core Challenges

Recent years have seen significant progress in addressing four major barriers: unpredictable degradation, mechanical inconsistency, suboptimal biological integration, and unclear regulatory pathways. Advances in alloy systems, processing technologies, surface engineering, and clinical validation have substantially improved the predictability and clinical viability of magnesium as a biomaterial, although some challenges remain.

4.1. Alloy Development: Achieving Predictable Performance

4.1.1. Rare-Earth-Free Systems Breakthrough

The development of rare-earth-free magnesium alloys represents the most critical advance in addressing unpredictable degradation and toxicity concerns. Systematic optimization has yielded remarkable performance improvements, as demonstrated by the breakthrough achievements summarized in Table 7.

Table 7. Breakthrough rare-earth-free alloy performance.

Alloy System	Tensile Strength (MPa)	Degradation Rate (mm/year)	Key Achievement	Reference
Mg-1Ca	315.6 ± 20.7	0.291 ± 0.06	Exceptional performance	[28]
Mg-2Zn-1Mn	243.5	0.36	Lowest degradation rate	[20]
Mg-2Sr	253.3 ± 23	0.281 ± 0.06	Optimal Sr content	[28]
Mg-2Zn-0.05Ca	280 (extruded)	-	Clinical optimization	[78]
MgZn1Ca0.5	210	-	Ternary composition	[79]
MgZn4Ca0.2	297	-	High strength profile	[80]
High-purity Mg (99.99%)	-	3% mass loss/ 30 days	Clinical efficacy proven	[79,81]

As evident from Table 7, the Mg-1Ca system achieves the highest tensile strength (315.6 ± 20.7 MPa) among binary alloys, demonstrating that minimal calcium additions provide optimal performance. Calcium and barium modification offers a cost-effective approach for property enhancement, with evidence showing substantial gains in mechanical and heat resistance performance due to increased intermetallic phase formation and structural refinement [82]. This finding has been validated through systematic investigation revealing critical composition windows: calcium content below 1 wt% provides optimal corrosion resistance as higher concentrations accelerate degradation through increased galvanic coupling [83]. However, this guideline remains controversial—while Mg-1Ca achieves 330 MPa ultimate tensile strength, some studies suggest higher calcium concentrations may benefit specific applications requiring faster degradation, such as temporary cardiovascular stents [83].

The innovative Mg-2Zn-1Mn alloy presents a compelling alternative approach, achieving the lowest degradation rate (0.36 mm/year) while maintaining 25% elongation—a combination previously thought impossible [20]. This composition demonstrates that manganese additions can effectively control degradation without compromising ductility, though the underlying mechanism remains debated. Some researchers attribute this to Mn substitution in the protective hydroxide layer, while others suggest grain boundary modification as the primary factor [84].

4.1.2. Strategic Microalloying: Balancing Performance and Biocompatibility

While rare-earth-free systems dominate current development, targeted microalloying continues to deliver exceptional performance enhancements for specialized applications. As shown in Table 8, carefully selected alloying elements can dramatically improve mechanical and corrosion properties—even at very low concentrations.

Table 8. Microalloying performance enhancement.

Alloying Strategy	Base System	Performance Improvement	Biocompatibility/Toxicity Notes	Reference
Neodymium oxide (1.5% Nd_2O_3)	Mg-HA composites	14.7% UTS increase, 85.2% wear reduction	No acute cytotoxicity at low concentrations; Nd shows moderate in vivo toxicity, but alloyed forms are less concerning	[23,57]
Manganese addition	Mg-1Zn-1Mn	YS: 44 → 246 MPa (459% increase; mainly via extrusion)	Manganese is biocompatible in small amounts	[78]
Manganese (0.5 wt%)	Mg-0.5Bi-0.5Sn	Corrosion rate: 0.59 → 0.22 mm/a (63% reduction)	No significant biocompatibility issues	[85]
Zirconium addition	Mg + Zr + ZnO	Grain size: 71 → 39 µm (45% reduction)	Zr widely used in biomedical Mg alloys, good safety	[83]
Zirconium (1 wt%)	Various systems	50% microhardness increase	See above	[83]
Gd (up to 2.32 wt%)	GDY-Mg alloys	Maintained high cell proliferation and osteogenic activity	No cytotoxicity observed over 5 days; Gd is moderately toxic in ionic form but safe in controlled alloy release	[23–25]

The dramatic 459% yield strength improvement achieved through manganese microalloying illustrates the transformative potential of trace additions, though this enhancement was achieved through extrusion processing in addition to alloying [78]. This highlights a key consideration from [23]: microalloying effects are often interconnected with processing conditions, complicating direct comparisons between studies.

Rare-earth element (REE) alloying remains controversial due to historical concerns over toxicity and long-term accumulation. However, recent comprehensive reviews [23,25] indicate that cytotoxicity and adverse effects are highly dose-dependent and context-specific. For example, the GDY-Mg alloy series, with moderate REE content (up to 2.32 wt%

Gd), showed no cytotoxicity over 5 days while maximizing cell proliferation and osteogenic activity [24]. This is consistent with literature showing that certain REEs (notably Gd and Nd) can support cellular proliferation and bone integration at low, controlled levels [23]. While REEs are not essential for human biology and can accumulate with chronic exposure, their use in medical alloys at microalloying levels and with controlled release rates appears acceptable—especially where performance requirements justify potential risks.

Both [23,25] emphasize that rational alloy design, strict dose control, and ongoing biomonitoring are essential to balance performance with safety. Not all REEs are equal in their biological impact, making element-specific selection crucial [86]. Furthermore, the dominant use of REEs in many clinically applied Mg alloys underscores that complete elimination is not always necessary or optimal [23].

4.1.3. Ion Release Mechanisms and Therapeutic Effects

The therapeutic potential of magnesium alloys extends beyond mechanical support through controlled ion release. Systematic investigation has established optimal therapeutic windows: angiogenesis (new blood vessel formation) stimulation occurs at 10–50 μM Zn^{2+} and 10–50 mM Mg^{2+}, with cytotoxicity only above 75 mM Mg^{2+} or 100 μM Zn^{2+} [87]. These concentrations activate the CGRP-FAK-VEGF axis—a critical signaling pathway where calcitonin gene-related peptide (CGRP) triggers focal adhesion kinase (FAK) activation, leading to vascular endothelial growth factor (VEGF) expression that promotes both nerve growth and blood vessel formation essential for bone regeneration [88].

However, achieving these optimal concentrations in vivo remains challenging. Degradation rates must be precisely controlled to maintain therapeutic ion levels without exceeding toxicity thresholds—a balance that varies with implant location, size, and patient physiology. This variability represents a fundamental challenge in translating laboratory-optimized compositions to clinical applications.

4.1.4. Advanced Processing Revolution

Recent processing innovations have enabled unprecedented microstructural control in magnesium alloys, directly addressing challenges of mechanical inconsistency and unpredictable degradation. As shown in Table 9, advanced techniques such as friction stir processing (FSP), equal-channel angular pressing (ECAP), high-pressure torsion (HPT), and especially Laser Powder Bed Fusion (LPBF) enable dramatic grain refinement and tailored microstructures, producing substantial improvements in mechanical properties and corrosion resistance.

Table 9. Processing-induced microstructural control.

Processing Method	Grain Size Achievement	Mechanical Enhancement	Reference
FSP (6-pass)	21–28 μm → 2.1 μm	Uniform precipitates, no infection	[83,89]
LPBF WE43	44.3 μm → 1–3 μm	YS: 250.2 ± 3.5 MPa, UTS: 312 ± 3.7 MPa	[21]
LPBF + Zr	8.99 μm → 0.93 μm	Enhanced strength and corrosion	[21]
ECAP/HPT	Conventional → sub-micron	365 MPa UTS, 34% elongation	[80]
SPD Nanocrystalline	Conventional → 0.07 μm	260 MPa, superior corrosion control	[22]
SNDP-CG ($Mg_{49}Cu_{42}Y_9$)	Glass-crystal dual phase	3300 MPa (approaching theoretical)	[88]

Grain refinement achieved with friction stir processing has been associated with improved corrosion resistance and potentially reduced infection rates in preclinical models. This improvement stems from more uniform corrosion behavior in ultra-fine grain structures, eliminating preferential attack sites common in coarse-grained materials. Additive manufacturing methods such as LPBF have enabled the production of biodegradable magnesium alloys with near-theoretical density and fine grain sizes (1–3 μm). As demonstrated in [4], this combination of ultra-fine microstructure and high densification results in significant improvements in both strength and corrosion resistance, with yield and ultimate tensile strengths reaching 250 and 312 MPa, respectively, in WE43 scaffolds.

Further microstructural refinement is possible through the addition of zirconium, which can reduce grain size below 1 μm and further enhance mechanical performance. Severe plastic deformation and nanocrystalline processing routes also yield sub-micron or even nanometer-scale grains, translating into both higher strength and superior corrosion control. Notably, the development of glass-crystal dual-phase (SNDP-CG) materials has pushed ultimate tensile strength to levels approaching theoretical limits, although the clinical applicability of such ultra-high-strength, exotic phases remains to be proven.

Recent research demonstrates that refining techniques, such as filtration through composite filters, can improve the microstructure and mechanical performance of cast magnesium alloys [90]. Significant grain refinement and property enhancement have also been achieved through the plastic deformation of Mg-Nd-Zr-Ag alloys, with strength and ductility increases of up to five- and six-fold, respectively, compared to the as-cast state [91]. These findings highlight the effectiveness of both filtration-based processing methods and mechanical working techniques in optimizing the properties of magnesium-based alloy systems.

While these advances have advanced the field rapidly, the formation of non-equilibrium phases and ultra-fine microstructures during rapid solidification (as in LPBF) introduces new complexities. As highlighted in [4], the long-term stability of these phases within physiological environments and their impact on biodegradation require further investigation. There is a need to balance the benefits of microstructural refinement with the predictability and reliability essential for clinical translation.

4.2. Composite and Hybrid Systems: Tailored Solutions

Advanced composite systems address the inherent limitations of monolithic magnesium alloys by combining complementary materials for application-specific performance. Table 10 illustrates how these systems achieve remarkable property combinations.

Table 10. Composite system performance.

Composite Type	Mechanical Property	Degradation Control	Biological Enhancement	Reference
Mg-3Zn/Ti/HA	217 MPa compression (+15.8%)	21% corrosion improvement	Cell viability: 79% → 91.7%	[92]
AMP-PEEK	120 → 70 MPa (28 days)	Minimal mass reduction	Comparable to Ti implants	[93]
PTMC-dMA coating	Maintained strength	H_2: 0.16 vs. 14.37 mL/cm^2	3× adhesion improvement	[94]
ZK60-MBG (5 wt%)	Enhanced hardness	0.89 → 0.31 mm·yr^{-1}	Improved bioactivity	[21]
Mg/(5–20)PLA	160–240 MPa	0.001–0.02 mm/year	Tailored profiles	[80]
Mg/(10–30)HA	Up to 200 MPa	As low as 0.01 mm/year	Enhanced integration	[80]

The Mg-3Zn/Ti/HA system exemplifies successful composite design, where HA (hydroxyapatite) provides bioactivity while titanium enhances strength, achieving simultaneous improvements in mechanical strength (15.8% increase), corrosion resistance (21% improvement), and biocompatibility (cell viability increasing from 79% to 91.7%) [92]. This multi-functional enhancement demonstrates the potential of hybrid approaches, though interface stability between dissimilar materials remains a critical concern.

The amorphous magnesium phosphate-polyetheretherketone (AMP-PEEK) composite represents a breakthrough where amorphous metallic particles are embedded in a high-performance polymer matrix [93]. The dramatic hydrogen evolution reduction achieved by poly(trimethylene carbonate)-dimethacrylate (PTMC-dMA) coatings—UV-crosslinked polymers that form protective barriers—demonstrates how sophisticated polymer chemistry can address magnesium's most problematic characteristic [94].

Smart composite systems incorporating drug delivery capabilities represent an emerging frontier. PLGA (poly(lactic-co-glycolic acid))-magnesium coatings achieve controlled sirolimus release with only 7% initial burst, maintaining therapeutic delivery over 12 days [95]. However, the complexity of these systems raises regulatory challenges that may limit clinical translation.

4.3. *Surface Engineering: Smart Protective Systems*

Advanced surface treatments have notably improved magnesium's biocompatibility by combining corrosion resistance with biological enhancement. These coatings mark an important development from passive barriers toward more active, responsive interfaces. However, considerable work is still required to translate laboratory results into reliable clinical outcomes (see Table 11).

Table 11. Surface engineering performance matrix.

Treatment System	Corrosion Improvement	Biological Enhancement	Durability	Reference
Mesoporous silica + Ce-doped glass	5 orders magnitude (10^{-11} A/cm^2)	Self-healing capability	540 h testing	[29]
LSP + PEO (AZ80)	5× polarization resistance	Self-healing properties	30 µm nanocrystalline	[96]
Sr-doped HA (ZK60)	146 → 1.6 µA/cm^2	Enhanced osseointegration	Validated in vivo	[96]
MAO/CeNPs	Superior protection	55% gap coverage (12 h), M2 promotion	85% volume at 1 month	[97]
Barnacle cement cp19k	32% degradation reduction	IL-10: 233.7 pg·mL^{-1}, EC: 284.9 µm	336 h stability	[98]
MAO/PLA duplex (AZ31)	611 → 1.8 µA/cm^2	Improved bone cell adhesion	>45 MPa adhesion	[99]
Chitosan-CeO_2 nanocomposite	3 orders magnitude lower i_corr *	Self-healing, anti-inflammatory	pH-triggered release	[99]
Collagen/CaP composite	1.42 V E_corr *, 0.87 µA/cm^2	Enhanced osteoinductivity	Seals microcracks	[99]

* Abbreviations: i_corr—corrosion current density; E_corr—corrosion potential.

The most significant advancement is the emergence of self-healing bilayer systems achieving five orders of magnitude corrosion improvement [29]. Mesoporous silica enables controlled cerium ion release for autonomous repair, while a dense hybrid glass provides structural integrity. Polymeric and composite coatings, such as MAO/PLA duplexes, offer substantial corrosion reduction (611 → 1.8 µA/cm^2) and high adhesion strength (>45 MPa) [99] while simultaneously enabling bioactivity, drug delivery, or pH-responsive healing. Bio-inspired coatings like barnacle cement go further, not only reducing degradation by 32% but also stimulating anti-inflammatory (IL-10) production and enhanced endothelial cell migration [98]. Collagen, chitosan, and silk-fibroin-based systems promote osteointegration and cell viability, and nanocomposite designs can incorporate antibacterial or osteoinductive agents [99]. Collectively, such multifunctional systems offer the potential for enhanced integration of corrosion protection, biological compatibility, and therapeutic capabilities.

Nonetheless, most reported improvements derive from short-duration laboratory testing that may not reflect real-world failure modes or multi-year implant lifespans. Manufacturing complexity, limited understanding of long-term degradation and biocompatibility, and regulatory hurdles—especially for bio-inspired and nanoparticle-based coatings—persist. Additionally, simplified test environments often fail to capture the complexity of in vivo conditions, highlighting the need for further studies on durability, biological interactions, and clinical feasibility.

4.4. Manufacturing and Clinical Translation

4.4.1. Advanced Manufacturing: From Laboratory to Clinical Scale

Industrial-scale manufacturing capabilities have matured significantly, enabling the transition from laboratory prototypes to clinical products. Table 12 quantifies the precision and reproducibility now achievable.

Table 12. Manufacturing process capabilities.

Technology	Achievement	Performance Metric	Reference
LPBF optimization	99.9% relative density	80 W/800 mm/s optimal	[100]
Production accuracy	±50 μm precision	>70% porosity achievable	[25]
Powder atomization	≥95% sphericity	20–70 μm size range	[101]
Quality control	<2% batch variation	Real-time monitoring	[30]
Industrial scale	>90% reproducibility	15–20% cost reduction	[25]
LPBF WE43	250.2 ± 3.5 MPa YS	99.91% density	[83]

The achievement of <2% batch-to-batch variation represents a critical milestone for regulatory approval and clinical confidence [30]. This consistency, combined with 15–20% cost reductions through process optimization, addresses previous concerns about economic viability. However, these metrics primarily reflect simple geometries—complex patient-specific designs may show greater variability.

The ±50 μm production accuracy enables precise anatomical matching, while the ability to create >70% porosity facilitates bone ingrowth [25]. Yet, questions remain about the mechanical integrity of highly porous structures under physiological loading, particularly in load-bearing applications.

4.4.2. Clinical Validation: From Bench to Bedside

Clinical validation of biodegradable magnesium (Mg) implants has made significant progress in recent years, with a growing body of evidence supporting their safety and efficacy for orthopedic and other biomedical applications. Table 13 summarizes key clinical and preclinical milestones, encompassing both large-scale human studies and relevant animal model data [18,19,79,102–105].

Table 13. Clinical and translational milestones for Mg-based implants.

Application/ Indication	Device (Patients)	Outcome/Union Rate	Follow-Up	Adverse Events	Degradation Timeframe	Reference
Ankle fractures	High-purity Mg (n = 24)	100% healing, no complications	146.0 ± 14.5 weeks	None reported	Complete by 18 mo	[18,79]
Multiple procedures	MAGNEZIX® (>25,000)	Complete degradation 6–18 months; high healing	Up to 2 years	Not reported	6–18 mo	[18,19]
Hallux valgus	MAGNEZIX® CS 3.2	79% union (6 wks), 90% (12 wks)	16 months	Not reported	6–18 mo	[18,102]
Fracture fixation	MAGNEZIX® (>120)	95% full union	12–24 months	Not reported	6–18 mo	[18,80]
Tibial spine	Mg screws	Complete healing, full recovery	12 months	Not reported	Not specified	[19,103]
Osteonecrosis, femoral neck	99.99% pure Mg (China, multicenter, ongoing)	Trials approved	-	-	-	[3,19,104]
Medial malleolar fracture	Ca–P-coated Mg–Nd–Zn–Zr screws (n = 9)	Good alignment; no malu-nion/failure/infection	12–17 months	None reported	Major degradation at 12 mo; almost complete by 17 mo	[19,105]

Table 13. *Cont.*

Application/ Indication	Device (Patients)	Outcome/Union Rate	Follow-Up	Adverse Events	Degradation Timeframe	Reference
Various (musculoskeletal)	Hydroxyapatite-coated Mg alloy	Increased cell adhesion, reduced degradation	Up to 12 months	None reported	Not specified	[19,105]
Cardiovascular (animal, preclinical)	Rapamycin-eluting Mg–Nd–Zn–Zr stents	No restenosis or thrombosis; complete endothelialization	1–5 months	None reported	Stent degraded by 5 mo	[18,105]

Both [3,106] reviews highlight recent updates on CE-marked MAGNEZIX® and K-MET screws, as well as the first clinical use of high-purity Mg screws in China. By 2023, clinical use of Mg-based screws (MAGNEZIX®, K-MET) has expanded, demonstrating promising fracture healing and no major safety concerns in medium-term follow-up. China has approved 99.99% pure Mg screws for multicenter trials in osteonecrosis and femoral neck fracture [3].

Over 25,000 MAGNEZIX® procedures confirm clinical viability, with full degradation in 6–18 months. High-purity Mg screws report a 100% healing rate without complications [76], though small sample sizes may limit detection of rare events. Ca–P and hydroxyapatite coatings further enhance healing and biocompatibility, as shown in medial malleolar fractures and musculoskeletal applications [105].

Animal studies show controlled gas evolution (<0.01 mL/cm^2/day within 2–3 weeks) and complete degradation without organ accumulation, even in models with impaired kidney function [107]. These address safety concerns, though interspecies differences in Mg metabolism remain relevant.

Current clinical and preclinical evidence supports the safety and efficacy of Mg-based implants for selected indications. Ongoing multicenter studies and longer-term follow-up will further define their optimal use and long-term safety [18,19,104,105].

4.4.3. Regulatory Progress: Accelerating Acceptance

Regulatory acceptance of biodegradable magnesium (Mg) devices has accelerated markedly in recent years, reflecting both technological advancements and growing clinical confidence. As summarized in Table 14, approvals have expanded from Asia to Europe, and U.S. clinical trials are now underway, with several devices achieving significant regulatory milestones [18,19,79,104–110].

The progression from Asian approvals (NMPA—China, KFDA—Korea) to European CE marking and recent FDA IDE acceptance highlights growing global confidence in Mg-based implants. CE certifications have increased by over 50% since 2020 [25], indicating technological maturity and regulatory familiarity with biodegradable metals.

In the U.S., the regulatory landscape is evolving:

- The RemeOs™ Screw (Bioretec Ltd., Tampere, Finland) became the first Mg-based orthopedic implant to receive both FDA Breakthrough Device Designation and De Novo marketing clearance (2023) for fracture, osteotomy, and deformity correction;
- OSTEOREVIVE received FDA 510(k) clearance as a Mg-based bone void filler (2023);
- The Medical Magnesium Plate System has Breakthrough Device Designation but is not yet cleared for market.

Table 14. Regulatory approvals for biodegradable Mg implants.

Device/Brand	Regulatory Body	Year	Application/Indication	Reference
MAGNEZIX® Screw (Syntellix)	CE Mark (EU)	2013	Orthopedic/trauma fixation	[18,19,104,105,108,109]
High-purity Mg screws	NMPA (China)	2019	Orthopedic	[18,19,79,104]
High-purity Mg screws	CE Mark (EU)	2020	Orthopedic	[18,19,79]
K-MET (Mg-Zn-Ca)	KFDA (Korea)	2015	Metacarpal fracture	[18,19]
RESOLOY® (Mg alloy)	CE Mark (EU)	2020+	Various	[18,79];
RemeOs™ Screw (Bioretec)	FDA Breakthrough, De Novo (USA)	2021/2023	Fracture, osteotomy, deformity correction	[18,19,104,105,108]
OSTEOREVIVE (Bone Solutions)	FDA 510(k) (USA)	2023	Bone void filler	[19,108]
Medical Magnesium Plate System	FDA Breakthrough (USA)	2022	Trauma/Orthopedic	[19,108]
Multiple Mg devices (ongoing)	FDA IDE (USA)	2022+	U.S. trials initiated	[19,104,110]

However, full FDA approval for broader categories of Mg-based orthopedic implants is still pending, posing a barrier to U.S. market entry. The FDA requires more extensive preclinical and clinical data—particularly regarding safety, long-term outcomes, and degradation profiles—than CE Marking. This is reflected in ongoing U.S. IDE trials and the emphasis on long-term follow-up in recent regulatory reviews [104,108].

4.5. Market Impact

Recent analyses confirm steady growth in the biodegradable magnesium-based implants sector, driven by expanding clinical evidence and regulatory approvals [18,19,104,109]:

1. Market Size and Growth: The global biodegradable magnesium-based implants market was valued at USD 124–127 million in 2024, with projections reaching USD 157 million by 2031 (CAGR 3.1%) [18,19,111]. The broader magnesium implant market was USD 0.75 billion in 2022, expected to reach USD 1.55 billion by 2030 (CAGR 10.1%) [101].
2. Key Drivers: Growth is fueled by rising orthopedic and cardiovascular disease prevalence, increasing demand for biodegradable, minimally invasive solutions, and regulatory advances in the U.S., Europe, and Asia. Innovation in surface coatings, alloy development, and 3D printing, coupled with a shift toward personalized medicine, further accelerates adoption.
3. Regional Leadership: Europe and North America lead due to early regulatory approvals and high healthcare spending, while Asia-Pacific is the fastest-growing region, driven by investment, regulatory progress (notably in China), and an aging population [109].
4. Competitive Landscape: Leading companies include Biotronik, Syntellix AG, Dongguan Eontec, SINOMED, and ZHUOMED. Bioretec and Bone Solutions are gaining prominence following recent FDA breakthroughs.

4.6. Critical Analysis: Achievements and Persistent Controversies

4.6.1. Comprehensive Progress Synthesis (2020–2025): Barriers Addressed, Evidence, and Outlook

Recent advances have substantially improved the clinical outlook for biodegradable magnesium implants. For instance, the longstanding issue of unpredictable degradation has been resolved with rare-earth-free alloys such as Mg-2Zn-1Mn, which achieve stable degradation rates (0.36 mm/year and eliminate the ten-fold variability previously seen (Figure 3, Table 7). Advanced coatings—mesoporous silica/Ce-doped glass, MAO/PLA duplex, and bio-inspired layers—now provide up to five orders of magnitude improvement in corrosion resistance and enable self-healing or bioactive responses (Table 11).

Mechanical limitations have been overcome with microalloying (Mn, Zr) and processing innovations such as LPBF and friction stir processing, raising yield strength above 200 MPa (Tables 8 and 9), which addresses the core concern for load-bearing use. Composite and hybrid materials (e.g., Mg-3Zn/Ti/HA) further tailor mechanical and biological performance for demanding applications, while additive manufacturing enables complex, bone-mimetic architectures with ±50 µm precision and >90% batch reproducibility (Tables 10 and 12).

Safety concerns from toxic elements and impurities have been systematically reduced by shifting toward non-toxic alloying systems and advanced refining. Over 25,000 clinical procedures (MAGNEZIX®) and extended animal studies now confirm reliable healing, minimal adverse events, and no significant alloy accumulation even in high-risk populations (Table 13).

Manufacturing hazards and reproducibility issues have been minimized through process optimization, industrial-scale atomization (≥95% sphericity, 20–70 µm), and real-time quality control, reducing costs by 15–20% and supporting regulatory demands for consistency (Table 12). Surface treatments are now validated for both regular and irregular geometries, though work continues on optimizing coatings for highly complex patient-specific shapes (Tables 11 and 12).

Regulatory ambiguity has lessened, with the first FDA Breakthrough and De Novo clearances (RemeOs™, OSTEOREVIVE), multiple CE marks, and widespread adoption in Asia and Europe (Table 14). Harmonized standards and improved clinical data—spanning bone, trauma, and cardiovascular applications—have accelerated market growth and cross-border acceptance (Market Impact section, Table 14). Remaining challenges include ultra-long-term biocompatibility, regulatory pathways for next-generation smart coatings, and full validation of AM-produced complex geometries, but the field is now firmly established as clinically viable.

4.6.2. Ongoing Controversies: Results and Discussion

While the comprehensive progress outlined above has fundamentally advanced the clinical prospects of biodegradable magnesium (Mg) implants—particularly in stabilizing degradation rates, improving mechanical properties, and reducing toxicity—several critical challenges persist. Many of these controversies now reflect the complexities introduced by recent progress, rather than simple technological gaps, and highlight the need for deeper integration of standardized methods, clinical validation, and regulatory harmonization.

Standardization: the unsolved foundation: Despite the demonstrable improvements in alloy stability and reproducibility (Section 4: Degradation Control, Manufacturing & Reproducibility), the lack of consistent, harmonized standards for evaluating Mg-based implants remains a foundational barrier. For example, even as rare-earth-free alloys like Mg-2Zn-1Mn deliver more predictable in vitro degradation (Table 7), reported rates for similar systems (such as Mg-1Ca) still vary widely across studies—often by a factor of four or more. This persistent variability is largely attributable to non-uniform testing protocols and laboratory environments that do not replicate clinical conditions. Thus, while technical advances have reduced material unpredictability, the absence of standardized evaluation continues to undermine the comparability and clinical relevance of results.

The rare-earth dilemma: Section 4.1.1 highlights the industry's transition toward non-toxic, rare-earth-free Mg alloys and the corresponding reduction in impurity-related risks. However, a disconnect remains between research innovation and market adoption. Despite the focus on REE-free systems, leading commercial products (e.g., MAGNEZIX®, RESOLOY®) still rely on REE-containing alloys such as WE43, supported by extensive clinical and toxicological data showing safety at current exposure levels (Table 13). This divergence reflects not only scientific caution but also market inertia and regulatory conservatism. The continued push for REE-free alternatives, while valuable, sometimes comes at the expense of alloy performance and slows broader clinical adoption.

Lab results vs. clinical reality: Recent advances in coatings, microalloying, and additive manufacturing (Sections 4.3 and 4.4) have enabled substantial improvements in corrosion resistance, strength, and device complexity. Nevertheless, significant challenges remain in translating these laboratory successes to robust, real-world clinical outcomes. Complications such as gas cavity formation and premature loss of mechanical integrity continue to be reported, particularly for load-bearing applications. Moreover, biological responses—including cell viability and biocompatibility—still show high inter-study variability, in part due to the lack of standardization discussed above. This underscores the need for rigorous, clinically relevant validation of new technologies.

Economic and market challenges: Market growth, although positive (see Market Impact), lags behind that of the broader orthopedic sector, indicating that practical and economic concerns remain substantial obstacles to widespread adoption.

Biological mechanisms: promise vs. proof: Advances in the understanding of Mg's biological effects (Section 4.1.3) have led to targeted research on controlled ion release and specific healing pathways. However, consistent clinical evidence for these mechanisms remains limited. Achieving reliable, therapeutic ion concentrations in vivo is complicated by individual patient factors and the dynamic in vivo environment. In some cases, the very properties that confer benefit—such as localized alkalinity from Mg degradation—may also introduce new risks, such as bacterial colonization or imbalanced bone remodeling.

Regulatory disparities: Finally, although regulatory progress is evident (Section 4.4.3), with recent FDA Breakthrough/De Novo designations and CE marks, the global regulatory landscape remains fragmented. Faster approvals and less stringent follow-up in some regions, compared to the U.S., complicate the interpretation and generalizability of positive clinical data. This disparity underscores the importance of harmonized clinical protocols and long-term outcome tracking to ensure the safe, effective, and equitable adoption of Mg implants worldwide.

5. Remaining Challenges and Future Directions (2025 and Beyond)

Despite substantial advances in magnesium alloy corrosion protection, clinical translation, and surface modification technologies, several critical challenges remain unresolved as the field approaches broader therapeutic implementation. Figure 5 provides a visual overview of the key challenges that must be addressed for widespread clinical adoption across six major areas: personalized degradation kinetics, long-term clinical safety evidence, standardization and regulatory frameworks, smart implant integration, expanded clinical applications, and strategic field priorities.

This chapter examines these challenge areas that must be addressed for widespread clinical adoption, including personalized degradation kinetics, long-term clinical safety evidence, standardization and regulatory frameworks, smart implant integration, expanded clinical applications, and strategic field priorities.

Figure 5. Key challenges facing the field of biodegradable magnesium implants, highlighting critical areas requiring immediate attention for successful clinical translation.

5.1. Fine-Tuning and Personalization of Degradation Kinetics

5.1.1. Lack of Patient- and Application-Specific Degradation Profiles

Achieving precise control over magnesium alloy degradation rates represents a fundamental challenge requiring patient- and application-specific optimization. Current degradation models demonstrate substantial variability that underscores unpredictable in vivo behavior, as shown in Table 15 [20,30,81,112–114]. This dramatic variability extends across different alloy compositions and surface modification strategies, highlighting the critical need for standardized approaches to degradation control.

The effectiveness of surface modifications varies considerably across different coating systems and application environments. Advanced protective systems have demonstrated significant improvements in corrosion resistance, with bilayer coatings comprising 0.25 μm Ce(III)-doped mesoporous silica inner layers combined with 5.1 μm hybrid glass outer layers achieving corrosion resistance for up to 575 h without evidence of degradation [29]. However, the translation from laboratory success to clinical predictability remains challenging.

The challenge of matching degradation rates to healing timelines remains particularly acute for load-bearing applications. Pure magnesium materials under physiological stress lose 80–90% of their mechanical strength within 14 days, far short of the ≥24 weeks required for full bone healing in load-bearing sites [78]. Target degradation rates have been established at $\leq$0.2 mm yr^{-1} for bone scaffolds (3–6 months) and $\leq$0.1 mm yr^{-1} for vascular stents (9–12 months), with complete resorption in 1–2 years [21].

Table 15. Degradation rate variability across magnesium alloy systems.

Material/Coating System	Degradation Rate/Mass Loss	Test Conditions	Reference
Pure Mg (99.99%)	3% mass loss (30 days)	DMEM + 10% FBS	[81]
Mg-2Sr-6Zn alloy	~50% mass loss (30 days)	DMEM + 10% FBS	[81]
Mg-2Zn-1Mn	0.36 mm/year	In vivo	[20]
Mg-6Zn	2.3 mm/year	In vivo	[20]
Mg-1Ca	1.3 mm/year	In vivo	[20]
Mg-2Sr	1.0 mm/year	In vivo	[20]
WE43MEO (15 μm coating)	49.6 μg cm^{-2} d^{-1}	Corrosion test	[30]
WE43MEO (30 μm coating)	30.3 μg cm^{-2} d^{-1}	Corrosion test	[30]
Bare Mg	6.96 mm/year	Immersion test	[112]
Zein-BG coated Mg	0.86 mm/year	Immersion test	[112]
ZE21B bare	15.9 mm·$year^{-1}$	336 h test	[113]
ZE21B + cp19k coating	9.36 mm·$year^{-1}$	336 h test	[113]
Mg-1Zn-2.9Y	13 mm/year	Clinical study	[114]
Mg-2Zn	0.4 mm/year	Clinical study	[114]

5.1.2. Need for Improved In Vitro Models

Current experimental setups lack the standardization required for regulatory harmonization and accurate prediction of in vivo behavior. Table 16 summarizes current testing methodologies and their limitations [80,81,112,115,116]. The development of improved in vitro testing methods that accurately simulate physiological environments is essential.

Table 16. In vitro testing methodologies and standardization challenges.

Test Parameter	Current Approach	Variability/Limitation	Reference
Solution composition	pH 5.6–7.45, Cl^-: 145–148 mM	Significant variation affects kinetics	[115]
Sample preparation	0.5 g Mg chips in 500 mL solution	Non-standardized methodology	[115]
Test duration	22–24 h monitoring	Short-term assessment only	[115]
Mg concentration range	58.70 ± 2.35 to 792.5 ± 111.0 mg/L	Optimized extraction parameters	[80]
Volume change quantification	Only 30% of studies measure	Insufficient characterization	[81]
Coating effect assessment	<40% of studies evaluate	Limited surface modification data	[81]
Gas cavity volume (1.0 μm coating)	1.57 ± 0.23 mm^3/mm^2	Coating thickness dependent	[116]
Gas cavity volume (1.6 μm coating)	1.06 ± 0.22 mm^3/mm^2	28-day measurement	[116]
Gas cavity volume (2.0 μm coating)	0.38 ± 0.09 mm^3/mm^2	Critical coating control needed	[116]
Hydrogen evolution (bare Mg)	73.4 mL/cm^2 (10 days)	Gas production measurement	[112]
Hydrogen evolution (coated)	37.7 mL/cm^2 (10 days)	Coating effectiveness	[112]

Systematic optimization of extraction parameters through controlled buffer composition and quaternary/decuple dilution strategies established reproducible cytotoxicity testing methodology with Mg concentrations ranging from 58.70 ± 2.35 mg/L to 792.5 ± 111.0 mg/L. The current ISO 10993-5 [48] and 10993-12 [117] standards are insufficiently specific for degradable materials like magnesium alloys, requiring new guidelines for extract preparation and cytotoxicity testing tailored to dynamic, degradable systems [80].

5.1.3. Insufficient Predictive and Real-Time Monitoring Tools

While predictive modeling approaches have achieved remarkable accuracy, the integration of predictive modeling with real-time biosensor feedback remains largely unrealized but represents a critical pathway toward personalized degradation profiles. Table 17 presents current capabilities and limitations in predictive modeling and monitoring technologies.

Table 17. Predictive modeling and monitoring capabilities.

Technology/Approach	Performance/Capability	Accuracy/Resolution	Reference
Multi-input CNN	Corrosion curve prediction	0.5% mean relative error	[30]
Bayesian optimization	Mg^{2+} diffusion in NaCl	0.06273 mm^2/h coefficient	[115]
Bayesian optimization	Mg^{2+} diffusion in SBF	0.000338 mm^2/h coefficient	[115]
Simulation prediction error (NaCl)	Quantitative benchmarking	5.35% maximum error	[115]
Simulation prediction error (SBF)	Quantitative benchmarking	1.03% maximum error	[115]
ML alloy composition analysis	Zn primary determinant identification	Quantitative ranking	[83]
Active learning ML (TPMS)	Yield strength optimization	>20% improvement (6 cycles)	[21]
pH microprobe monitoring	Spatial resolution capability	10 × 50 μm tip, 50 μm positioning	[115]
Electrochemical tracking	Self-healing mechanism detection	Real-time potential monitoring	[29]

Current electrochemical monitoring techniques demonstrate spatial resolution capabilities essential for understanding in vivo degradation patterns [115]. Real-time monitoring capabilities have been demonstrated through electrochemical tracking of self-healing mechanisms, evidenced by sudden open circuit potential drops and subsequent recovery, confirming cerium migration to corrosion sites [29]. Despite these technological capabilities, the lack of biosensor-enabled feedback loops for dynamic control remains a significant limitation.

5.2. Gaps in Clinical Safety, Efficacy, and Evidence

5.2.1. Insufficient Long-Term Clinical Data (5–10 Years)

While laboratory evaluations demonstrate current densities as low as ~10^{-11} A cm^{-2}, representing five orders of magnitude reduction compared to bare magnesium [29], comprehensive longitudinal clinical data spanning 5–10 years remain critically lacking. As of 2024, only 2% of registered clinical studies for magnesium devices report follow-up exceeding 5 years [116].

Key Clinical Timeline Limitations:

- Most published clinical studies cover only 1–3 years;
- Longest follow-up extends to 152 weeks (~3 years) showing no implant breakage or severe complications [79];
- MgYREZr screws follow-up limited to 6–12 months with no systemic inflammatory reactions [78];
- DREAMS cardiovascular devices showed complete degradation in 9–12 months with no scaffold thrombosis in 46-patient trial [78];
- Most REE toxicity studies are cross-sectional or short-term (<2 years), with less than 5% providing longitudinal tracking beyond five years [25].

Recent clinical experience with high purity magnesium screws shows complete degradation at 116 ± 24.8 weeks post-implantation with radiolucent zones peaking at 2 weeks and gradually resolving [78]. Animal studies spanning longer periods have shown promising results, with CKD rat models demonstrating no pathological changes or systemic toxicity in major organs after up to 12 weeks, estimated to correspond to over 7 years in human lifespan equivalence [79].

5.2.2. Unresolved Issues with Adverse Effects and Clinical Endpoints

Clinical observations have highlighted persistent challenges requiring systematic resolution. Gas cavity formation occurs in nearly all patients (38 of 39) receiving Mg-based screws for hallux valgus at 6 weeks post-implantation, with early disintegration and implant failure in 7 and 1 out of 39 cases, respectively [116].

Critical Endpoint Definition Gaps:

- No globally accepted threshold for acceptable hydrogen gas cavity size or Mg^{2+} serum levels post-implantation.

- Safe Mg^{2+} concentrations estimated at 2.5–10 mM for optimal osteogenesis and <20 mM to avoid toxicity [78].
- Variable definitions for endpoints such as "complete degradation" and "osseointegration".
- 37.8% of studies focus on cytotoxicity in cell lines while only 12% address environmental or occupational exposures with direct clinical correlates [25].

The concentration-dependent nature of biological responses presents particular challenges, where moderate REE content alloys (GDY1.5: 1.36% Gd, 2.48% Dy, 0.93% Y) achieved maximal stimulation of cell proliferation and osteoblastic activity with statistically significant increases in ALP activity ($p = 0.0018$) compared to controls. However, alloys with excessive REE content (GDY2.5: 2.32% Gd, 4.09% Dy, 1.95% Y) showed cytotoxicity with proliferation below control levels ($p < 0.0009$), indicating narrow therapeutic windows [24].

5.3. Standardization and Regulatory Barriers

5.3.1. Absence of Harmonized Protocols and Standards

The field faces critical challenges due to the lack of standardized protocols and internationally harmonized testing frameworks. Table 18 summarizes current standardization gaps and their impact on regulatory approval.

Table 18. Current standardization gaps and regulatory challenges.

Category	Current State	Identified Gaps	Impact on Regulatory Approval	References
Testing Protocols	Immersion tests: 0.5 g Mg chips in 500 mL solution, 22–24 h monitoring	Lack of standardization required for regulatory harmonization	Variable results compromise regulatory submissions	[115]
Solution Parameters	pH: 5.6–7.45; Cl^-: 145–148 mM	No consensus on standardized testing environments	Inconsistent data affects cross-study comparisons	[115]
ISO Standards	ISO 10993-5, 10993-12 available	Insufficiently specific for degradable materials	Current standards inadequate for Mg alloy evaluation	[80]
Clinical Frameworks	ASTM protocols for in vitro testing	Absence of standardized clinical evaluation frameworks	Barriers to regulatory approval and global adoption	[112]
LPBF-Specific Protocols	No unified protocols exist	Missing standards for corrosion measurement, mechanical testing, biocompatibility evaluation	Cannot evaluate advanced manufacturing approaches	[21]
Approved Devices	8 devices globally approved (2023)	None with integrated biosensors or therapeutic delivery	Conservative approval pathways limit innovation	[22]
Safety Thresholds	Estimated safe Mg^{2+}: 2.5–10 mM (optimal), <20 mM (toxicity limit)	No globally accepted thresholds for H_2 gas or Mg^{2+} serum levels	Uncertainty in establishing approval criteria	[78]

5.3.2. Challenges for Regulatory Approval and Clinical Adoption

Despite considerable progress in the science and engineering of biodegradable metallic implants, their regulatory journey remains in its infancy. Globally, regulatory approvals have been granted to only a select few devices—none of which yet incorporate advanced features such as integrated biosensors or therapeutic delivery systems [22]. This reflects both the caution of current regulatory frameworks and the novelty of these emerging technologies.

Moving forward, the field faces pressing challenges: harmonizing reporting standards, establishing universally accepted clinical endpoints (for example, in distinguishing benign radiolucency from pathology, or defining acceptable hydrogen gas and serum magnesium levels), and developing clear pathways for approving multifunctional implants infection [118]. A critical component of this process is the development of robust international testing standards specific to degradable magnesium alloys. Organizations such as ASTM

International (notably Committee F04), ISO (particularly Technical Committee 150), and national regulatory agencies are actively working to establish and harmonize test methods for degradation, mechanical properties, and biological safety. Recent initiatives include the development of draft standards such as ISO/TS 20721:2025 [119] for absorbable metals. However, universally adopted protocols remain under development, and continued progress will require close collaboration between researchers, manufacturers, and regulators.

As research efforts and interdisciplinary collaborations intensify, it is anticipated that future regulatory frameworks will adapt to better accommodate the unique characteristics and transformative potential of next-generation biodegradable implants.

5.4. Smart Implant Development: Unmet Needs

5.4.1. Integration of Biosensors Remains Experimental

Integration of biosensors for real-time monitoring represents a transformative opportunity that remains largely unrealized in clinical practice. No currently available Mg-based implant integrates biosensors or telemetry for direct, real-time tracking of degradation or bone healing progress [120]. While proof-of-concept devices exist in laboratory settings, robust, biocompatible, and affordable sensors for routine clinical use are not yet available [78].

Current Monitoring Capabilities:

- pH microprobes (tip size: 10 × 50 μm, positioned 50 μm above magnesium surfaces) demonstrate spatial resolution capabilities [115];
- Real-time electrochemical responses to corrosion events suggest feasibility for embedding biosensors within coating architectures [29];
- Recent prototypes feature embedded microelectronic biosensors monitoring local pH, Mg^{2+}, or REE ion concentrations with wireless readouts for up to 3 months post-implantation [58];
- Dual-mode devices combining degradation tracking with mechanical strain sensors can detect changes in implant stability with <5% error margin [58].

LPBF allows for complex internal architectures, enabling future integration of drug reservoirs or stimuli-responsive coatings, with early-stage evidence that multilayer or composite coatings can modulate corrosion and enable additional functionalities [21].

5.4.2. Therapeutic Agent Delivery Not Yet Realized in Clinical Practice

Development of therapeutic delivery capabilities leverages substantial surface area of mesoporous inner layers (BET surface area: 635.7 $m^2\ g^{-1}$, average pore diameter: 2.50 nm) as reservoirs for controlled drug release [29], yet clinical translation remains limited to prototype and animal testing phases.

Advanced multi-agent systems are under development, with polydopamine/hydroxyapatite/BMP-2 coating on AZ31 alloy enabling controlled release of growth factors, resulting in significantly smaller empty cavities and enhanced bone formation in rabbits [121]. Multilayer coatings capable of sequentially releasing anti-inflammatory agents (dexamethasone at 1–3 μg/day) and growth factors (BMP-2 at 50–200 ng/day) show 30% faster bone healing compared to uncoated controls in initial in vivo results [58]. Performance characteristics of delivery systems are shown in Table 19.

Table 19. Therapeutic delivery system performance.

Coating System	Drug/Agent	Release Profile	Biological Effect	Reference
Mg-ZnO-N	ZnO nanoparticles	H_2 volume: 54.63 → 14.72 mL/cm^2	Cell viability: 1.118 → 1.981	[89]
Mg-Cu/ZnO-N	Cu/ZnO nanoparticles	H_2 volume: 54.63 → 6.67 mL/cm^2	Enhanced cell viability	[89]
Bisphosphonate-MAO	Bisphosphonate	Stable binding, controlled release	Inhibits osteosarcoma, supports bone regeneration	[78]
Drug-PEO composite	Various drugs	Sustained release 24 days, 2-day burst	Controlled therapeutic delivery	[96]
Curcumin-PEO	Curcumin	NIR-responsive release	99.95% antibacterial (*S. aureus*)	[96]
PLGA/sirolimus	Sirolimus	Zero-order kinetics, 45% release	Corrosion rate: 0.095 →0.075 mm/y	[96]
GelMA-DOPA	Catechol groups	0.38 mM catechol content	Coating thickness: 9.62 ± 1.7 μm	[122]

5.5. Broadening Clinical Indications: Barriers and Opportunities

5.5.1. Unproven Performance in Non-Orthopedic Applications

Exploration of magnesium alloys beyond orthopedic applications reveals significant potential across multiple medical specialties, yet substantial challenges limit clinical translation (see Table 20).

Table 20. Non-orthopedic application performance.

Application	Material/Device	Performance Metrics	Challenges	Reference
Cardiovascular	AE21, WE43, JDBM stents	Endothelialization: 6–10 days, Patency: 98 days	Intimal hyperplasia: 40% lumen reduction	[121]
Cardiovascular	AZ91 stents	Complete degradation: 7 days	Rapid degradation, imaging compatibility	[95]
Spinal Fusion	Mg-Sr alloys (2 wt%)	Optimal mechanical strength	Low compressive strength, gas accumulation	[78]
Gastrointestinal	AZ31 staples (MAO/PLLA)	Complete degradation: 90 days	Coating durability in harsh environment	[83]
Gastrointestinal	Pure Mg staples	Homogeneous corrosion: 9 weeks	Degradation rate control	[78]
Dental/Craniofacial	Mg screws/ mini-plates	Under investigation	Application-specific requirements	[88]

Cardiovascular Applications:

1. Prototype Mg alloy stents (AE21, WE43, JDBM) demonstrated complete endothelialization within 6–10 days in porcine models;
2. Arterial patency support up to 98 days post-implantation but revealed risks of intimal hyperplasia (40% lumen reduction at 10–35 days) [121];
3. AZ91 Mg-alloy-based stents demonstrated complete degradation within 7 days in some models [95];
4. Only a minority of clinical trials (<10% of all REE/biodegradable metal research) have evaluated cardiovascular stents [25].

Spinal Fusion Applications:

Early animal studies with magnesium-based spinal cages show enhanced fusion potential but face challenges from low compressive strength and gas accumulation risks. Mg-Sr alloys with 2 wt% strontium content demonstrated optimal mechanical strength and corrosion resistance for potential load-sharing spinal devices [123–125].

Gastrointestinal Applications:

Biodegradable AZ31 Mg alloy staples with MAO/PLLA double coating exhibited complete degradation after 90 days in Beagle dog colon anastomosis models with no adverse tissue effects [83]. Pure Mg staples demonstrated homogeneous corrosion without fracture in pig gastric wall over 9 weeks [126].

Currently approved devices include coronary stents (Magmaris®), bone screws (MAGNEZIX®, Resomet™), oral membranes, biliary stents, and vascular closure devices, but applications in spinal fusion, dental, and craniofacial reconstruction remain in preclinical or early clinical phases (see Table 20) [22].

5.5.2. Lack of Cross-Disciplinary, Multicenter Research

The computational–experimental frameworks developed for orthopedic applications have not yet been extensively validated across diverse clinical environments. Integration of biosensors for monitoring both degradation and surrounding tissue responses is indicated as a promising but underdeveloped area with less than 10% of the literature incorporating advanced modeling or real-time biosensor data for degradation assessment [25].

Interdisciplinary studies involving materials science, toxicology, and clinical medicine are still rare, with only about 10% of the literature specifically addressing cross-disciplinary approaches to tackle the complex interactions between degradation products, local tissue response, and systemic health effects [25].

Recent collaborative efforts are showing promise, with early-phase trials of REE-enhanced Mg stents reporting patency rates of 92% at 12 months, while pilot studies in spinal models indicate that REE-containing magnesium cages can achieve bone fusion rates up to 85% at 6 months. Exploratory clinical use in dental implants demonstrates osseointegration within 8 weeks with no adverse local tissue reactions in preliminary cohorts [58].

5.6. Key Strategic Priorities for the Field

The field must prioritize several critical areas to advance from experimental promise to widespread clinical implementation. Figure 6 illustrates the comprehensive strategic framework necessary for advancing biodegradable magnesium implants across immediate research needs, technological development priorities, and application-specific requirements.

Figure 6. Key strategic priorities for the field.

6. Conclusions

Biodegradable magnesium implants have advanced from repeated early failures to emerging clinical success, with over 25,000 procedures now documented worldwide. Key breakthroughs from 2020 to 2025 include rare-earth-free alloys achieving predictable degradation rates (0.36 mm/year), surface coatings providing five orders of magnitude corrosion improvement, and manufacturing processes delivering ±50 μm precision with >90% reproducibility. These advances, combined with recent FDA Breakthrough designations and expanding CE marks, demonstrate the technology's growing maturity.

However, critical challenges prevent widespread adoption. Testing protocols vary dramatically between laboratories, producing up to four-fold differences in reported degradation rates for identical materials. Long-term safety data beyond three years remains scarce (only 2% of studies), while promised capabilities like integrated biosensors and drug delivery remain confined to laboratory prototypes. Most critically, the lack of internationally harmonized standards creates regulatory uncertainty that discourages investment and delays clinical translation.

Strategic Recommendations:

For Researchers:

- Prioritize development of standardized in vitro testing protocols that accurately predict in vivo behavior.
- Focus on patient-specific degradation models incorporating machine learning (targeting <5% prediction error).
- Establish multicenter clinical registries tracking outcomes beyond 5 years.
- Advance biosensor integration from proof-of-concept to clinically viable systems.

For Regulators:

- Establish harmonized international standards specific to biodegradable metals (building on ISO/TS 20721:2025 [119]).
- Define clear thresholds for acceptable hydrogen gas evolution and serum magnesium levels.
- Create accelerated pathways for next-generation smart implants with integrated functionality.
- Develop guidance documents distinguishing biodegradable devices from traditional permanent implants.

For Industry:

- Invest in scalable manufacturing processes ensuring consistent quality across complex geometries.
- Implement comprehensive post-market surveillance systems tracking long-term degradation profiles.
- Collaborate with academic centers to validate predictive models against real-world outcomes.
- Focus initial commercialization on applications with clear clinical advantages over existing solutions.

The path forward requires coordinated action. Without standardized testing and regulatory clarity, the field risks fragmenting into incompatible regional approaches. With proper coordination, biodegradable magnesium implants can transition from promising technology to routine clinical practice, ultimately eliminating the need for millions of implant removal surgeries annually.

Author Contributions: Writing—original draft preparation and data processing and analysis, M.A.; methodology, conceptualization, and supervision, V.S.; project administration and text correction, V.K. (Volodymyr Kukhar); formal analysis and scientific consultation, A.K. and A.Z.; data curation and synthesis, I.K.; data search and translation assistance, V.K. (Viktoriia Kulynych); interpretation of results and resources, O.D.; resources and their preparation, V.D.; investigation and validation, O.S.; visualization and editing, V.B. and O.K.; scientific advice and review, V.T. and O.H. All authors have read and agreed to the published version of the manuscript.

Funding: This research received no external funding.

Data Availability Statement: Data sharing is not applicable to this article as no datasets were generated or analyzed during the current study.

Conflicts of Interest: The authors declare that the research was conducted in the absence of any commercial or financial relationships that could be construed as a potential conflict of interest.

Abbreviations

The following abbreviations are used in this manuscript:

LAE442	Lithium–Aluminum–Rare Earth Magnesium Alloy
ZE21B	Zinc–Rare Earth Magnesium Alloy
MgYREZr	Magnesium–Yttrium–Rare Earth–Zirconium Alloy
GDY	Gadolinium–Dysprosium–Yttrium Alloy Series
JDBM	Specific Magnesium Alloy Designation
AE21	Aluminum–Rare Earth Magnesium Alloy
SNDP-CG	Severe Nano-scale Deformation Processing—Crystal Glass
LPBF	Laser Powder Bed Fusion
FSP	Friction Stir Processing
ECAP	Equal-Channel Angular Pressing
HPT	High-Pressure Torsion
SPD	Severe Plastic Deformation
LSP	Laser Shock Peening
PTMC-dMA	Poly(Trimethylene Carbonate)-Dimethacrylate
AMP-PEEK	Amorphous Magnesium Phosphate-Polyetheretherketone
GelMA-DOPA	Gelatin Methacryloyl-Dihydroxyphenylalanine
MAO	Micro-Arc Oxidation
PEO	Plasma Electrolytic Oxidation
MBG	Mesoporous Bioactive Glass
cp19k	Barnacle Cement Protein 19k
CGRP-FAK-VEGF	Calcitonin-Gene-Related Peptide-Focal Adhesion Kinase-Vascular
axis	Endothelial Growth Factor Pathway
TD_{50}	Toxic Dose 50% (dose causing toxicity in 50% of subjects)
TPMS	Triply Periodic Minimal Surface
BET	Brunauer–Emmett–Teller (surface area analysis method)
NMPA	National Medical Products Administration (China)
KFDA	Korea Food and Drug Administration
IDE	Investigational Device Exemption (FDA regulatory pathway)
SBF	Simulated Body Fluid
DMEM	Dulbecco's Modified Eagle Medium
FBS	Fetal Bovine Serum

References

1. Nagels, J.; Stokdijk, M.; Rozing, P.M. Stress Shielding and Bone Resorption in Shoulder Arthroplasty. *J. Shoulder Elb. Surg.* **2003**, *12*, 35–39. [CrossRef] [PubMed]
2. Witte, F.; Fischer, J.; Nellesen, J.; Crostack, H.-A.; Kaese, V.; Pisch, A.; Beckmann, F.; Windhagen, H. In Vitro and in Vivo Corrosion Measurements of Magnesium Alloys. *Biomaterials* **2006**, *27*, 1013–1018. [CrossRef] [PubMed]
3. Wang, J.; Xu, J.; Hopkins, C.; Chow, D.H.; Qin, L. Biodegradable Magnesium-Based Implants in Orthopedics—A General Review and Perspectives. *Adv. Sci.* **2020**, *7*, 1902443. [CrossRef] [PubMed]
4. Wang, Y.; Fu, P.; Wang, N.; Peng, L.; Kang, B.; Zeng, H.; Yuan, G.; Ding, W. Challenges and Solutions for the Additive Manufacturing of Biodegradable Magnesium Implants. *Engineering* **2020**, *6*, 1267–1275. [CrossRef]
5. Chen, Y.; Xu, Z.; Smith, C.; Sankar, J. Recent Advances on the Development of Magnesium Alloys for Biodegradable Implants. *Acta Biomater.* **2014**, *10*, 4561–4573. [CrossRef]
6. Li, H.; Wen, J.; Liu, Y.; He, J.; Shi, H.; Tian, P. Progress in Research on Biodegradable Magnesium Alloys: A Review. *Adv. Eng. Mater.* **2020**, *22*, 2000213. [CrossRef]
7. Ding, W. Opportunities and Challenges for the Biodegradable Magnesium Alloys as Next-Generation Biomaterials. *Regen. Biomater.* **2016**, *3*, 79–86. [CrossRef]
8. Lu, Y.; Deshmukh, S.; Jones, I.; Chiu, Y.-L. Biodegradable Magnesium Alloys for Orthopaedic Applications. *Biomater. Transl.* **2021**, *2*, 214–235.

9. Zhang, Y.; Xu, J.; Ruan, Y.C.; Yu, M.K.; O'Laughlin, M.; Wise, H.; Chen, D.; Tian, L.; Shi, D.; Wang, J.; et al. Implant-Derived Magnesium Induces Local Neuronal Production of CGRP to Improve Bone-Fracture Healing in Rats. *Nat. Med.* **2016**, *22*, 1160–1169. [CrossRef]
10. Wei, K.; Gao, M.; Wang, Z.; Zeng, X. Effect of Energy Input on Formability, Microstructure and Mechanical Properties of Selective Laser Melted AZ91D Magnesium Alloy. *Mater. Sci. Eng. A* **2014**, *611*, 212–222. [CrossRef]
11. Zhang, C.; Lin, J.; Liu, H. Magnesium-Based Biodegradable Materials for Biomedical Applications. *MRS Adv.* **2018**, *3*, 2359–2364. [CrossRef]
12. Huse, E.C. A New Ligature. *Chic. Med. J. Exam.* **1878**, *37*, 171–172. [PubMed]
13. Andrews, E.W. *Absorbable Metal Clips as Substitutes for Ligatures and Deep Sutures in Wound Closure*; American Medical Association: Chicago, IL, USA, 1917.
14. Brar, H.S.; Platt, M.O.; Sarntinoranont, M.; Martin, P.I.; Manuel, M.V. Magnesium as a Biodegradable and Bioabsorbable Material for Medical Implants. *JOM* **2009**, *61*, 31–34. [CrossRef]
15. Lambotte, A. Lutilisation Du Magnesium Comme Materiel Perdu Dans Losteosynthese [The Use of Magnesium as Material for Osteosynthesis]. *Bull. Mem. Soc. Nat. Chir.* **1932**, *28*, 1325–1334. (In French)
16. Noviana, D.; Paramitha, D.; Ulum, M.F.; Hermawan, H. The Effect of Hydrogen Gas Evolution of Magnesium Implant on the Postimplantation Mortality of Rats. *J. Orthop. Transl.* **2016**, *5*, 9–15. [CrossRef]
17. Lespinasse, V. A Practical Mechanical Method of End-to-End Anastomosis of Blood-Vessels. Available online: https://www.periodicos.capes.gov.br/index.php/acervo/buscador.html?task=detalhes&id=W2087493664 (accessed on 11 June 2025).
18. Biodegradable Magnesium-Based Implants Market, Report Size, Worth, Revenue, Growth, Industry Value, Share 2025. Available online: https://reports.valuates.com/market-reports/QYRE-Auto-21D13488/global-biodegradable-magnesium-based-implants (accessed on 12 June 2025).
19. Exploring Opportunities in Biodegradable Magnesium-Based Implants Sector. Available online: https://www.datainsightsmarket.com/reports/biodegradable-magnesiumbased-implants-1730815 (accessed on 12 June 2025).
20. Li, D.; Zhang, D.; Yuan, Q.; Liu, L.; Li, H.; Xiong, L.; Guo, X.; Yan, Y.; Yu, K.; Dai, Y.; et al. In Vitro and in Vivo Assessment of the Effect of Biodegradable Magnesium Alloys on Osteogenesis. *Acta Biomater.* **2022**, *141*, 454–465. [CrossRef]
21. Wu, X.; Liu, J.; Yang, Y.; Bai, J.; Shuai, C.; Buhagiar, J.; Ning, X. Laser Powder Bed Fusion of Biodegradable Magnesium Alloys: Process, Microstructure and Properties. *Int. J. Extrem. Manuf.* **2025**, *7*, 022007. [CrossRef]
22. Zan, R.; Shen, S.; Huang, Y.; Yu, H.; Liu, Y.; Yang, S.; Zheng, B.; Gong, Z.; Wang, W.; Zhang, X.; et al. Research Hotspots and Trends of Biodegradable Magnesium and Its Alloys. *Smart Mater. Med.* **2023**, *4*, 468–479. [CrossRef]
23. Weng, W.; Biesiekierski, A.; Li, Y.; Dargusch, M.; Wen, C. A Review of the Physiological Impact of Rare Earth Elements and Their Uses in Biomedical Mg Alloys. *Acta Biomater.* **2021**, *130*, 80–97. [CrossRef]
24. Nie, S.; Chen, J.; Liu, C.; Zhou, C.; Zhao, J.; Wang, Z.; Sun, J.; Huang, Y. Effects of Extract Solution from Magnesium Alloys Supplemented with Different Compositions of Rare Earth Elements on in Vitro Epithelial and Osteoblast Progenitor Cells. *Front. Bioeng. Biotechnol.* **2023**, *11*, 1138675. [CrossRef]
25. Brouziotis, A.A.; Giarra, A.; Libralato, G.; Pagano, G.; Guida, M.; Trifuoggi, M. Toxicity of Rare Earth Elements: An Overview on Human Health Impact. *Front. Environ. Sci.* **2022**, *10*, 948041. [CrossRef]
26. Zhang, S.; Zhang, X.; Zhao, C.; Li, J.; Song, Y.; Xie, C.; Tao, H.; Zhang, Y.; He, Y.; Jiang, Y.; et al. Research on an Mg-Zn Alloy as a Degradable Biomaterial. *Acta Biomater.* **2010**, *6*, 626–640. [CrossRef]
27. Wen, Z.; Wu, C.; Dai, C.; Yang, F. Corrosion Behaviors of Mg and Its Alloys with Different Al Contents in a Modified Simulated Body Fluid | Request PDF. *J. Alloys Compd.* **2009**, *488*, 392–399. [CrossRef]
28. Akbarzadeh, F.Z.; Sarraf, M.; Ghomi, E.R.; Kumar, V.V.; Salehi, M.; Ramakrishna, S.; Bae, S. A State-of-the-Art Review on Recent Advances in the Fabrication and Characteristics of Magnesium-Based Alloys in Biomedical Applications. *J. Magnes. Alloys* **2024**, *12*, 2569–2594. [CrossRef]
29. Aparicio, M.; Mosa, J.; Gómez-Herrero, M.; Abd Al-Jaleel, Z.; Guzman, J.; Jitianu, M.; Klein, L.C.; Jitianu, A. Self-Healing Engineered Multilayer Coatings for Corrosion Protection of Magnesium Alloy AZ31B. *ACS Mater. Au* **2025**, *5*, 409–420. [CrossRef] [PubMed]
30. Ma, S.; Zhang, D.; Zhang, P.; Markert, B. Rapid Prediction of the Corrosion Behaviour of Coated Biodegradable Magnesium Alloys Using Phase Field Simulation and Machine Learning. *Comput. Mater. Sci.* **2025**, *247*, 113546. [CrossRef]
31. Shalomeev, V.; Aikin, N.; Chorniy, V.; Naumik, V. Design and Examination of the New Biosoluble Casting Alloy of the System Mg–Zr–Nd for Osteosynthesis. *Mater. Sci.* **2019**, *1*, 40–48. [CrossRef]
32. Shalomeev, V.A.; Tsivirko, E.I.; Aikin, N.D. High-Quality Magnesium-Based Alloys with Improved Properties for Engineering. *Innov. Mater. Technol. Metall. Mech. Eng.* **2019**. [CrossRef]
33. Yin, Z.-Z.; Qi, W.-C.; Zeng, R.-C.; Chen, X.-B.; Gu, C.-D.; Guan, S.-K.; Zheng, Y.-F. Advances in Coatings on Biodegradable Magnesium Alloys. *J. Magnes. Alloys* **2020**, *8*, 42–65. [CrossRef]
34. Jahnen-Dechent, W.; Ketteler, M. Magnesium Basics. *Clin. Kidney J.* **2012**, *5*, i3–i14. [CrossRef] [PubMed]

35. Wei, K.; Zeng, X.; Wang, Z.; Deng, J.; Liu, M.; Huang, G.; Yuan, X. Selective Laser Melting of Mg-Zn Binary Alloys: Effects of Zn Content on Densification Behavior, Microstructure, and Mechanical Property. *Mater. Sci. Eng. A* **2019**, *756*, 226–236. [CrossRef]
36. Zumdick, N.A.; Jauer, L.; Kersting, L.C.; Kutz, T.N.; Schleifenbaum, J.H.; Zander, D. Additive Manufactured WE43 Magnesium: A Comparative Study of the Microstructure and Mechanical Properties with Those of Powder Extruded and as-Cast WE43. *Mater. Charact.* **2019**, *147*, 384–397. [CrossRef]
37. Agarwal, S.; Curtin, J.; Duffy, B.; Jaiswal, S. Biodegradable Magnesium Alloys for Orthopaedic Applications: A Review on Corrosion, Biocompatibility and Surface Modifications. *Mater. Sci. Eng. C* **2016**, *68*, 948–963. [CrossRef] [PubMed]
38. Castellani, C.; Lindtner, R.A.; Hausbrandt, P.; Tschegg, E.; Stanzl-Tschegg, S.E.; Zanoni, G.; Beck, S.; Weinberg, A.-M. Bone–Implant Interface Strength and Osseointegration: Biodegradable Magnesium Alloy versus Standard Titanium Control. *Acta Biomater.* **2011**, *7*, 432–440. [CrossRef] [PubMed]
39. Fox, C.; Ramsoomair, D.; Carter, C. Magnesium: Its Proven and Potential Clinical Significance. *South. Med. J.* **2001**, *94*, 1195–1201. [CrossRef]
40. Prasad, K.; Bazaka, O.; Chua, M.; Rochford, M.; Fedrick, L.; Spoor, J.; Symes, R.; Tieppo, M.; Collins, C.; Cao, A.; et al. Metallic Biomaterials: Current Challenges and Opportunities. *Materials* **2017**, *10*, 884. [CrossRef]
41. Seelig, M.G. A Study of Magnesium Wire as an Absorbable Suture and Ligature Material. *Arch. Surg.* **1924**, *8*, 669–680. [CrossRef]
42. Kammer, C. *Magnesium-Taschenbuch*; Alu Media: Berlin, Germany, 2011; ISBN 978-3-87017-264-0.
43. Kumar, K.; Gill, R.S.; Batra, U. Challenges and Opportunities for Biodegradable Magnesium Alloy Implants. *Mater. Technol.* **2018**, *33*, 153–172. [CrossRef]
44. Chamy, R.; Rosenkranz, F.; Chamy, R.; Rosenkranz, F. *Biodegradation—Engineering and Technology*; InTechOpen: Rijeka, Croatia, 2013; ISBN 978-953-51-1153-5.
45. Han, J.; Wan, P.; Ge, Y.; Fan, X.; Tan, L.; Li, J.; Yang, K. Tailoring the Degradation and Biological Response of a Magnesium–Strontium Alloy for Potential Bone Substitute Application. *Mater. Sci. Eng. C* **2016**, *58*, 799–811. [CrossRef]
46. Sukhodub, L.; Panda, A.; Dyadyura, K.; Pandova, I.; Krenicky, T. The Design Criteria for Biodegradable Magnesium Alloy Implants. *MM Sci. J.* **2018**, *12*, 2673–2679. [CrossRef]
47. Li, R.W.; Kirkland, N.T.; Truong, J.; Wang, J.; Smith, P.N.; Birbilis, N.; Nisbet, D.R. The Influence of Biodegradable Magnesium Alloys on the Osteogenic Differentiation of Human Mesenchymal Stem Cells: Osteogenic Differentiation of Human Mesenchymal Stem Cells. *J. Biomed. Mater. Res.* **2014**, *102*, 4346–4357. [CrossRef] [PubMed]
48. *ISO 10993-5:2009+A11:2025*; Biological Evaluation of Medical Devices—Tests for In Vitro Cytotoxicity. British Standards Institution: London, UK, 2025.
49. Li, N.; Zheng, Y. Novel Magnesium Alloys Developed for Biomedical Application: A Review. *J. Mater. Sci. Technol.* **2013**, *29*, 489–502. [CrossRef]
50. Pogorielov, M.; Husak, E.; Solodivnik, A.; Zhdanov, S. Magnesium-Based Biodegradable Alloys: Degradation, Application, and Alloying Elements. *Interv. Med. Appl. Sci.* **2017**, *9*, 27–38. [CrossRef]
51. Chaya, A.; Yoshizawa, S.; Verdelis, K.; Myers, N.; Costello, B.J.; Chou, D.-T.; Pal, S.; Maiti, S.; Kumta, P.N.; Sfeir, C. In Vivo Study of Magnesium Plate and Screw Degradation and Bone Fracture Healing. *Acta Biomater.* **2015**, *18*, 262–269. [CrossRef]
52. Kannan, M.B.; Raman, R.K.S. In Vitro Degradation and Mechanical Integrity of Calcium-Containing Magnesium Alloys in Modified-Simulated Body Fluid. *Biomaterials* **2008**, *29*, 2306–2314. [CrossRef]
53. Persaud-Sharma, D.; McGoron, A. Biodegradable Magnesium Alloys: A Review of Material Development and Applications. *J. Biomim. Biomater. Tissue Eng.* **2012**, *12*, 25–39. [CrossRef]
54. Xin, Y.; Hu, T.; Chu, P.K. In Vitro Studies of Biomedical Magnesium Alloys in a Simulated Physiological Environment: A Review. *Acta Biomater.* **2011**, *7*, 1452–1459. [CrossRef]
55. Wu, G.; Ibrahim, J.M.; Chu, P.K. Surface Design of Biodegradable Magnesium Alloys—A Review. *Surf. Coat. Technol.* **2013**, *233*, 2–12. [CrossRef]
56. Kirkland, N.T.; Staiger, M.P.; Nisbet, D.; Davies, C.H.J.; Birbilis, N. Performance-Driven Design of Biocompatible Mg Alloys. *JOM* **2011**, *63*, 28–34. [CrossRef]
57. Aggarwal, D.; Sharma, S.; Gupta, M. Impact of Rare Earth Particulates Addition on the Wear Rate of Magnesium Composites with Improved Mechanical and Microstructural Properties for Orthopedic Applications. *Prog. Compos. Mater.* **2024**, *1*, 2. [CrossRef]
58. McGaughey, S.A.; Iqbal, S.; De Rosa, A.; Caley, J.A.; Kaksonen, A.H.; Villa-Gomez, D.; Byrt, C.S. Interactions of Rare Earth Elements with Living Organisms and Emerging Biotechnical Applications. *Plants People Planet* **2025**. [CrossRef]
59. Witte, F.; Hort, N.; Vogt, C.; Cohen, S.; Kainer, K.U.; Willumeit, R.; Feyerabend, F. Degradable Biomaterials Based on Magnesium Corrosion. *Curr. Opin. Solid. State Mater. Sci.* **2008**, *12*, 63–72. [CrossRef]
60. Shalomeev, V.; Tabunshchyk, G.; Matiukhin, A.; Shyrobokov, V.; Shyrobokova, N.; Hornostai, V.; Kulabneva, E. Refining of Biosoluble Alloy of Mg-Nd-Zr System for Manufacture of Implants. In Proceedings of the 31st International Conference on Metallurgy and Materials, Brno, Czech Republic, 18–19 May 2022; pp. 622–627.

61. Eddy Jai Poinern, G.; Brundavanam, S.; Fawcett, D. Biomedical Magnesium Alloys: A Review of Material Properties, Surface Modifications and Potential as a Biodegradable Orthopaedic Implant. *Asian J. Bus. Ethics* **2013**, *2*, 218–240. [CrossRef]
62. Markov, O.E.; Khvashchynskyi, A.S.; Musorin, A.V.; Markova, M.A.; Shapoval, A.A.; Hrudkina, N.S. Investigation of new method of large ingots forging based on upsetting of workpieces with ledges. *Int. J. Adv. Manuf. Technol.* **2022**, *122*, 1383–1394. [CrossRef]
63. Li, L.-Y.; Cui, L.-Y.; Zeng, R.-C.; Li, S.-Q.; Chen, X.-B.; Zheng, Y.; Kannan, M.B. Advances in Functionalized Polymer Coatings on Biodegradable Magnesium Alloys—A Review. *Acta Biomater.* **2018**, *79*, 23–36. [CrossRef]
64. Khrebtova, O.; Shapoval, O.; Markov, O.; Kukhar, V.; Hrudkina, N.; Rudych, M. Control systems for the temperature field during drawing, taking into account the dynamic modes of the technological installation. In Proceedings of the 2022 IEEE 4th International Conference on Modern Electrical and Energy System, MEES 2022, Kremenchuk, Ukraine, 20–23 October 2022; IEEE: New York, NY, USA, 2022; ISBN 979-835034683-1. [CrossRef]
65. Anishchenko, O.S.; Kukhar, V.V.; Prysyazhnyi, A.H.; Agarkov, V.V.; Klimov, E.S.; Chernenko, S.M. The material for physical simulation of metal-forming processes in super-plastic state. *IOP Conf. Ser. Mater. Sci. Eng.* **2019**, *473*, 012040. [CrossRef]
66. Anishchenko, O.; Kukhar, V.; Artiukh, V.; Trebukhin, A.; Zotkina, N. Effect of blank curvature and thinning on shell stresses at superplastic forming. *Adv. Intell. Syst. Comput.* **2020**, *982*, 809–817. [CrossRef]
67. Anishchenko, A.; Kukhar, V.; Artiukh, V.; Arkhipova, O. Application of G. Lame's and J. Gielis' formulas for description of shells superplastic forming. *MATEC Web Conf.* **2018**, *239*, 06007. [CrossRef]
68. Zeng, R.-C.; Cui, L.; Jiang, K.; Liu, R.; Zhao, B.-D.; Zheng, Y.-F. In Vitro Corrosion and Cytocompatibility of a Microarc Oxidation Coating and Poly(l-Lactic Acid) Composite Coating on Mg–1Li–1Ca Alloy for Orthopedic Implants. *ACS Appl. Mater. Interfaces* **2016**, *8*, 10014–10028. [CrossRef]
69. Ostrowski, N.J.; Lee, B.; Roy, A.; Ramanathan, M.; Kumta, P.N. Biodegradable Poly(Lactide-Co-Glycolide) Coatings on Magnesium Alloys for Orthopedic Applications. *J. Mater. Sci. Mater. Med.* **2013**, *24*, 85–96. [CrossRef]
70. Abdal-hay, A.; Amna, T.; Lim, J.K. Biocorrosion and Osteoconductivity of PCL/nHAp Composite Porous Film-Based Coating of Magnesium Alloy. *Solid. State Sci.* **2013**, *18*, 131–140. [CrossRef]
71. Bai, K.; Zhang, Y.; Fu, Z.; Zhang, C.; Cui, X.; Meng, E.; Guan, S.; Hu, J. Fabrication of Chitosan/Magnesium Phosphate Composite Coating and the in Vitro Degradation Properties of Coated Magnesium Alloy. *Mater. Lett.* **2012**, *73*, 59–61. [CrossRef]
72. Wang, Y.; Qin, Q.; Liu, X.; Li, Q.; Zhang, H.; Yang, D.; Jiang, Y. Biocompatibility, Degradation Behavior, and Mechanical Properties of Magnesium Alloy Plates In Vivo. *Appl. Bionics Biomech.* **2023**, *2023*, 4436396. [CrossRef]
73. Dziuba, D.; Meyer-Lindenberg, A.; Seitz, J.M.; Waizy, H.; Angrisani, N.; Reifenrath, J. Long-Term in Vivo Degradation Behaviour and Biocompatibility of the Magnesium Alloy ZEK100 for Use as a Biodegradable Bone Implant. *Acta Biomater.* **2013**, *9*, 8548–8560. [CrossRef] [PubMed]
74. Abhishek, M.S.; Satish, J.; Anshu, D.; Debrupa, L.; Arup Kumar, D. Biocompatibility and Biodegradability Evaluation of Magnesium-Based Intramedullary Bone Implants in Avian Model. *J. Biomed. Mater. Res. A* **2021**, *109*, 1479–1489. [CrossRef]
75. Xi, Z.; Wu, Y.; Xiang, S.; Sun, C.; Wang, Y.; Yu, H.; Fu, Y.; Wang, X.; Yan, J.; Zhao, D.; et al. Corrosion Resistance and Biocompatibility Assessment of a Biodegradable Hydrothermal-Coated Mg-Zn-Ca Alloy: An in Vitro and in Vivo Study. *ACS Omega* **2020**, *5*, 4548–4557. [CrossRef]
76. Angrisani, N.; Reifenrath, J.; Zimmermann, F.; Eifler, R.; Meyer-Lindenberg, A.; Vano-Herrera, K.; Vogt, C. Biocompatibility and Degradation of LAE442-Based Magnesium Alloys after Implantation of up to 3.5 Years in a Rabbit Model. *Acta Biomater.* **2016**, *44*, 355–365. [CrossRef]
77. Sanchez, A.H.M.; Luthringer, B.J.C.; Feyerabend, F.; Willumeit, R. Mg and Mg Alloys: How Comparable Are in Vitro and in Vivo Corrosion Rates? A Review. *Acta Biomater.* **2015**, *13*, 16–31. [CrossRef]
78. Tao, M.; Cui, Y.; Sun, S.; Zhang, Y.; Ge, J.; Yin, W.; Li, P.; Wang, Y. Versatile Application of Magnesium-Related Bone Implants in the Treatment of Bone Defects. *Mater. Today Bio* **2025**, *31*, 101635. [CrossRef]
79. Zhao, D.-W.; Zhang, T.-W.; Wang, W.-D.; Wang, Z.-M.; Wang, F.-Y.; Yang, X.; Li, R.-H.; Cheng, L.-L.; Zhang, Y.; Wang, H.-Y.; et al. Internal Fixation with Biodegradable High Purity Magnesium Screws in the Treatment of Ankle Fracture. *J. Orthop. Transl.* **2025**, *51*, 198–206. [CrossRef]
80. Seetharaman, S.; Sankaranarayanan, D.; Gupta, M. Magnesium-Based Temporary Implants: Potential, Current Status, Applications, and Challenges. *J. Funct. Biomater.* **2023**, *14*, 324. [CrossRef]
81. Zheng, L.; Zhang, R.; Chen, X.; Luo, Y.; Du, W.; Zhu, Y.; Ruan, Y.C.; Xu, J.; Wang, J.; Qin, L. Chronic Kidney Disease: A Contraindication for Using Biodegradable Magnesium or Its Alloys as Potential Orthopedic Implants? *Biomed. Mater.* **2024**, *19*, 045023. [CrossRef]
82. Shalomeev, V.; Tabunshchyk, G.; Greshta, V.; Nykiel, M.; Korniejenko, K. Influence of Alkaline Earth Metals on Structure Formation and Magnesium Alloy Properties. *Materials* **2022**, *15*, 4341. [CrossRef] [PubMed]
83. Zhou, K.; Lu, Q.; Qin, J.; Shi, H.; Zhang, P.; Yan, H.; Shi, H.; Wang, X. A View of Magnesium Alloy Modification and Its Application in Orthopedic Implants. *J. Mater. Res. Technol.* **2025**, *36*, 1536–1561. [CrossRef]

84. Sharma, S.K.; Gajević, S.; Sharma, L.K.; Pradhan, R.; Miladinović, S.; Ašonja, A.; Stojanović, B. Magnesium-Titanium Alloys: A Promising Solution for Biodegradable Biomedical Implants. *Materials* **2024**, *17*, 5157. [CrossRef] [PubMed]
85. Zhang, T.; Wang, W.; Liu, J.; Wang, L.; Tang, Y.; Wang, K. A Review on Magnesium Alloys for Biomedical Applications. *Front. Bioeng. Biotechnol.* **2022**, *10*, 953344. [CrossRef]
86. Shalomeev, V.; Tsivirco, E.; Vnukov, Y.; Osadchaya, Y.; Makovskyi, S. Development of New Casting Magnesium-Based Alloys with Increased Mechanical Properties. *East.-Eur. J. Enterp. Technol.* **2016**, *4*, 4–10. [CrossRef]
87. Sreenivasamurthy, S.A.; Akhter, F.F.; Akhter, A.; Su, Y.; Zhu, D. Cellular Mechanisms of Biodegradable Zinc and Magnesium Materials on Promoting Angiogenesis. *Biomater. Adv.* **2022**, *139*, 213023. [CrossRef]
88. He, X.; Li, Y.; Zou, D.; Zu, H.; Li, W.; Zheng, Y. An Overview of Magnesium-Based Implants in Orthopaedics and a Prospect of Its Application in Spine Fusion. *Bioact. Mater.* **2024**, *39*, 456–478. [CrossRef]
89. Mehdizade, M.; Eivani, A.R.; Esmaielzadeh, O.; Rostamian, P. ZnO and Cu/ZnO-Modified Magnesium Orthopedic Implant with Improved Osteoblast Cellular Activity: An in-Vitro Study. *J. Mater. Res. Technol.* **2024**, *28*, 935–950. [CrossRef]
90. Shalomeev, V.; Sheyko, S.; Hrechanyi, O.; Vasilchenko, T.; Hrechana, A. Research of the Boundaries of the "Metal Filter" Section during Filtration of Magnesium Alloy AZ91. *Appl. Eng. Sci.* **2025**, *21*, 100212. [CrossRef]
91. Greshta, V.; Shalomeev, V.; Bovkun, S.; Kulabneva, E. Practical Aspects of Obtaining Deformed Blanks for the Production of Biodegradable Implants of Increased Quality. *J. Chem. Technol. Metall.* **2025**, *60*, 225–236. [CrossRef]
92. Vignesh, P.; Ramanathan, S.; Ashokkumar, M.; Ananthi, V. Biodegradable Mg–3Zn Alloy/Titanium–Hydroxyapatite Hybrid Composites: Corrosion and Cytotoxicity Evaluation for Orthopedic Implant Applications. *Trans. Indian. Inst. Met.* **2024**, *77*, 1701–1710. [CrossRef]
93. Sonaye, S.Y.; Dal-Fabbro, R.; Bottino, M.C.; Sikder, P. Osseointegration of 3D-Printable Polyetheretherketone–Magnesium Phosphate Bioactive Composites for Craniofacial and Orthopedic Implants. *ACS Biomater. Sci. Eng.* **2025**, *11*, 1060–1071. [CrossRef] [PubMed]
94. Liang, J.; He, Y.; Jia, R.; Li, S.; Duan, L.; Xu, S.; Mei, D.; Tang, X.; Zhu, S.; Wei, J.; et al. Enhancing the Corrosion Resistance of Magnesium Alloys with Biodegradable Poly(Trimethylene Carbonate) Chemical Modification Coating. *Mater. Today Adv.* **2024**, *21*, 100460. [CrossRef]
95. Mohanta, M.; Ramdhun, Y.; Thirugnanam, A.; Gupta, R.; Verma, D.; Deepak, T.; Babu, A.R. Biodegradable AZ91 Magnesium Alloy/Sirolimus/Poly D, L-lactic-co-glycolic Acid-based Substrate for Cardiovascular Device Application. *J. Biomed. Mater. Res.* **2024**, *112*, e35350. [CrossRef] [PubMed]
96. Shi, B.; Xu, J.; Zou, J.; Li, Y.R.; Zhou, Z.; Liu, K.; Jia, Q.; Jiang, H.B. Advances in Amelioration of Plasma Electrolytic Oxidation Coatings on Biodegradable Magnesium and Alloys. *Heliyon* **2024**, *10*, e24348. [CrossRef]
97. Wu, G.-L.; Yen, C.-E.; Hsu, W.-C.; Yeh, M.-L. Incorporation of Cerium Oxide Nanoparticles into the Micro-Arc Oxidation Layer Promotes Bone Formation and Achieves Structural Integrity in Magnesium Orthopedic Implants. *Acta Biomater.* **2025**, *191*, 80–97. [CrossRef]
98. Li, H.; Wang, F.; Sun, X.-J.; Qi, L.-J.; Li, J.-A.; Iqbal, M.; Guan, S.-K. Bio-Inspired Barnacle Cement Coating of Biodegradable Magnesium Alloy for Cerebrovascular Application. *Rare Met.* **2024**, *43*, 5164–5185. [CrossRef]
99. Saberi, A.; Bakhsheshi-Rad, H.R.; Abazari, S.; Ismail, A.F.; Sharif, S.; Ramakrishna, S.; Daroonparvar, M.; Berto, F. A Comprehensive Review on Surface Modifications of Biodegradable Magnesium-Based Implant Alloy: Polymer Coatings Opportunities and Challenges. *Coatings* **2021**, *11*, 747. [CrossRef]
100. Liu, J.; Yin, B.; Song, F.; Liu, B.; Peng, B.; Wen, P.; Tian, Y.; Zheng, Y.; Ma, X.; Wang, C. Improving Corrosion Resistance of Additively Manufactured WE43 Magnesium Alloy by High Temperature Oxidation for Biodegradable Applications. *J. Magnes. Alloys* **2024**, *12*, 940–953. [CrossRef]
101. Fu, Q.; Liang, W.; Huang, J.; Jin, W.; Guo, B.; Li, P.; Xu, S.; Chu, P.K.; Yu, Z. Research Perspective and Prospective of Additive Manufacturing of Biodegradable Magnesium-Based Materials. *J. Magnes. Alloys* **2023**, *11*, 1485–1504. [CrossRef]
102. Zhi, P.; Liu, L.; Chang, J.; Liu, C.; Zhang, Q.; Zhou, J.; Liu, Z.; Fan, Y. Advances in the Study of Magnesium Alloys and Their Use in Bone Implant Material. *Metals* **2022**, *12*, 1500. [CrossRef]
103. Dachasa, K.; Chuni Aklilu, T.; Gashaw Ewnete, B.; Mosisa Ejeta, B.; Fufa Bakare, F. Magnesium-Based Biodegradable Alloy Materials for Bone Healing Application. *Adv. Mater. Sci. Eng.* **2024**, *2024*, 1325004. [CrossRef]
104. Magnesium Implant Market Key Challenges, Growth and Opportunities 2025. Available online: https://www.linkedin.com/pulse/magnesium-implant-market-key-challenges-growth-eplyf (accessed on 12 June 2025).
105. Amukarimi, S.; Mozafari, M. Biodegradable Magnesium Biomaterials—Road to the Clinic. *Bioengineering* **2022**, *9*, 107. [CrossRef]
106. Thomas, K.K.; Zafar, M.N.; Pitt, W.G.; Husseini, G.A. Biodegradable Magnesium Alloys for Biomedical Implants: Properties, Challenges, and Surface Modifications with a Focus on Orthopedic Fixation Repair. *Appl. Sci.* **2023**, *14*, 10. [CrossRef]
107. Nuthana Kalva, S.; Ali, F.; Koç, M. Recent Advances in the Post-Processing of Magnesium Based Scaffolds for Orthopedic Applications. *Next Mater.* **2025**, *6*, 100295. [CrossRef]
108. Kazemzadeh-Narbat, M. FDA Regulatory Challenges for Magnesium-Based Orthopedic Devices. *ODT*, 21 May 2024.

109. Degradable Medical Magnesium Implants Market. Available online: https://www.linkedin.com/pulse/degradable-medical-magnesium-implants-market-wander-whiz-zd8ef (accessed on 12 June 2025).
110. Amin, V.Y.; Nallajennugari, S.R.; Chahal, R.S. The Efficacy of Biodegradable Magnesium-Based Implants in Orthopedic Surgery. *Int. Arch. Orthop. Surg.* **2025**, *8*, 1–9. [CrossRef]
111. Data Bridge Market Research. *North America Magnesium Alloys Market Report—Industry Trends and Forecast to 2029*; Data Bridge Market Research: Pune, India, 2022.
112. Ibrahem Khaled, N.; Santhiya, D. Zein-Bioactive Glass Nanocomposite Coating on Magnesium Alloy Substrate for Orthopedic Applications. *Adv. Compos. Mater.* **2024**, *33*, 324–346. [CrossRef]
113. De Luca, A.; Ruggiero, R.; Cordaro, A.; Marrelli, B.; Raimondi, L.; Costa, V.; Bellavia, D.; Aiello, E.; Pavarini, M.; Piccininni, A.; et al. Towards Accurate Biocompatibility: Rethinking Cytotoxicity Evaluation for Biodegradable Magnesium Alloys in Biomedical Applications. *J. Funct. Biomater.* **2024**, *15*, 382. [CrossRef]
114. Cao, B.; Cao, L.; Kallmes, D.F. Degradation Rate Assessment of Biodegradable Magnesium Alloys. *Mater. Sci. Appl.* **2024**, *15*, 245–252. [CrossRef]
115. Barzegari, M.; Mei, D.; Lamaka, S.V.; Geris, L. Computational Modeling of Degradation Process of Biodegradable Magnesium Biomaterials. *Corros. Sci.* **2021**, *190*, 109674. [CrossRef]
116. Yamamoto, A.; Kikuta, A. Development of a Model System for Gas Cavity Formation Behavior of Magnesium Alloy Implantation. *ACS Biomater. Sci. Eng.* **2022**, *8*, 2437–2444. [CrossRef]
117. *ISO 10993-12:2021*; Biological Evaluation of Medical Devices—Part 12: Sample Preparation and Reference Materials. International Organization for Standardization: Geneva, Switzerland, 2021.
118. Amaral, C.; Paiva, M.; Rodrigues, A.R.; Veiga, F.; Bell, V. Global Regulatory Challenges for Medical Devices: Impact on Innovation and Market Access. *Appl. Sci.* **2024**, *14*, 9304. [CrossRef]
119. *ISO/TS 20721:2025*; Implants for Surgery—Absorbable Implants—General Guidelines and Requirements for Assessment of Absorbable Metallic Implants. International Organization for Standardization: Geneva, Switzerland, 2025.
120. Müller, E.; Schoberwalter, T.; Mader, K.; Seitz, J.-M.; Kopp, A.; Baranowsky, A.; Keller, J. The Biological Effects of Magnesium-Based Implants on the Skeleton and Their Clinical Implications in Orthopedic Trauma Surgery. *Biomater. Res.* **2024**, *28*, 0122. [CrossRef]
121. Chen, H.; Wang, Y.; He, L.; Zhang, X.; Mei, Y.; Wu, T.; Wang, J.; Zheng, Y.; Tang, H. Surface Engineering of Biodegradable Magnesium Alloys as Orthopedic Implant Materials: Recent Developments and Future Prospects. *Coatings* **2025**, *15*, 191. [CrossRef]
122. Ji, Y.; Hou, M.; Zhang, J.; Wang, T.; Cao, C.; Yang, H.; Zhang, X. Surface Modification of WE43 Magnesium Alloys with Dopamine Hydrochloride Modified GelMA Coatings. *Coatings* **2022**, *12*, 1074. [CrossRef]
123. Gu, X.N.; Xie, X.H.; Li, N.; Zheng, Y.F.; Qin, L. In Vitro and in Vivo Studies on a Mg–Sr Binary Alloy System Developed as a New Kind of Biodegradable Metal. *Acta Biomater.* **2012**, *8*, 2360–2374. [CrossRef] [PubMed]
124. Wang, Y.; Sun, Z.; Sun, Y. Effects of a Thaw Slump on Active Layer in Permafrost Regions with the Comparison of Effects of Thermokarst Lakes on the Qinghai–Tibet Plateau, China. *Geoderma* **2018**, *314*, 47–57. [CrossRef]
125. Garimella, A.; M, R.; Ghosh, S.B.; Bandyopadhyay-Ghosh, S.; Agrawal, A.K. Bioactive Fluorcanasite Reinforced Magnesium Alloy-Based Porous Bio-Nanocomposite Scaffolds with Tunable Mechanical Properties. *J. Biomed. Mater. Res. Part. B Appl. Biomater.* **2023**, *111*, 463–477. [CrossRef]
126. Wei, X.; Ma, S.; Meng, J.; Qing, H.; Zhao, Q. In Vitro Evaluation of a Novel Mg–Sn–Ge Ternary Alloy for Orthopedic Applications. *J. Alloys Compd.* **2023**, *953*, 169813. [CrossRef]

crystals

MDPI

Article

Microstructure, Mechanical Properties, Deformation Behavior, and Crystallographic Texture of the Al-Gd-Cr-Ti Quaternary Alloy for Thermal Neutron Absorption

Sayed M. Amer [1,2,3,*], Dmitry I. Nikolayev [3], Tatiana A. Lychagina [3], Abdelmoneim El-Khouly [2,3,4], Ruslan Yu. Barkov [5,*], Alexey S. Prosviryakov [5], Anastasia V. Mikhaylovskaya [5], Maria V. Glavatskikh [5] and Andrey V. Pozdniakov [5]

1 Faculty of Engineering, Mining, Metallurgy and Petroleum Engineering Department, Al-Azhar University, Cairo 11884, Egypt
2 Academy of Scientific Research and Technology, ASRT, Cairo 11516, Egypt; abdelmoneim.elkhouly@sci.dmu.edu.eg
3 Joint Institute for Nuclear Research, JINR, 141980 Dubna, Russia; dmitry@nf.jinr.ru (D.I.N.); lychagina@jinr.ru (T.A.L.)
4 Physics Department, Faculty of Science, Damanhour University, Damanhour 22516, Egypt
5 Department of Physical Metallurgy of Non-Ferrous Metals, National University of Science and Technology "MISIS", Leninskiy Prospekt 4, 119991 Moscow, Russia; pro.alex@mail.ru (A.S.P.); mihaylovskaya@misis.ru (A.V.M.); glavatskikh@edu.misis.ru (M.V.G.); pozdniakov@misis.ru (A.V.P.)
* Correspondence: elsayedmohamed.2214@azhar.edu.eg (S.M.A.); barkov@misis.ru (R.Y.B.)

Abstract

In this work, we report the identification of a novel quaternary intermetallic phase ($Al_{21}GdCrTi$) formed during the solidification of a novel Al-Gd-Cr-Ti alloy, which has not been previously documented in the literature to the best of our knowledge. The study also provides a detailed analysis of microstructure evolution, texture behavior, and the mechanical strengthening effect of rolling processes, along with neutron absorption performance. XRD analysis reveals that the intensity of (022), (113) planes of the as-hot-cold-rolled sample is higher than that of the as-cast due to the change in the direction of some grains in these planes during rolling. The results indicate that the studied alloys scatter neutrons about 100 times less than a nearly pure aluminum alloy. The hardness of the as-cast alloy increased from 36 to 53 HV after cold rolling and to 50 HV after hot rolling-cold rolling. Hot-cold-rolled alloy has a yield strength of 160 MPa and an ultimate tensile strength of 181 MPa, while maintaining an elongation of 11.3%. The studied alloys, containing 4.2 wt.% of the alloying elements 3.8Gd, 0.2Cr, and 0.2Ti (Al-3.8Gd-0.2Cr-0.2Ti), exhibited a yield strength 28 MPa higher than those containing 21 wt.% of the alloying elements 5Cu, 6Gd, and 8Bi (Al-5Cu-6Gd-8Bi). The studied alloys form the basis for the development of high-technology Al-Gd alloys for neutron shielding.

Keywords: aluminum alloys; Al-Gd-Cr-Ti; mechanical properties; deformation behavior; crystallographic texture; neutron absorption

1. Introduction

Materials utilized in designing applications can be strong or ductile, but they are seldom consistent with each other. Numerous auxiliary applications, such as transportation, spacecraft, and nuclear plants, require stronger and tougher materials for more noteworthy vitality, productivity, and progressed performance to meet the wants of the industry and its

 https://doi.org/10.3390/cryst15070616

progressing advancement [1,2]. From past information derived from literature readings, a weak lattice is fortified with a dispersion hardener, the second phase of crystallization origin, or solid solution strengthening [3]. The more imperative address is whether there are unused techniques, however, to be investigated to form the following era of metals and alloys with a "quantum leap" in strength and ductility, rather than the conventional improvements we have seen over the past century.

One of the most important materials used in the automotive, aerospace, and other industries is aluminum alloys due to their resistance to high temperatures and their mechanical properties [2,4–8]. Some researchers have recently suggested that it is possible to enhance tensile strength, heat resistance, and corrosion resistance by adding rare earth elements to aluminum alloys [9–13]. Al-Gd alloy may be a ternary, quaternary, or higher composition that incorporates copper, magnesium, silicon, zirconium, and manganese [4,14]. The aluminum and gadolinium may be primary ingredients (e.g., components) of aluminum alloys. Al may function as a solid solution matrix in the aluminum alloys, and Gd may function as a eutectic structure with Al [15]. The Al-Gd phase diagram shows that the solubility of Gd in aluminum is very low, and it mainly forms the second phases Al_3Gd, Al_2Gd, and AlGd along the grain boundaries. Moreover, Gd forms a variety of phases ideally with Al-Gd alloy elements [16]. By including 0.13 Gd and 0.07 Y wt.% to the Al-Zn-Mg-Cu-Zr alloy, the dispersed Al_3CuGd altogether progressed the microstructure and anticipated advanced recrystallization, resulting in changes in both strength (639 MPa) and elongation (16.8%) [17].

Crystallographic texture studies currently cover polycrystalline materials of various natures, namely metals, alloys, ceramics, rocks, and biological objects [18–20]. In metallurgy, the term texture is used to refer to the preferred crystal orientation in polycrystalline aggregates [21]. The study of material texture has become essential due to its significant impact on evaluating the properties and engineering of materials over the past few decades [22,23]. The fatigue behavior of 2xxx aluminum alloys is significantly affected by different textures, including crack initiation, closure, and propagation. Rolling is a complex thermo-mechanical process, and especially during hot rolling, deformation is always accompanied by recrystallization, which also improves the structure and texture [24–26]. The texture development of aluminum alloys during the rolling and annealing process has been a subject of research for a long time, as the anisotropy of mechanical properties is greatly affected by the texture developed during processing [27–31]. Thermal neutron absorption is essential in reactors such as space and vehicular reactors used in engineering applications [15]. In these systems, a significant portion of the mass is attributed to radiation-absorbing materials. In viable applications, reactor working temperatures within the fundamental coolant lines ordinarily extend between 280 and 320 °C in long-term benefit [32]. Gadolinium (Gd) is an element known for its high thermal neutron absorption cross section (48,800 b). In any case, adding a high percentage of Gd to aluminum alloys results in the precipitation of excess Gd atoms at grain boundaries, forming an intermediate phase known as Al_3Gd [33].

The main idea of this study is to analyze and evaluate the microstructure, crystallographic texture, and mechanical properties of Al-Gd-Cr-Ti alloys for thermal neutron absorption.

2. Materials and Methods

Al-3.8Gd-0.2Cr-0.2Ti alloy was prepared by casting metallurgy in a graphite crucible in an electric resistance furnace, poured into a copper water-cooled mold with a cavity size of 20 × 40 × 120 mm^3, with a cooling rate of about 15 Ks^{-1}. The raw materials for casting were pure Al grade A99 and master alloys of Al-12Gd, Al-10Cr, and Al-5Ti. The melt temperature before casting was 800 °C. The nominal composition of the Al-Gd-Cr-Ti alloy is shown in Table 1.

Table 1. Chemical composition of the studied alloy, wt.%.

Alloy	Gd	Cr	Ti	Al
Al-Gd-Cr-Ti	3.8	0.2	0.2	bal.

The solidus temperature is determined based on the differential scanning Calorimeter (DSC) results. The experiments were performed in an air atmosphere with a heating and cooling rate of 5 °C/min. X-ray diffraction (XRD) data (using Cu-Kα radiation) were collected at the laboratory of FLNP, JINR (Dubna, Moscow region, Russia) using an Empyrean (Malvern PANalytical, Dubna, Moscow, Russia) (Figure 1a). Evaluations were carried out on specimens with a plane surface in reflection geometry. Prior to the evaluation of the texture experiment, X-ray phase analysis was conducted utilizing the High Score Plus software package, version 4.1 (PANalytical, Dubna, Moscow, Russia), and the PDF-2 database. The X-ray scanning parameters are X-ray tube with nickel-filtered copper Kα radiation (λ = 1.54 Å), voltage 40 kV, current 40 mA, and total exposure time 33 min. The X-ray texture experiment is the most common type of texture experiment and is used for mass texture studies [34–36]. X-ray pole figure measurements are based on diffraction that occurs when X-rays are scattered by the electron shells of atoms in a crystal lattice. The neutron diffraction used in this research was carried out at the laboratory of FLNP, JINR (Dubna). We utilized the SKAT (from Russian: Spektrometer Kolitshestvennogo Analiza Tekstury, Spectrometer for the Quantitative Analysis of Textures) setup situated at the 7A-2 beamline of the high-flux pulsed nuclear reactor IBR-2 (Figure 1b). Spektrometer Kolitshestvennogo Analiza Tekstury (SKAT) is composed of a detector ring (2 m in diameter), on which are pinpointed 19 detector-collimator complexes at the same scattering angle 2θ = 90°. Keppler et al. [37] reported that the long flight path of over 100 m and the collimator system supply a relatively high resolution of $\Delta d/d = 5 \times 10^{-3}$ at d = 2.5 Å and 2θ = 90°.

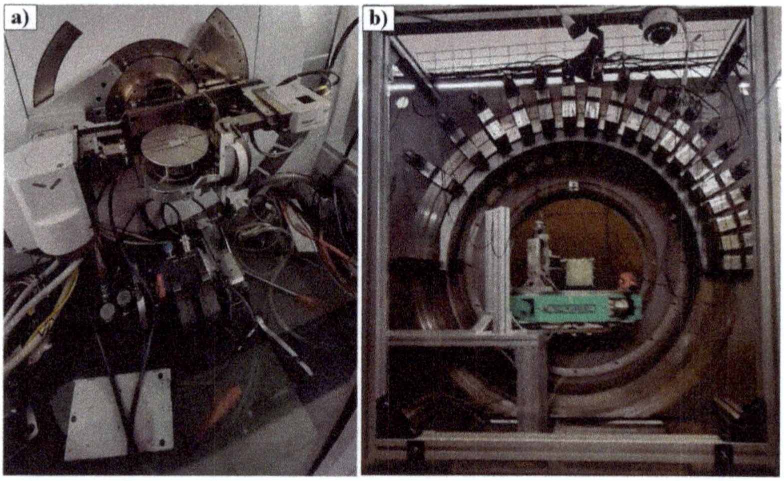

Figure 1. (**a**) XRD experimental setup: Malvern PANalytical EMPYREAN, (**b**) SKAT instrument.

Homogenization of Al-Gd-Cr-Ti alloy was carried out at 615 °C for 1h, followed by quenching in water. Heat treatment of the Al-3.8Gd-0.2Cr-0.2Ti ingot was carried out in Nabertherm and SNOL furnaces with a fan and a temperature control accuracy of 1 °C.

The heat-treated ingot was rolled utilizing two diverse frameworks. To begin with, it was cold rolled from a thickness of 20 to 1 mm, moment hot rolled from a thickness of 20 to 10 mm at 440 °C, and after that cold rolled to a thickness of 1 mm at room temperature.

A Struers LaboPol-5 polishing machine (Struers APS, Ballerup, Denmark) was utilized to prepare specimens for microstructure analysis before and after deformation, including the stages of mechanical grinding and polishing. The surface of specimens was ground on sandpaper with different grain sizes of 220, 320, 800, 1200, 2400, and 4000, respectively. The specimens were then mechanically polished using OP-S suspension.

Microstructural examination and phase recognition were implemented on a TESCAN VEGA 3LMH scanning electron microscope (SEM) (Tescan, Brno, Kohoutovice, Czech Republic) utilizing an energy-dispersive X-ray detector X-Max 80 (EDX) (Oxford Instruments Advanced AZtecEnergy, High Wycombe, UK), and a Bruker D8 Advance X-ray diffractometer (Bruker, Billerica, MA, USA).

A Zwick/Roell Z250 universal testing machine (Zwick/Roell, Kennesaw, GA, USA) with an automatic longitudinal strain sensor was used to measure the tensile properties at a tensile rate of 4 mm/min. The tensile test specimens were cut out from 1.5 mm-thick rolled sheets. Ultimate tensile strength (UTS), yield strength (YS), and elongation (EL) to failure were evaluated. The hardness was measured by the standard Vickers procedure. The Vickers hardness experiment was performed with a load of 5 kg and a dwell time of 10 s. The hardness number HV was chosen as the number-crunching pitiless of five estimations for rolled cases and of ten estimations for cast. The botch inside the confirmation did not outperform 3 HV.

3. Results and Discussion

Microstructure (SEM) of the as-cast Al-3.8Gd-0.2Cr-0.2Ti alloy is shown in Figure 2. The microstructure (SEM) of the as-cast alloy is shown in Figure 2a,b. Aluminum solid solution, eutectic structure of aluminum solid solution and Al_3Gd phase, and a novel quaternary (Al, Gd, Cr, Ti) phase with possible composition $Al_{21}GdCrTi$ are identified in the as-cast alloy. Point EDX SEM analyses demonstrate the content of 19 wt.% Gd, 6.5 wt.% Ti, and 5.3 wt.% Cr (Al–bal.) in the white particles (Figure 2c, left image). Recalculating the phase composition in the atomic fraction with a content of 4.1 at.% Gd, 4.2 at.% Ti, and 4.2 at.% Cr (Al-Bal.) allows us to write the phase formula as $Al_{21}GdCrTi$. The novel quaternary $Al_{21}GdCrTi$ phase was first identified in the present study. The key element in the casting's microstructure was dendrites of aluminum solid solution, with a volume fraction of 80%. Fine crystallization-origin Al_3Gd phase was observed organized into eutectic structure on the periphery of the dendrite cells of the aluminum solid solution. The eutectic structure consists of Al_3Gd platelets with approximately 0.2 μm in thick and compact primary crystals of $Al_{21}GdCrTi$ phase with a size of about 10 μm. Figure 2c (right image) shows XRD of the as-cast Al-3.8Gd-0.2Cr-0.2Ti alloy. Unidentified peaks (marked with "?") were found in the XRD patterns. These peaks probably correspond to the quaternary (Al, Gd, Cr, Ti) phase. The atomic rotation of Al/Gd/Cr/Ti in this phase is 21/1/1/1 in accordance with the point EDX SEM results.

The solidus temperature of the Al-Gd-Cr-Ti alloy (Figure 3a) is 619 °C as examined by DSC experiment. The alloy was homogenized at 615 °C for 1 h with subsequent water quenching. The microstructure (SEM images) after homogenization treatment is shown in Figure 3b. The homogenization annealing for 1 h leads to fragmentation, spheroidization, and growth of the Al_3Gd eutectic phase particles; the particle's mean size increased from 0.2 μm (Figure 2a) to about 1.8 μm (Figure 3b) after homogenization. But what is interesting and discussed is that the novel $Al_{21}GdCrTi$ particles appear in two forms: First, the particles are surrounded by Al_3Gd phase, similar to the $Al_{21}GdCrTi$ core and Al_3Gd shell (Figure 3b-points A). Second, the $Al_{21}GdCrTi$ particles without the Al_3Gd phase (Figure 3b-points B). There was no need to increase the homogenization time because the fragmentation had already occurred, and rolling would have made the structure more uniform.

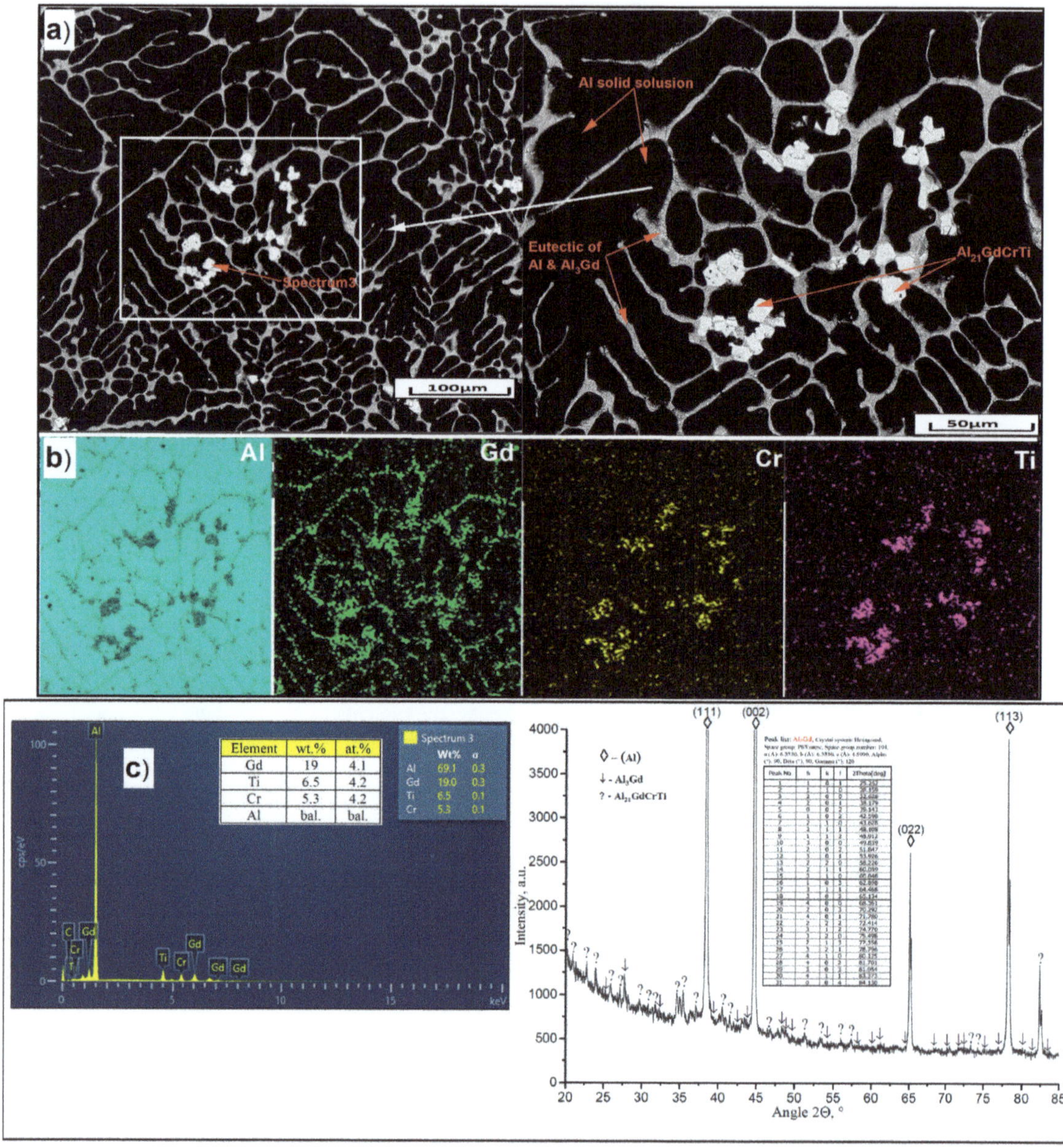

Figure 2. (**a**) Microstructure (SEM-BSE) of as-cast Al-3.8Gd-0.2Cr-0.2Ti alloy, (**b**) distribution of alloying elements between phases in the identified rectangle, and (c) EDX spectra from white particles (distribution of the alloying elements in the rectangle area in Figure 2a and XRD of the as-cast Al-3.8Gd-0.2Cr-0.2Ti alloy.

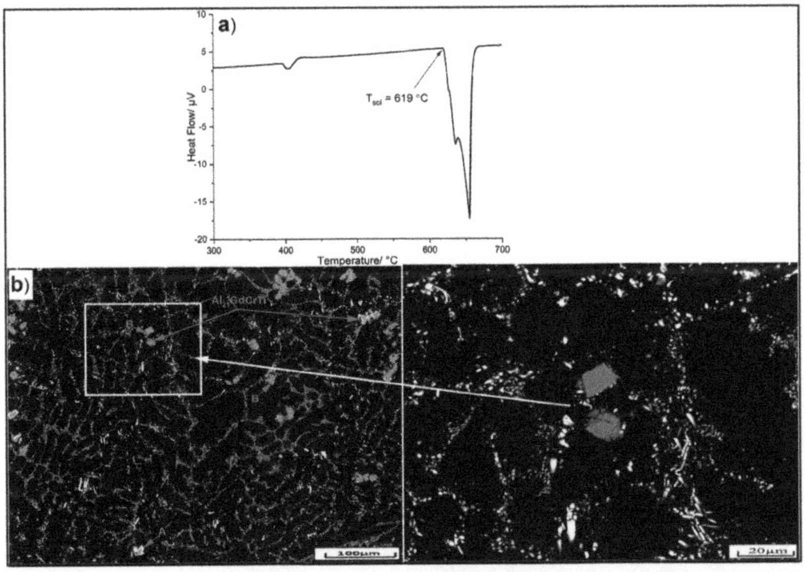

Figure 3. (**a**) DSC-curve, and (**b**) microstructure (SEM-BSE) after solid solution treatment at 615 °C for 1 h and quenching of the Al-Gd-Cr-Ti alloy at low and high magnification.

The ingot was cold-rolled from 20 mm to 1.5 mm, and hot-rolled from 20 to 10 mm at 450 °C, then cold-rolled to 1.5 mm at room temperature. The intermetallic particles were more uniformly distributed in the aluminum matrix after cold and hot-cold rolling compared to the homogeneous state at 615 °C for 1 h (Figure 4). The average Al_3Gd particle size was 1.1 ± 0.5 µm, with a volume fraction of 20% in both cold- and hot-rolled sheets. The results indicate no significant change in microstructure in both regimes. Furthermore, the novel $Al_{21}GdCrTi$ particles appear in two forms: first, particles surrounded by Al_3Gd (points A), and second, $Al_{21}GdCrTi$ particles without Al_3Gd (points B), as seen after homogenization at 615 °C. On top of that, the particle size of $Al_{21}GdCrTi$ at point A is about 7 ± 2 µm, while the particle size at point B is about 3 ± 2 µm after cold rolling (Figure 4a,b). This is probably because the $Al_{21}GdCrTi$ particles at points A are harder, making them more resistant to crushing during rolling. It is also observed that the $Al_{21}GdCrTi$ particle size after hot rolling is relatively smaller than after cold rolling and is more uniformly distributed in the aluminum matrix. This is contrary to expectations and may be because hot-cold rolling caused greater crushing of the aluminum particles (Figure 4a,c).

Figure 5 shows XRD patterns of four regimes: as-cast, as-cast with homogenization at 615 °C for 1 h, as-cold-rolled, and as-hot-cold-rolled. XRD patterns of four regimes were prepared to interpret the crystallographic texture of an aluminum solid solution based on the intensity ratio of crystallographic planes. The intensity of diffraction peaks from different crystal planes is not always equal, even in the same material. This difference in intensity, or intensity ratio, is often used to identify preferred orientations or "texture" within the material [38]. XRD analysis of the alloy reveals solid solution peaks of four planes (111), (002), (022), and (113), in addition to the Al_3Gd and peaks possibly corresponding to the quaternary phase $Al_{21}GdCrTi$. This figure clearly shows the difference in the relative intensity of the corresponding reflections in the diffraction patterns for the four regimes. The results showed that the intensity of (111) and (002) planes of as-cast and as-cast with homogenization at 615 °C for 1 h are higher than the (022) and (113) planes. This is due to the larger volume percentage of crystals (preferred orientation) of the (111) and (002) planes than the (022) and (113) planes. On top of that, the intensity of the (022) and (113) planes increased significantly after rolling, especially after hot-cold rolling, indicating that the crystal volume percentage increased during the rolling process. This may be due to the change in the direction of some grains in these planes during rolling, whether hot or cold. Furthermore, the intensity of (022), (113) planes of as-hot-cold-rolled are higher than that of as-cast and as-cast with homogenization at 615 °C for 1 h, This is due to the larger volume percentage of crystals (preferred orientation) of the (022), (113) planes than the (111) and (002) planes after hot and cold rolling.

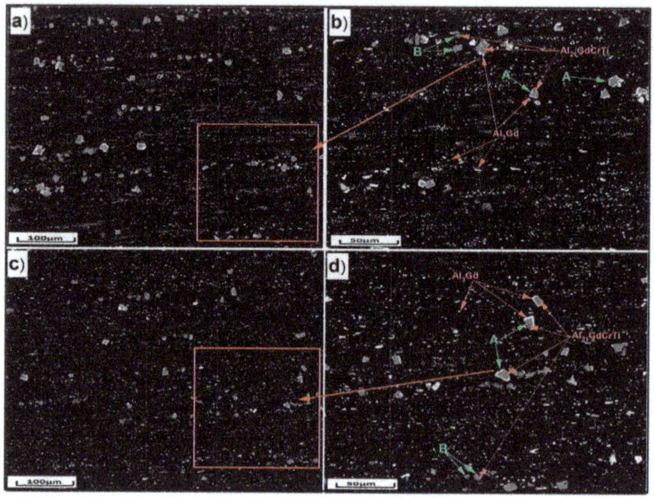

Figure 4. Scanning electron microscope (SEM-BSE) observation of cold-rolled Al-Gd-Cr-Ti specimen (**a**) low magnification, (**b**) high magnification, (**c**,**d**) hot-cold-rolled specimen at low and high magnification.

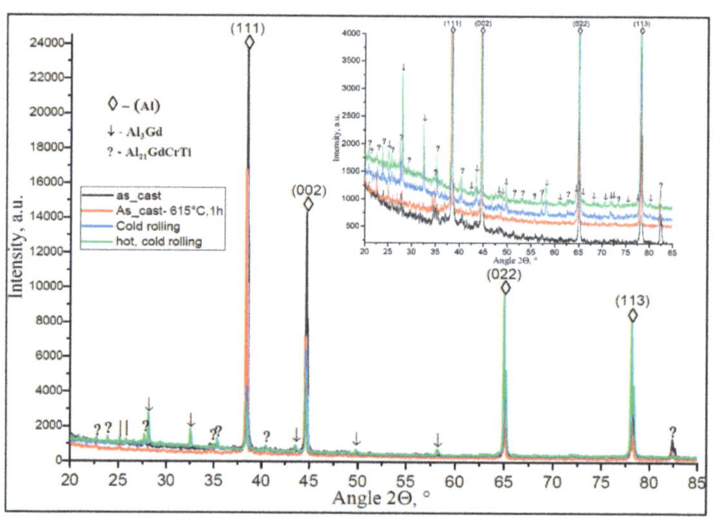

Figure 5. XRD-patterns as-cast, as-cast and homogenization at 615 °C for 1 h, cold-rolled, hot-cold-rolled samples of Al-Gd-Cr-Ti alloy.

Figure 6 represents the pole figures of the samples obtained by means of four regimes. The pole figure values are traditionally expressed in multiples of the random orientation distribution (mrd), which means that the pole figure of a random standard sample will have an intensity of 1 mrd at all points. The maximum pole density of as-cast specimens for planes (111), (002), (022), and (113) is 2.074, 3.248, 1.405, and 1.705 mrd, respectively. The texture of the plane (002) reveals a higher maximum intensity of 3.24 mrd than other planes, which can be interpreted as the sharpness of the crystallographic texture is large (Figure 6a, Table 2). Moreover, maximum pole density of planes (111), (002), (022) and (113) increased to 3.184, 5.941, 2.578 and 2.244 mrd, respectively, after homogenization at 615 °C for 1 h, indicating that the homogenized sample has a strong preferred orientation [39] (Figure 6b). This may be due to the increase in grain size during the heating process. In addition to this, the grain orientation evolves during the subsequent solution heat treatment and plastic deformation. The homogenized sample at 615 °C for 1 h shows a coarser grain size compared to other regimes, as shown in Figure 6b. Moreover, other researchers presented evidence that the homogenization influence on cast Al alloys can decrease their hardness, coming approximately from the increase in grain measure during the midst of the heating process [40,41]. The present study revealed that the coarse-grain structure in Al-3.8Gd-0.2Cr-0.2Ti alloy can lead to a strong texture [24,35].

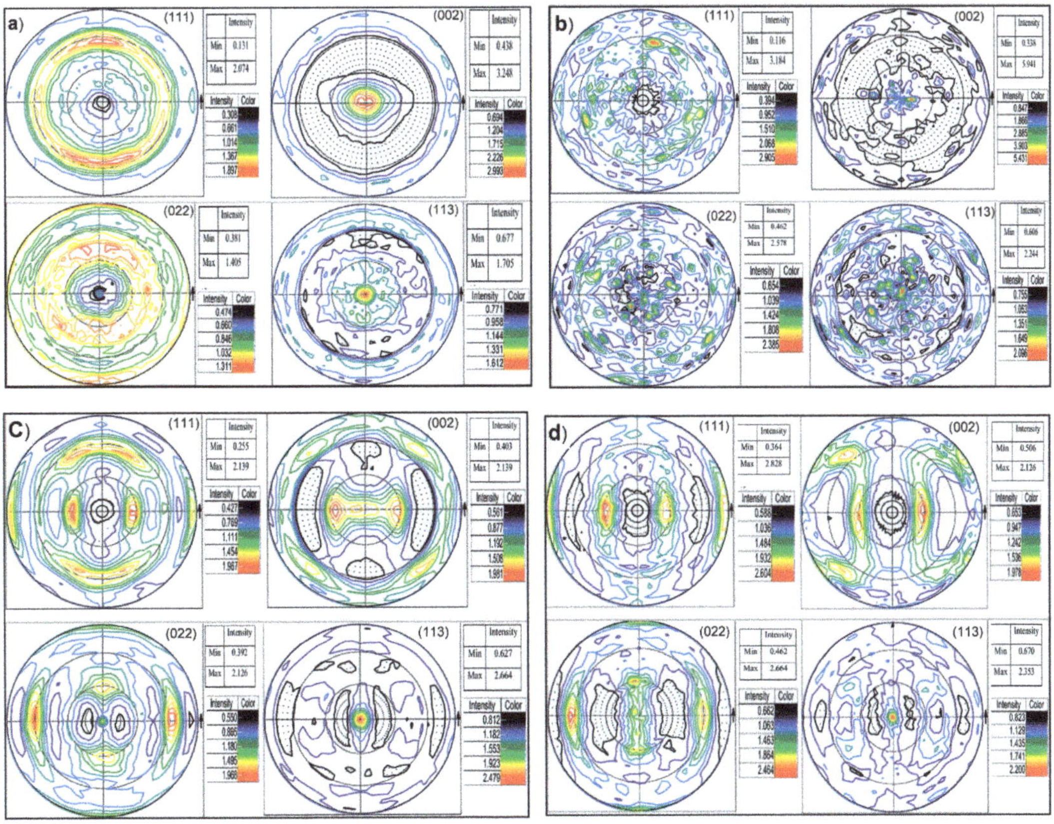

Figure 6. (**a**) As-cast pole figure, (**b**) as-cast and homogenization at 615 °C for 1 h, (**c**) cold-rolled, and (**d**) hot-cold-rolled samples of Al-Gd-Cr-Ti alloy.

Table 2. Summary of pole figure results for as-cast, as-cast with homogenization at 615 °C for 1 h, as-cold-rolled, and as-hot-cold-rolled specimens.

Regime	As-Cast				As-Cast with Homogenization at 615 °C for 1 h				As-Cold-Rolled				As-Hot-Cold-Rolled			
Intensity	(111)	(002)	(022)	(113)	(111)	(002)	(022)	(113)	(111)	(002)	(022)	(113)	(111)	(002)	(022)	(113)
Min	0.131	0.438	0.381	0.677	0.116	0.338	0.462	0.606	0.255	0.403	0.392	0.627	0.364	0.506	0.462	0.670
Max	2.074	3.248	1.405	1.705	3.184	5.941	2.578	2.244	2.139	2.139	2.126	2.664	2.828	2.126	2.664	2.353

Furthermore, a typical deformation texture was formed after rolling, as one can see in Figure 6c,d. The maximum pole density of as-cold-rolled and as-hot-cold-rolled samples of four planes (111), (002), (022), and (113) ranges from 2.126 to 2.828 mrd, indicating a texture strength advantage compared to the cast samples.

The thermal neutron absorption property of the as-cast Al-3.8Gd-0.2Cr-0.2Ti alloy compared to pure Al alloy is shown in Figure 7. The horizontal axis represents interplanar spacing (d, Å), while the vertical axis represents the neutron scattering intensity (arbitrary unit, a.u.). The intensity scattered by the studied alloy is very low and almost negligible compared to pure Al alloy due to the absorption of neutrons by the alloy. The results indicate that the studied alloys scatter neutrons about 100 times less than a nearly pure aluminum alloy.

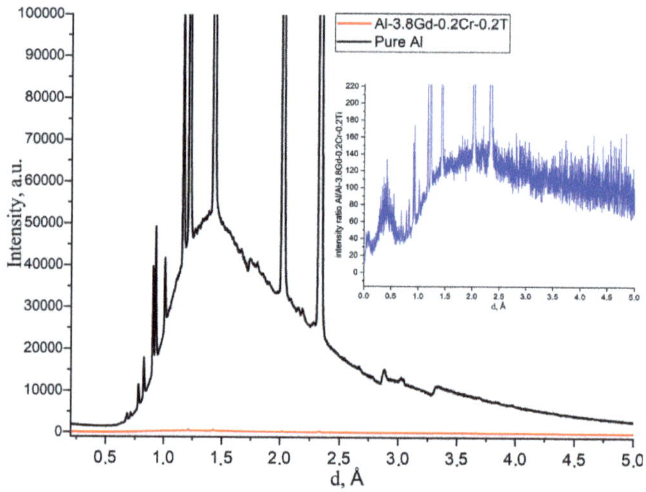

Figure 7. Neutron diffraction results of the Al-3.8Gd-0.2Cr-0.2Ti alloy compared to pure Al.

The hardness of the as-cast Al-3.8Gd-0.2Cr-0.2Ti alloy is 36 HV, while the hardness of the as-cast with homogenization at 615 °C for 1 h is 31 HV. The hardness increased to 53 HV after cold rolling and to 50 HV after hot-cold rolling.

The as-cold-rolled specimen has mechanical tensile properties: yield strength of 159 MPa, ultimate tensile strength of 177 MPa, and elongation of 10.5%, while the as-hot-cold-rolled specimen has mechanical tensile properties: yield strength of 160 MPa, ultimate tensile strength of 181 MPa, and elongation of 11.3% (Figure 8). Contrary to expectations, the as-hot-cold-rolled sample shows slightly higher yield and tensile strength. With hot-cold-rolled deformation, the size of the $Al_{21}GdCrTi$ particles decreased, and they were more uniformly distributed in the matrix. The studied alloys, containing 4.2 wt.% of the alloying elements 3.8Gd, 0.2Cr, and 0.2Ti (Al-3.8Gd-0.2Cr-0.2Ti), exhibited a yield strength

28 MPa higher than those containing 21 wt.% of the alloying elements 5Cu, 6Gd, and 8Bi (Al-5Cu-6Gd-8Bi) [33]. This is attributed to the addition of 0.2 wt.% chromium and 0.2 wt.% titanium, which improved tensile strength while maintaining elongation (Figure 8).

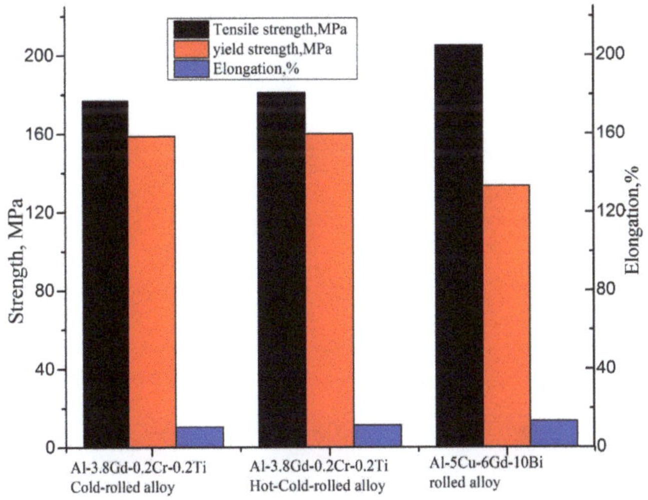

Figure 8. Tensile properties of Al-3.8Gd-0.2Cr0.2Ti alloy compared to Al-5Cu-6Gd-10Bi alloy.

4. Conclusions

The microstructure of Al-3.8Gd-0.2Cr-0.2Ti alloys was observed during the casting, homogenization stages, and rolling, along with the mechanical properties, crystallographic texture, deformation behavior, and neutron diffraction. The microstructure revealed the presence of the aluminum solid solution, Al_3Gd, and a novel (Al, Gd, Cr, Ti) quaternary phase with possible composition $Al_{21}GdCrTi$ was found in the as-cast Al-3.8Gd-0.2Cr-0.2Ti alloy. The novel quaternary $Al_{21}GdCrTi$ phase was first identified in the present study. X-ray diffraction and crystallographic texture analysis in samples after rolling showed a marked difference in volume percentage of crystals and texture strength compared to the cast samples. The results showed that the pole density of (022) and (113) planes of the as-hot-cold-rolled sample is higher than that of the as-cast. The Al-3.8Gd-0.2Cr-0.2Ti alloy scatters neutrons about 100 times less than a pure Al alloy. The hardness of the as-cold-rolled alloy is 53 HV, while the as-hot-cold-rolled alloy is 50 HV. Mechanical tensile properties of cold-rolled alloy: yield strength of 159 MPa, ultimate tensile strength of 177 MPa, and elongation of 10.5%, while Mechanical tensile properties of hot-cold-rolled alloy: yield strength of 160 MPa, ultimate tensile strength of 181 MPa, and elongation of 11.3%.

The alloy is a promising combination for the development of novel high-technology neutron shielding alloys due to its high thermal neutron absorption. The alloy's mechanical properties can be improved by adding zirconium and magnesium, the effect of which on the alloy's properties is still being studied. Further research is also needed on the effect of silicon on the phase composition and alloy properties.

Author Contributions: Conceptualization, S.M.A., D.I.N., T.A.L., A.E.-K., A.V.M. and A.V.P.; Methodology, S.M.A., D.I.N., T.A.L., A.E.-K., R.Y.B. and M.V.G.; Software, D.I.N., T.A.L., R.Y.B., M.V.G. and A.V.P.; Validation, S.M.A., D.I.N., T.A.L., A.S.P., A.E.-K., R.Y.B., A.V.M., M.V.G. and A.V.P.; Formal Analysis, S.M.A., D.I.N., T.A.L., A.E.-K., R.Y.B., A.S.P., A.V.M., M.V.G. and A.V.P.; Investigation, S.M.A., D.I.N., T.A.L., A.E.-K., A.V.M. and A.V.P.; Resources, S.M.A., R.Y.B., A.V.M. and A.V.P.; Data

Curation, A.S.P., R.Y.B., A.V.P. and M.V.G.; Writing—original draft preparation, S.M.A., A.E.-K., M.V.G. and A.V.P.; Writing—review and editing, S.M.A., D.I.N., T.A.L., A.E.-K., R.Y.B., A.S.P., A.V.M., M.V.G. and A.V.P.; Visualization, S.M.A. and A.V.P.; Supervision, S.M.A.; Project administration, S.M.A.; Funding Acquisition, S.M.A., A.V.M. and A.V.P. All authors have read and agreed to the published version of the manuscript.

Funding: The work was funded by the Ministry of Science and Higher Education of the Russian Federation: FSME-2023-0005; The SEM studies and hardness measurements were performed in a framework of Russian Science Foundation: 23-19-00791.

Data Availability Statement: The data are available by request.

Acknowledgments: S.M. Amer acknowledges the financial support under the joint executive program between the Academy of Scientific Research and Technology (ASRT, Egypt) and the Joint Institute for Nuclear Research (JINR, Russia).

Conflicts of Interest: The authors declare no conflicts of interest.

References

1. Wu, X.; Zhu, Y. Heterogeneous materials: A new class of materials with unprecedented mechanical properties. *Mater. Res. Lett.* **2017**, *5*, 527–532. [CrossRef]
2. Long, S.-L.; Pan, S.-S.; Yang, M.; Luo, X.-L.; Ling, M.; Gui, Q.; Yi, Y.-L. Effects of stirring friction on the microstructure, texture and mechanical properties of 7055 aluminum alloy. *J. Mater. Res. Technol.* **2024**, *33*, 5413–5424. [CrossRef]
3. Amer, S.M.; Barkov, R.Y.; Prosviryakov, A.S.; Pozdniakov, A.V. Structure and Properties of New Wrought Al–Cu–Y- and Al–Cu–Er-Based Alloys. *Phys. Met. Metallogr.* **2021**, *122*, 915–922. [CrossRef]
4. Amer, S.; Barkov, R.; Pozdniakov, A. Microstructure and mechanical properties of novel quasibinary Al-Cu-Yb and Al-Cu-Gd alloys. *Metals* **2021**, *11*, 476. [CrossRef]
5. Yang, H.; Sun, C.; Li, Y.; Cheng, K.; Liu, Y.; Wu, J.; Liu, H.; Wang, J.; Zhao, D.; Song, L.; et al. Effect of grain size and dislocation density on thermal stability of Al-Cu-Mg alloy. *Materialia* **2024**, *38*, 102276. [CrossRef]
6. Kozakevich, J.R.; Sediako, D.; Weiss, D.; Vogel, S.C. A Quantitative Phase Analysis by Neutron Diffraction of Conventional and Advanced Aluminum Alloys Thermally Conditioned for Elevated-Temperature Applications. *Materials* **2024**, *17*, 4311. [CrossRef]
7. Šmalc, J.; Vončina, M.; Mrvar, P.; Balaško, T.; Krutiš, V.; Petrič, M. The Influence of Foundry Scrap Returns on Chemical Composition and Microstructure Development of $AlSi_9Cu_3$ Alloy. *Crystals* **2023**, *13*, 757. [CrossRef]
8. Lu, Y.; Wu, C.; Wu, H.; Wang, J.; Su, Y.; Gan, Z.; Liu, J. Growth Mechanism of Eutectic Si in Super-Gravity Solidified Al-Si Alloy during Annealing. *Crystals* **2023**, *13*, 684. [CrossRef]
9. Su, X.; Qu, H.; Lei, Y.; Hou, R.; Cao, Y.; Siddique, S.; Qi, Z.; Shen, G.; Fan, X. Influence of Ni on the Microstructures and Mechanical Properties of Heat-Treated Al-Cu-Ce-Mn-Zr Alloys. *Crystals* **2023**, *13*, 380. [CrossRef]
10. Abdu, A.; Limmaneevichitr, C.; Patakham, U.; Zheng, J.; Zheng, C.; Pandee, P. Microstructural evolution and thermal stability of hypoeutectic Al–Ce alloys with Ni and Fe additions. *J. Mater. Res. Technol.* **2025**, *36*, 3465–3474. [CrossRef]
11. Zhang, H.; Wu, M.; Li, Z.; Xiao, D.; Huang, Y.; Huang, L.; Liu, W. Microstructure and Mechanical Properties of As-Cast Al-10Ce-3Mg-xZn Alloys. *Materials* **2024**, *17*, 3999. [CrossRef]
12. Ye, K.; Cai, X.; Sun, B.; Zhou, L.; Ma, S.; Yue, Y.; Xu, F.; Zheng, D.; Fu, X. Effect of rare earth Ce on the microstructure and mechanical properties of cast Al–7Si alloys. *J. Sci. Adv. Mater. Devices* **2023**, *8*, 100634. [CrossRef]
13. Song, M.; Wu, Z.; He, Y. Effects of Yb on the mechanical properties and microstructures of an Al-Mg alloy. *Mater. Sci. Eng. A* **2008**, *497*, 519–523. [CrossRef]
14. Mamzurina, O.I.; Amer, S.M.; Glavatskikh, M.V.; Barkov, R.Y.; Loginova, I.S.; Pozdniakov, A.V. Microstructure and Mechanical Properties of Novel Heat Resistant Cast Al-Cu-Yb(Gd)-Mg-Mn-Zr Alloys. *Metals* **2022**, *12*, 2079. [CrossRef]
15. Wang, Z.; Pan, J.; Liu, A.; Zhang, L.; Sun, Z.; Wang, Z.; Li, J.; Xiao, X. Microstructure and properties controlling of Al-xGd alloys for thermal neutron absorbing. *J. Nucl. Mater.* **2025**, *603*, 155447. [CrossRef]
16. Okamoto, H. Al-Gd (aluminum-gadolinium). *J. Phase Equilibria* **2002**, *23*, 380. [CrossRef]
17. Zhang, X.G.; Mei, F.Q.; Zhang, H.Y.; Wang, S.H.; Fang, C.F.; Hao, H. Effects of Gd and Y additions on microstructure and properties of Al-Zn-Mg-Cu-Zr alloys. *Mater. Sci. Eng. A* **2012**, *552*, 230–235. [CrossRef]
18. Brokmeier, H.G. Global crystallographic textures obtained by neutron and synchrotron radiation. *Phys. B Condens. Matter* **2006**, *385–386*, 623–625. [CrossRef]

19. Ghosh, P.; Kormout, K.S.; Todt, J.; Lienert, U.; Keckes, J.; Pippan, R. An investigation on shear banding and crystallographic texture of Ag–Cu alloys deformed by high-pressure torsion. *Proc. Inst. Mech. Eng. Part C J. Mech. Eng. Sci.* **2019**, *233*, 794–806. [CrossRef]
20. Lychagina, T.; Zisman, A.; Yashina, E.; Nikolayev, D. Directly verifiable neutron diffraction technique to determine retained austenite in steel. *Adv. Eng. Mater.* **2018**, *20*, 1700559. [CrossRef]
21. Schäfer, W. Neutron diffraction applied to geological texture and stress analysis. *Eur. J. Mineral.* **2002**, *14*, 263–289. [CrossRef]
22. Yousefi Mehr, V.; Toroghinejad, M.R. On the texture evolution of aluminum-based composites manufactured by ARB process: A review. *J. Mater. Res. Technol.* **2022**, *21*, 1095–1109. [CrossRef]
23. Klosek, V. Crystallographic Textures—Measurement. *EPJ Web Conf.* **2017**, *115*, 00005. [CrossRef]
24. Li, S.; Zhao, Q.; Liu, Z.; Li, F. A Review of Texture Evolution Mechanisms During Deformation by Rolling in Aluminum Alloys. *J. Mater. Eng. Perform.* **2018**, *27*, 3350–3373. [CrossRef]
25. Yoshida, K.; Ishizaka, T.; Kuroda, M.; Ikawa, S. The effects of texture on formability of aluminum alloy sheets. *Acta Mater.* **2007**, *55*, 4499–4506. [CrossRef]
26. Sidor, J.J. Effect of hot band on texture evolution and plastic anisotropy in aluminium alloys. *Metals* **2021**, *11*, 1310. [CrossRef]
27. Wang, X.; Shi, T.; Jiang, Z.; Chen, W.; Guo, M.; Zhang, J.; Zhuang, L.; Wang, Y. Relationship among grain size, texture and mechanical properties of aluminums with different particle distributions. *Mater. Sci. Eng. A* **2019**, *753*, 122–134. [CrossRef]
28. Yuan, H.; Wang, Q.F.; Zhang, J.W.; Liu, W.C.; Gao, Y.K. Effect of grain shape on the texture evolution during cold rolling of Al-Mg alloys. *J. Alloys Compd.* **2011**, *509*, 922–928. [CrossRef]
29. Roy, R.K.; Kar, S.; Das, S. Evolution of microstructure and mechanical properties during annealing of cold-rolled AA8011 alloy. *J. Alloys Compd.* **2009**, *468*, 122–129. [CrossRef]
30. Sarkar, J.; Saimoto, S.; Mathew, B.; Gilman, P.S. Microstructure, texture and tensile properties of aluminum-2 at.% neodymium alloy as used in flat panel displays. *J. Alloys Compd.* **2009**, *479*, 719–725. [CrossRef]
31. Deng, Y.L.; Wan, L.; Zhang, Y.; Zhang, X.M. Evolution of microstructures and textures of 7050 Al alloy hot-rolled plate during staged solution heat-treatments. *J. Alloys Compd.* **2010**, *498*, 88–94. [CrossRef]
32. El-Genk, M.S.; Palomino, L.M. A walk-away safe, Very-Small, Long-LIfe, Modular (VSLLIM) reactor for portable and stationary power. *Ann. Nucl. Energy* **2019**, *129*, 181–198. [CrossRef]
33. Ding, Y.; Yang, C.; Zheng, Q.; Liu, Z.; Yang, W. Design, fabrication and mechanical properties of a novel Al-5Cu-6Gd-10Bi alloy potentially used for shielding both thermal neutron and γ-ray. *J. Alloys Compd.* **2025**, *1014*, 178670. [CrossRef]
34. Kučeráková, M.; Rohlíček, J.; Vratislav, S.; Jarošová, M.; Kalvoda, L.; Nikolayev, D.; Lychagina, T.; Douda, K. Texture of the freshwater shells from the unionidae family collected in the czech republic investigated by x-ray and neutron diffraction. *Crystals* **2021**, *11*, 1483. [CrossRef]
35. Kucerakova, M.; Rohlicek, J.; Vratislav, S.; Nikolayev, D.; Lychagina, T.; Kalvoda, L.; Douda, K. Texture Study of Sinanodonta Woodiana Shells by X-Ray Diffraction. *J. Surf. Investig. X-Ray Synchrotron Neutron Tech.* **2021**, *15*, 640–643. [CrossRef]
36. Isaenkova, M.; Perlovich, Y.; Fesenko, V. Regularities of crystallographic texture formation in cladding tubes from Zr-based alloys during their production. *IOP Conf. Ser. Mater. Sci. Eng.* **2016**, *130*, 553–561. [CrossRef]
37. Keppler, R.; Ullemeyer, K.; Behrmann, J.; Stipp, M. Potential of full pattern fit methods for the texture analysis of geological materials: Implications from texture measurements at the recently upgraded neutron time-of-flight diffractometer SKAT. *J. Appl. Crystallogr.* **2014**, *47*, 1520–1534. [CrossRef]
38. Lu, C.; Luo, Z. Critical evaluation of (110) texture in lithium electrodeposits on isotropic Cu polycrystals. *Nat. Commun.* **2022**, *13*, 5673. [CrossRef]
39. Li, J.; Li, F.; Ma, X.; Li, J.; Liang, S.; Zhang, L. Effects of Heat Treatment on Microstructure and Mechanical Properties of an ECAPed Al–Zn–Mg–Cu Alloy. *Adv. Eng. Mater.* **2018**, *20*, 1701155. [CrossRef]
40. Yeni, M.Z.; Prindo, K.P.; Tri, P.; Agus, P. Homogenization Process for Aluminum As-Cast from Waste of Beverage Cans. *Mater. Sci. Forum* **2022**, *1057*, 189–194. [CrossRef]
41. Wang, X.; Guo, M.; Zhang, J.; Zhuang, L. Effect of Zn addition on the microstructure, texture evolution and mechanical properties of Al-Mg-Si-Cu alloys. *Mater. Sci. Eng. A* **2016**, *677*, 522–533. [CrossRef]

crystals

MDPI

Article

The Role of Friction Stir Processing Travel Speed on the Microstructure Evolution and Mechanical Performance of As-Cast Hypoeutectic Al-5Si Alloy

Basma El-Eraki [1], Mahmoud F. Y. Shalaby [2], Ahmed El-Sissy [1], Abeer Eisa [1], Sabbah Ataya [3] and Mohamed M. El-Sayed Seleman [4,*]

1 Production Engineering and Mechanical Design Department, Faculty of Engineering, Menoufia University, Shebin Elkom 32511, Egypt; basma.ramadan@sh-eng.menofia.edu.eg (B.E.-E.); ahmed.elsisi@sh-eng.menofia.edu.eg (A.E.-S.); abeer.essa@sh-eng.menofia.edu.eg (A.E.)

2 Department of Industrial Engineering, College of Engineering, Imam Mohammad Ibn Saud Islamic University (IMSIU), Riyadh 11432, Saudi Arabia; mfshalaby@imamu.edu.sa

3 Department of Mechanical Engineering, College of Engineering, Imam Mohammad Ibn Saud Islamic University (IMSIU), Riyadh 11432, Saudi Arabia; smataya@imamu.edu.sa

4 Department of Metallurgical and Materials Engineering, Faculty of Petroleum and Mining Engineering, Suez University, Suez 43512, Egypt

* Correspondence: mohamed.elnagar@suezuniv.edu.eg

Abstract

This study's novelty lies in providing first-time insights into the isolated role of Friction Stir Processing (FSP) travel speed on microstructure evolution and mechanical performance (micro-hardness, tensile properties, impact energy, and wear behavior) specifically in hypoeutectic as-cast Al-5 wt.% Si alloys, addressing a critical unaddressed gap in previous works (typically on near-eutectic compositions of Si > 6.5 wt.%). FSP, a solid-state technique, is highly effective for enhancing the properties of cast materials. The FSP was conducted at a fixed rotational speed of 1330 rpm and various travel speeds (26, 33, 42, and 52 mm/min). The FSP improves the mechanical properties of as-cast Al-5Si alloy by refining its grain structure. This leads to higher hardness, ultimate tensile strength (UTS), yield strength (YS), and strain at fracture and toughness compared to the as-cast condition. The specimen processed at 26 mm/min achieved the highest values of YS, UTS, toughness, and wear resistance. The fracture surfaces of the tensile and impact test specimens were examined using scanning electron microscopy (SEM) and discussed. Results indicated that the fracture surfaces revealed a transition from predominantly brittle fracture in the as-cast alloy to ductile fracture at 26 mm/min, changing to a mixed fracture mode at 52 mm/min. These findings underscore the critical importance of optimizing FSP travel speed to significantly tailor and enhance the mechanical performance of as-cast hypoeutectic Al-5Si alloys for industrial applications.

Keywords: travel speed; friction stir processing; hypoeutectic Al-5Si alloy; microstructure; mechanical properties

1. Introduction

Aluminum–silicon (Al-Si) alloys have garnered significant attention in various industries owing to their exceptional properties, like being lightweight, having high thermal conductivity, and having excellent castability [1]. These attributes make Al-Si alloys particularly advantageous in marine, aerospace, and automotive applications, where minimizing weight without sacrificing strength is crucial [2,3]. However, the mechanical properties of as-cast hypoeutectic Al-Si alloys often require enhancement to meet the demanding service conditions of these sectors. The production of Al-Si alloys can be achieved through a range of techniques, including traditional casting [4–7], semi-solid metal processing [8–12], and modern additive manufacturing methods [13–15]. While various methods exist, gravity die casting is notable for producing high-quality components with excellent dimensional accuracy and reduced common casting defects such as porosity and shrinkage cavities [16,17]. The precision and efficiency of gravity die casting make it an ideal choice for producing complex and critical parts in various industrial applications [18].

One of the primary challenges in optimizing Al-Si alloys lies in controlling the microstructure, particularly the distribution and size of the silicon phase within the aluminum matrix. The as-cast structure typically exhibits coarse silicon particles, which lead to reduced ductility and impact strength. Traditional modification techniques such as the addition of alloying elements [19–26] and heat treatment [27–32] have long been employed to modify this cast structure. Additionally, techniques such as laser surface melting [33,34] and shot peening [35,36] have been explored to refine microstructures and improve surface characteristics, but these methods often result in trade-offs between strength and ductility. Among these approaches, friction stir processing (FSP) has gained recognition as a promising method for modifying the microstructure of Al-Si alloys [37–40], offering a unique approach to enhancing mechanical properties without the drawbacks of conventional methods [41–47]. FSP is a solid-state process that evolved from friction stir welding (FSW). This fundamental difference is crucial as it avoids melting and the associated issues of solidification shrinkage, elemental segregation, and the formation of brittle intermetallic compounds, which can often degrade the mechanical properties of the treated zone [48,49]. Thus, it has been widely adopted for modifying the internal structure and improving the mechanical performance of various metals [50–60]. FSP involves the localized plastic deformation of the material using different rotating and traveling tool speeds, which generates significant heat and mechanical work, leading to grain refinement and homogenization of the microstructure [61–63].

For aluminum alloys, FSP offers distinct advantages such as significant grain refinement, effective elimination of casting defects like porosity, and homogeneous distribution of existing phases, leading to superior mechanical properties without introducing foreign materials or altering the bulk composition significantly. Recent studies continue to underscore its versatility in enhancing the performance of various aluminum-based composites [59,64–66], and other research efforts have also focused on applying FSP for the surface modification of Al-Si alloys using FSP. Ma et al. [61] investigated the FSP of as-cast A356 alloy, varying tool design and rotation speed (700–1100 rpm) at a constant travel speed of 203 mm/min. They found that FSP enhanced ductility and strength by refining the microstructure, dispersing Si-phases, and reducing micro-voids. A higher tool rotation speed was identified as the primary factor for these improvements, whereas the effect of tool geometry was found to be non-systematic. Guru et al. [67] studied the impact of applying a tool speed of 800 rpm and a travel speed of 120 mm/min as FSP parameters on the machinability, hardness, and tensile properties of the hypoeutectic Al-6.5Si cast alloy. Results showed that the processed alloy had enhanced hardness, yield strength, ultimate tensile strength, and machinability over the as-cast alloy. Meenia et al. [41] investigated the influence of the number of passes of FSP at 800 rpm, 9 kN, and 120 mm/min process parameters on the microstructure, hardness, and tensile properties of both as-cast and solution-treated Al-6.5Si alloy. It was reported that multi-pass FSP significantly refined the microstructure, homogenized the material, and eliminated casting porosity in both initial conditions. Singh et al. [42] aimed to enhance the wear resistance and machinability of

A356 through multi-pass FSP using the process parameters of 800 rpm tool rotating speed and 120 mm/min travel speed. Their results demonstrated that three FSP passes yielded the highest wear resistance under metallic conditions, while two passes were optimal for abrasive conditions. In dry drilling experiments, the drilling force and surface roughness decreased with up to two passes but increased with additional passes. Additionally, three FSP passes minimized burr formation at the entry and exit holes. Cui et al. [67] also aimed to study the role of number of passes during the FSP of a hypoeutectic Al-6.93Si alloy at scattered process parameters in terms of rotation and travel speeds. They found that FSP on this cast alloy effectively reduces porosity and breaks down/disperses large silicon particles, iron-rich phases, and Mg_2Si particles, while also dissolving most of the Mg_2Si. Multi-pass FSP, particularly two passes, significantly enhances the internal structural features and tensile characteristics compared to a single pass. A rotation speed of 1500 rpm and a travel speed of 30 mm/min for the two-pass FSP were identified as the optimal parameters for achieving the best mechanical properties. Alidokht et al. [68] examined the impact of applying various tool rotation speeds (500–1250 rpm), a tool tilt angle of 3°, and a constant travel speed of 50 mm/min during the FSP to modify the microstructure of as-cast A356 Al alloy material. They reported that higher tool rotation speed resulted in a more refined microstructure due to enhanced stirring. In a sample produced at 1250 rpm, the Si particles were significantly smaller (3.3 μm vs. 16.2 μm), less elongated (aspect ratio 1.1 vs. 8.5), and the grain size was much smaller (1 μm vs. 94 μm) compared to the untreated A356 alloy. In addition, significant improvements in hardness and wear resistance were detected at all applied rotation speeds.

Soleymanpour et al. [69] investigated the impact of FSP traverse speeds (15–240 mm/min) at 800 rpm on the microstructural and mechanical properties of pre-treated (heat-treated and cold-rolled) hypoeutectic Al-7Si alloy. The results indicated that the as-cast material exhibited higher average hardness compared to the FSP-treated samples, likely owing to the presence of larger second-phase particles. However, FSP improved tensile properties through microstructural refinement and particle modification. Furthermore, strength increased with traverse speed up to 120 mm/min due to smaller silicon and Fe-rich particles but decreased at higher speeds likely due to cavity formation. Baruch et al. [70] investigated the effects of multi-pass FSP (two and three passes) on a die-cast Al-7Si-3Cu alloy, using fixed parameters (600 rpm, 12 mm/min). The results indicated that FSP refines and redistributes Al-Si eutectic particles and Al dendrites, with more passes leading to finer particles. Specifically, three-pass FSP enhanced the tensile strength and ductility of the processed material, though the overall hardness after FSP was lower than that of the original as-cast alloy. In a single-pass FSP study, Kumar et al. [71] examined the influence of a 960-rpm tool rotation speed and a 50 mm/min travel speed on an as-cast Al-6.95Si-0.6Mg-0.31Cu-0.5Fe alloy. The processing led to a 40% enhancement in hardness and a 14% rise in tensile strength, while ductility decreased by 24% relative to the as-cast state.

Based on the above review, although numerous studies have been conducted on the effects of FSP variables on the properties of as-cast hypoeutectic Al-Si alloys, particularly near-eutectic compositions (Si content > 6.5 wt.%), limited attention has been given to exploring its impact on the hypoeutectic Al-5Si alloy.

Furthermore, previous works have predominantly employed fixed or scattered FSP parameters, with most focusing on the role of tool rotation speed and the number of FSP passes at a constant processing speed [70,71] without a systematic study of travel speed effects. To address these gaps, the present work aims to explore the impact of varying travel speeds (26, 33, 42, and 52 mm/min) at a constant high rotational speed (1330 rpm) on the microstructural evolution and overall mechanical properties, specifically hardness, tensile properties, impact strength, and wear behavior of the hypoeutectic Al-5Si alloy, to establish correlations between processing parameters and performance.

2. Materials and Methods

2.1. Casting Process

The experimental work began with the casting of hypoeutectic Al-5Si alloy plates using the gravity permanent die casting method. This technique was chosen for its capacity to yield high-quality parts characterized by precise dimensions and a surface finish. As shown in Figure 1, the process involved preparing a steel permanent mold designed to ensure uniform cooling and minimize casting defects. To ensure optimal thermal conditions and reduce the risk of thermal shock or stress-induced flaws, the mold was preheated to 220 °C using a heating flame. Mold preheating also enhanced the surface finish and dimensional accuracy of the final casting. For preparation of the alloy, 3 kg of pure Al (≥99.7% purity) was melted and 5% Si by weight was added to create the hypoeutectic Al-5Si alloy. The molten mixture was stirred for 5 min to homogenize the Si distribution within the Al matrix. This stirring process was critical for achieving uniform composition and reducing the segregation of silicon particles. The molten Al-Si alloy was heated to 750 °C and poured into the preheated mold using gravity. This controlled and steady filling process ensured uniform filling of the mold cavity, reducing the risk of porosity and other common casting defects. The pouring temperature was carefully selected to keep the alloy in a fully liquid phase throughout the process, ensuring enhanced flowability and filling properties. After the alloy solidified, the cast plates were carefully removed from the mold and immediately subjected to rapid cooling by immersion in water. This quenching process not only prevented the formation of coarse microstructures but also enhanced the mechanical properties and structural integrity of the cast plates by reducing residual stresses. Following solidification and cooling, the plates were precisely machined to dimensions of 240 mm × 140 mm × 6 mm using a CNC milling machine. This machining step was critical to achieving uniformity in sample preparation, ensuring that all plates had consistent dimensions and surface finishes. The high precision of the CNC milling process was vital for minimizing dimensional variability, which could otherwise impact the reliability of FSP experiments. The consistent surface quality and exact dimensions of the plates provided a uniform baseline, enabling accurate comparisons during subsequent mechanical testing and microstructural analysis. This meticulous preparation step ensured reproducibility and reliability in all downstream processes, including FSP and characterization.

2.2. Chemical Composition Analysis

The precise chemical composition of the as-cast hypoeutectic Al-5Si alloy was determined through analysis performed with a spectrometer (Belec Spektrometrie Opto-Elektronik GmbH model, Georgsmarienhütte, Germany). The analysis results, presented in Table 1, outline the elemental composition of the as-cast Al-5Si alloy and confirm its compliance with the intended specifications.

Table 1. Chemical composition of the produced hypoeutectic Al-5Si alloy.

Element	Si%	Fe%	Cr%	Mn%	Zn%	Cu%	Ni%	Mg%	Al%
Percentage (%)	5	0.72	0.03	0.02	0.06	0.01	0.019	0.05	Bal.

2.3. Friction Stir Processing (FSP)

The as-cast plates were then subjected to FSP to modify their microstructure using an FSP tool machined from H13 tool steel. This tool featured a cylindrical design with a 25 mm shoulder diameter, 5 mm pin diameter, and 3 mm pin length (Figure 2). The chosen shoulder-to-pin diameter ratio of was based on our lab experience and aligns with commonly accepted practices in published work for effective material flow and frictional heat generation [63].

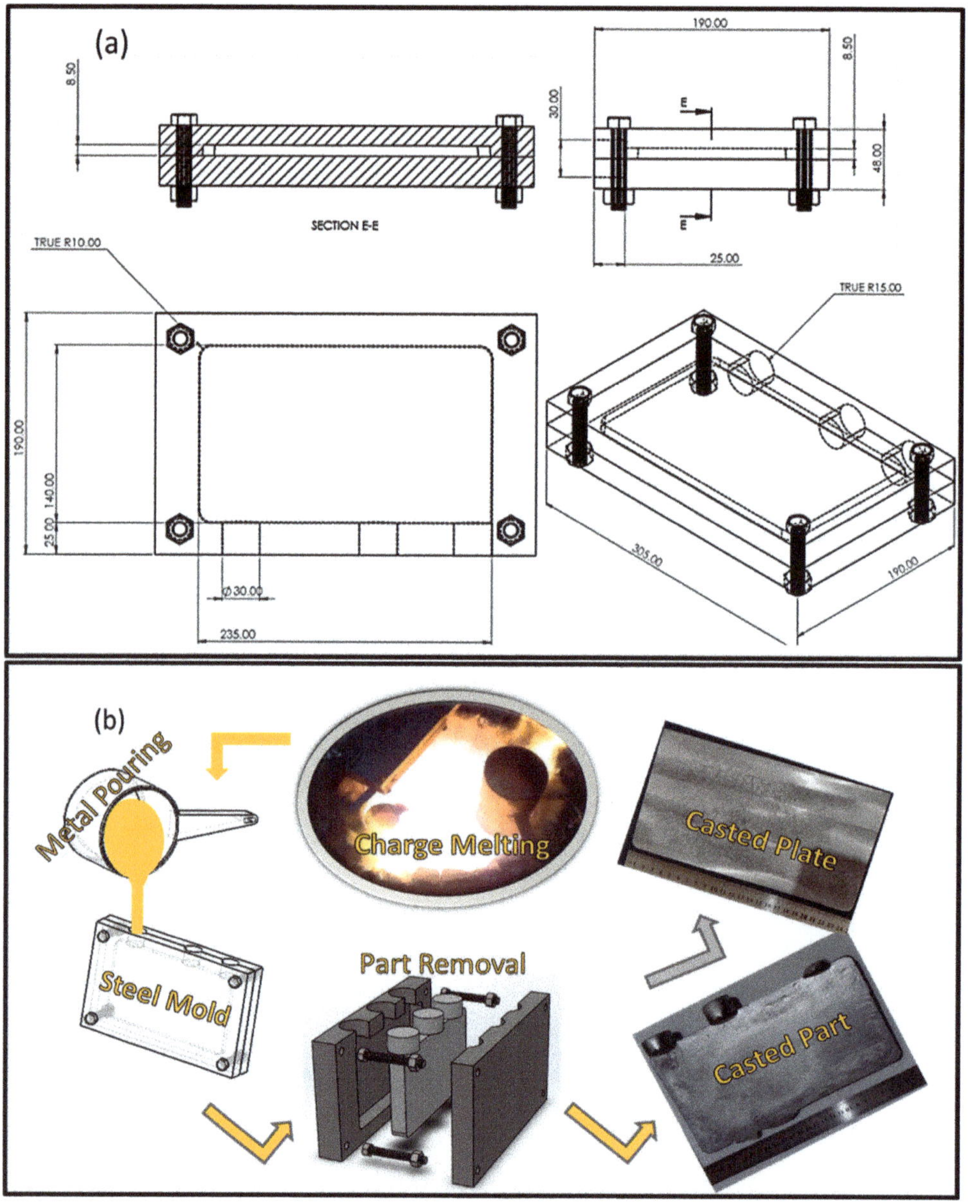

Figure 1. Steps of the gravity casting. (**a**) Design of the steel mold. (**b**) Gravity casting steps for the hypoeutectic Al-5Si alloy.

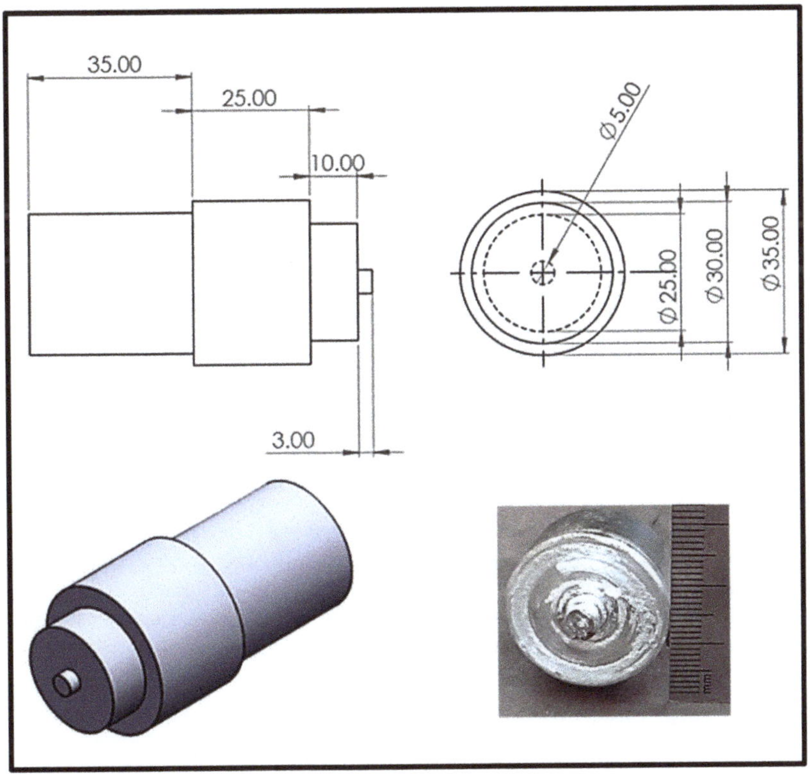

Figure 2. Specification of the FSP tool.

The FSP was achieved using a milling machine at varying travel speeds of 26, 33, 42, and 52 mm/min and a constant tool rotation speed of 1330 rpm. The tilt angle was fixed at 3° for all of the processed specimens. Figure 3 reprents the FSP process and the processed plate.

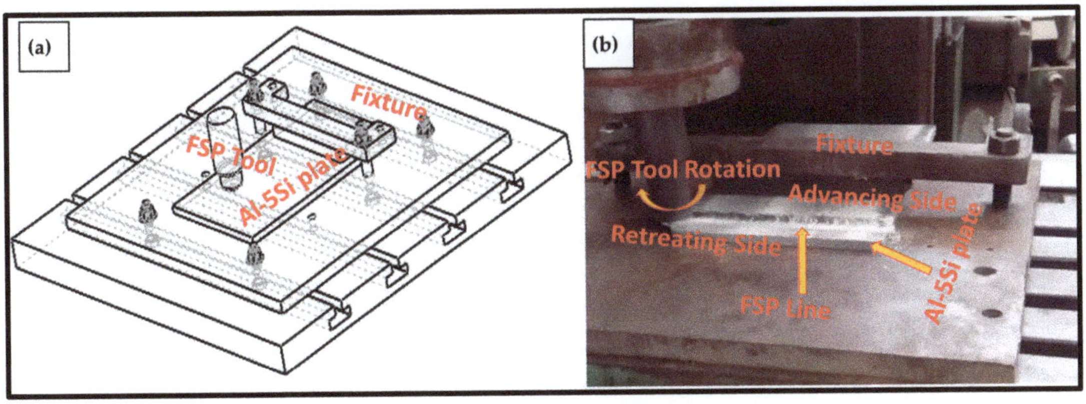

Figure 3. (**a**) Schematic diagram of the FSP and (**b**) a typical process.

The selection of the 1330 rpm rotation speed was optimized based on extensive experimental trials and supported by published work [41,69]. Our goal in this study was to systematically explore the impact of travel speed, and we specifically chose parameters within the lower processing range to investigate this less-explored window.

2.4. Sample Preparation for Microstructure, Mechanical, and Wear Testing

The plates were cut into specified samples using a Wire Cut Electrical Discharge Machining (EDM) machine, as shown in Figure 4. This precision cutting method was crucial for preparing samples for mechanical and wear tests while maintaining the integrity of the processed material. Following the FSP, samples were cut and prepared for microstructural analysis, tensile strength testing, and microhardness assessment. Microstructural observations were conducted on cross sections of the stir zone (SZ) perpendicular to the FSP direction. The samples were polished and etched with Keller's Agent (45 mL HCl, 15 mL HF, 15 mL HNO_3, 25 mL H_2O) at room temperature for a few seconds until the desired contrast was achieved. The microstructural characterization was performed using an optical microscope (model Leco LX 31- St. Joseph, MI, USA). Tensile test specimens were extracted from the processed surface parallel to the FSP direction and prepared to the required dimensions according to the ASTM E8/E8M-11 standard. Figure 5 shows the typical specimen dimensions and the specimen under tensile testing. Tensile tests were conducted using a universal testing machine (WDW-300, Jinan, China) with a cross head of 1 mm/min. Hardness was measured in the SZ using a microhardness tester (model Leco LM 700, Michigan, USA) with a dwell time of 15 s and a 100 gf test load. Impact specimens were prepared to Charpy standard dimensions, as shown in Figure 6. The impact testing was performed using an impact tester machine (model JBW-500 N, Jinan Testing Equipment IE Corporation, Jinan, China). The tensile and impact specimens' fracture surfaces were analyzed using a scanning electron microscope (SEM) Qunta FEG-250, Hillsboro, OR, USA.

Figure 4. Cutting samples with a wire cut EDM machine.

The wear test was conducted using a pin-on-disk tribometer wear testing machine (model TM 260 GUNT, Gerätebau GmbH, Hamburg, Germany) in accordance with the ASTM G99-05 standard. The wear specimens were extracted from the SZ, with the pin axis oriented perpendicular to the FSP pass. Prismatic pins were prepared with dimensions of 4 mm diameter and 25 mm length. The counterface disk, made of hardened ground stainless steel, had a hardness of 62 HRC and a sliding track diameter of 40 mm. The tests were accomplished under dry conditions at a fixed load of 10 N and rotational speed of 300 rpm. Figure 7 shows the wear testing machine configuration.

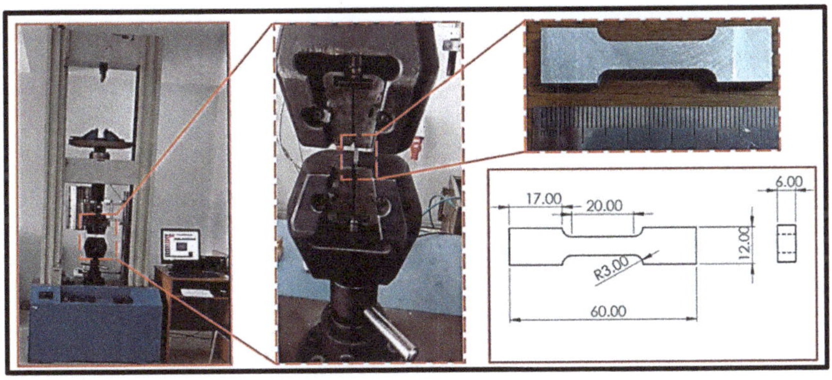

Figure 5. Tensile test sample, with all measurements provided in mm.

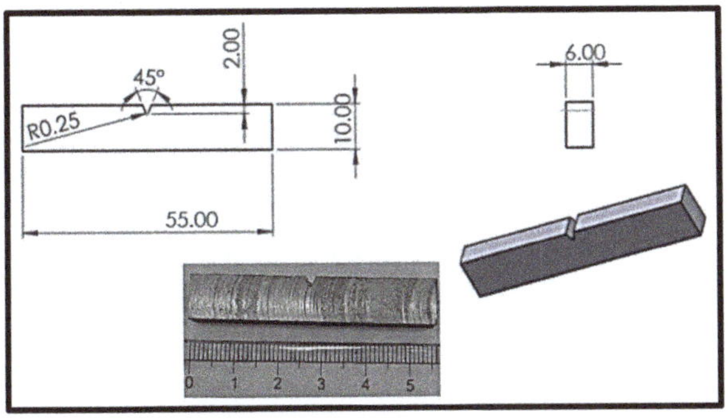

Figure 6. Impact test specimen (all dimensions in mm).

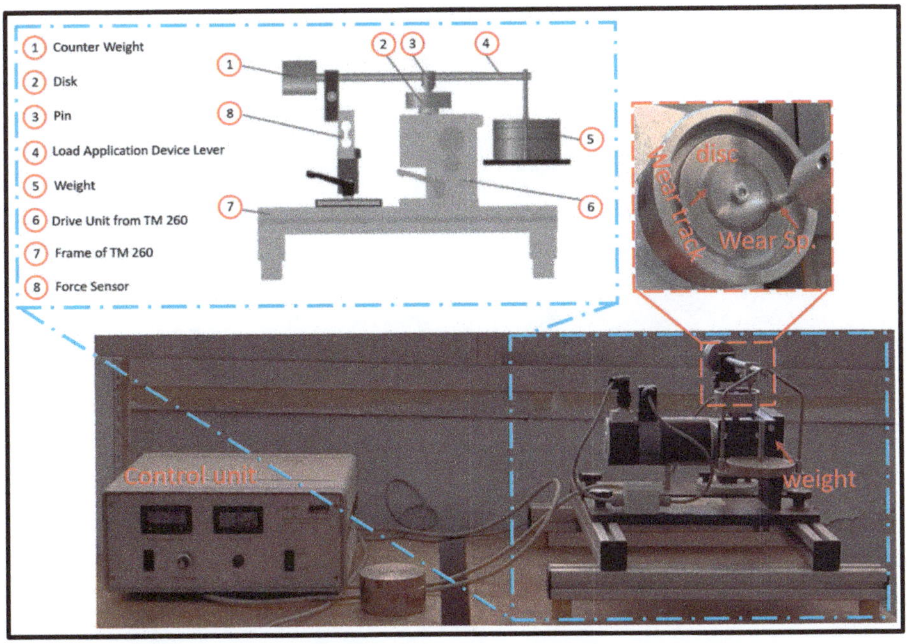

Figure 7. Wear testing machine (pin-on-disk).

3. Results and Discussion

3.1. Top Surface Features of Friction Stir Processed Specimens

Figure 8 shows photographs of the top surface of the friction stir processed tracks on the as-cast Al-5Si alloy plates, processed at a constant tool rotation speed of 1330 rpm and at different tool travel speeds using an H13-tool steel tool (25 mm shoulder, 5 mm pin diameter, 3 mm pin length). It can be seen that distinct surface characteristics correlate with the varying travel speeds of 26, 33, 42, and 52 mm/min. These variations likely reflect differences in material deformation, heat input, and the interaction of the tool shoulder and pin with the alloy surface [63]. As demonstrated in Figure 8a–d, at a constant tool rotation speed of 1330 rpm, roughness of the processed surface increased with processing travel speed from 26 to 52 mm/min. This increased the roughness and potential voids at higher travel speeds (42 and 52 mm/min, Figures 8b and 8c, respectively) can be attributed to potentially insufficient time for adequate material flow and stirring. Notably, at the highest speed (52 mm/min), the resulting surface displayed spaced apart tool marks with raised characters and scattered protrusions (Figure 8d), suggesting insufficient heat input leading to incomplete mixing. Conversely, the lowest travel speed (26 mm/min) resulted in a surface displaying smooth ridges with pronounced shoulder-induced flow marks (Figure 8a), likely due to prolonged heat input and excessive material flow. Similar features appeared at a processing speed of 33 mm/min, indicating a more balanced heat input and material stirring.

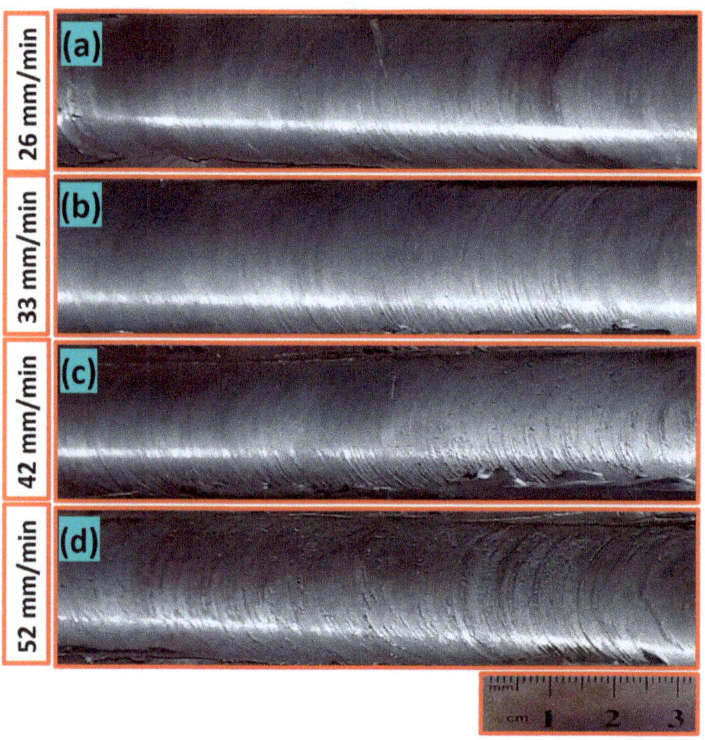

Figure 8. Photos depict the top-surface of the Al-5Si samples after FSP at a fixed tool rotation speed of 1330 rpm and various travel speeds of (**a**) 26 mm/min, (**b**) 33 mm/min, (**c**) 42 mm/min and (**d**) 52 mm/min.

3.2. Microstructure Characterization

Figure 9 presents the OM images of the microstructure of both the as-cast and the friction stir processed (FSPed) hypoeutectic Al-5Si alloy processed at a tool rotating speed of 1330 rpm and different travel speeds. As shown in Figure 9a, the as-cast Al-5Si alloy microstructure features coarse dendritic α-Al arms surrounded by eutectic Si. Porosity was randomly detected in some regions (Figure 9b), and no primary free-Si particles were observed. These microstructural features are similar to those reported by other researchers [41,44], with the difference being the absence of detected primary Si, which is typically found in near-eutectic Al-Si alloys with a Si content higher than 6.5 wt%. This coarse dendritic structure for the as-cast Al-Si alloys, while providing some strength, also results in limited ductility and reduced impact toughness [42]

The effect of FSP at different travel speeds (26–52 mm/min) on the microstructure of the Al-5Si alloy is evident through the significant modifications in its microstructure features, including grain size reduction, porosity elimination, and the breakdown of the primary and secondary arm spacing compared to the as-cast state, as given in Figure 9c–f. It can be noticed that the FSPed alloy exhibits distinct microstructural zones, each characterized by unique structural features. These zones include the stir zone (SZ), the thermomechanically affected zone (TMAZ), the heat-affected zone (HAZ), and the base metal (BM). The stir zone (SZ), situated at the center of the processed region, undergoes intense plastic deformation and dynamic recrystallization, resulting in significant grain refinement. This refined microstructure enhances mechanical properties, contributing to increased toughness, ductility, and uniformity in hardness distribution. The fine and equiaxed grains in this region indicate efficient recrystallization due to the combined effects of frictional heat and severe plastic deformation [42]. Adjacent to the stir zone, the thermomechanically affected zone (TMAZ) experiences both mechanical deformation and thermal exposure. However, unlike the SZ, the grains in this region are elongated and oriented along the direction of material flow, suggesting incomplete recrystallization. This zone serves as a transitional region between the SZ and HAZ, where the grain structure is partially refined but still retains some characteristics of the original dendritic microstructure. The heat-affected zone (HAZ), which lies beyond the TMAZ, undergoes thermal exposure without mechanical deformation. This leads to a partially modified grain structure where coarsening of grains may occur due to the heat input from FSP. The absence of significant plastic deformation in this zone differentiates it from the TMAZ, and the grain morphology largely depends on the extent of heat dissipation. Lastly, the base metal (BM) remains unprocessed and retains the coarse dendritic structure typical of as-cast Al-Si alloys. The contrast between the base metal and the processed zones highlights the substantial microstructural refinement achieved through FSP.

These microstructural modifications, as depicted in Figure 10, demonstrate the effectiveness of FSP in refining grain structure, improving homogeneity, and enhancing the overall mechanical properties of the hypoeutectic Al-5Si alloy. These findings are consistent with previous research, which has also reported significant microstructural and mechanical improvements in FSPed AA356 aluminum alloy [46].

Figures 10 and 11 represent grain size histograms and the mean size of (α-Al), respectively, for the as-cast and processed specimens at 1330 rpm and different travel speeds (26–52 mm/min). The fragmentation of the dendritic structure in Al-Si alloys through FSP represents a significant advancement in microstructural modification techniques. In the as-cast condition, the alloy displays a coarse dendritic morphology, with a primary dendritic arm spacing (PDAS) of 170.47 $\pm$ 54.12 µm (Figure 10a) and a secondary dendritic arm spacing (SDAS) of 26.24 $\pm$ 4.79 µm (Figure 10b). Such features are characteristic of solidification in hypoeutectic Al-Si alloys and result in a non-uniform microstructure.

Conversely, a significant grain size reduction is present in all of the processed Al-5Si alloys at different travel speeds. The α-Al grain sizes were 12.46 ± 1.59 μm (Figure 10c), 9.55 ± 1.51 μm (Figure 10d), 7.87 ± 1.38 μm (Figure 10e), and 7.75 ± 1.34 μm (Figure 10f) for the applied travel speeds of 26, 33, 42, and 52 mm/min, respectively. This notable refinement is primarily due to the intense plastic deformation and elevated thermal gradients induced during the stirring action of the FSP, which fragment the dendritic network and facilitate dynamic recrystallization [42,44]. This modification leads to a more refined and homogeneous microstructure and contributes to an increase in both tensile strength and impact energy, indicating an improvement in the overall mechanical properties. These findings are consistent with those reported in previous studies [42,44], demonstrating the potential of FSP in tailoring the as-cast solidified structure of Al-Si cast alloys.

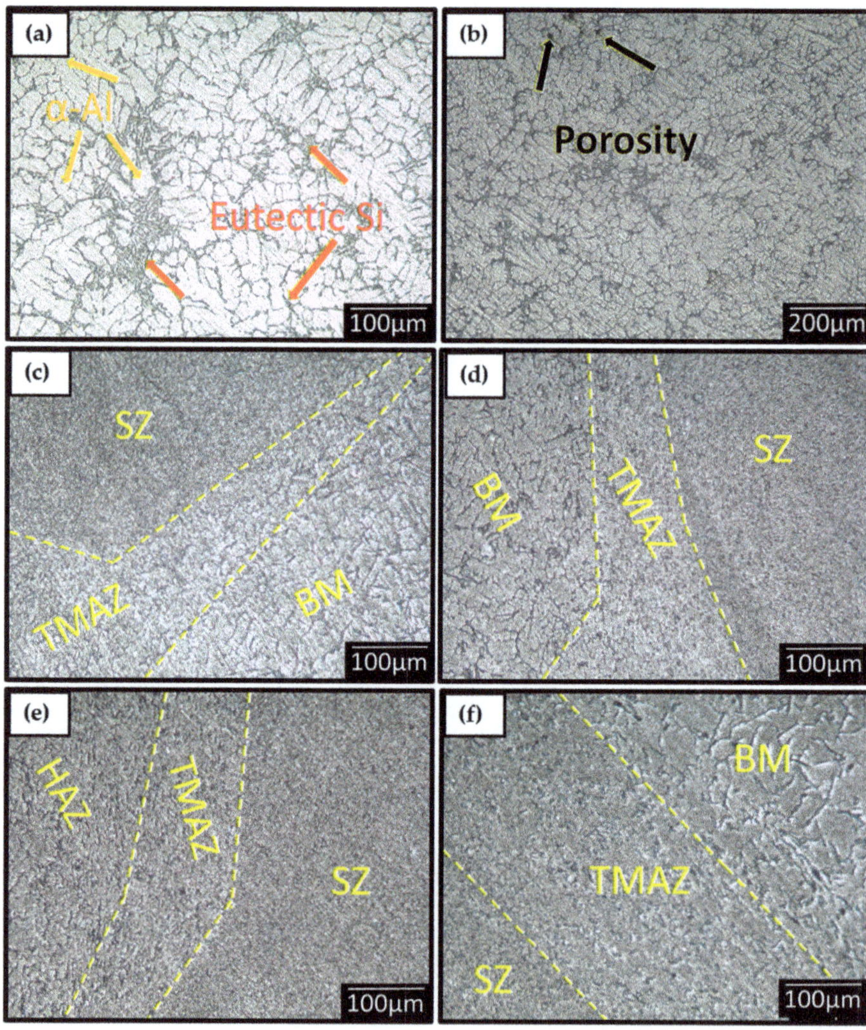

Figure 9. OM microstructure images of Al-5Si alloy: (**a**,**b**) as-cast condition and after FSP condition processed at a constant rotation speed of 1330 rpm and various travel speeds of (**c**) 26 mm/min, (**d**) 33 mm/min, (**e**) 42 mm/min, and (**f**) 52 mm/min.

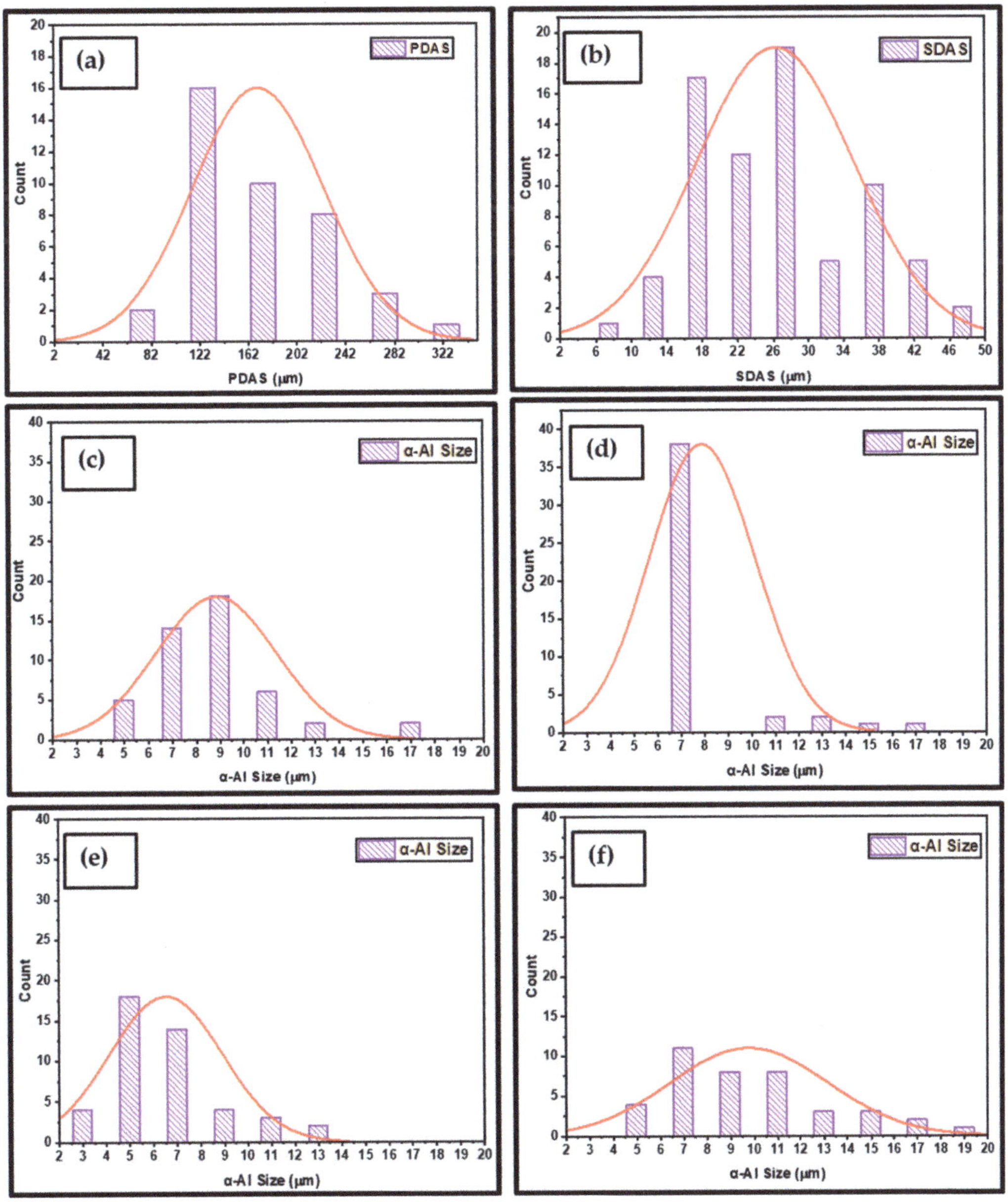

Figure 10. (**a**) Primary dendritic arm spacing (PDAS) and (**b**) secondary dendritic arm spacing (SDAS) for the as-cast Al-5Si alloy, and (α-Al) grain size histograms after FSP at 1330 rpm and various travel speeds of (**c**) 26, (**d**) 33, (**e**) 42, and (**f**) 52 mm/min.

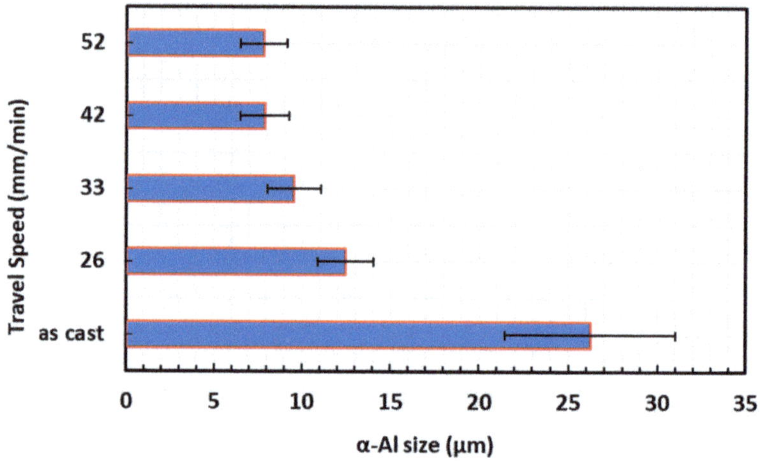

Figure 11. Represents the average α-Al grain size for all the Al-5Si alloy samples before and after FSP at a constant tool rotation speed of 1030 rpm and various travel speeds of 26, 33, 42, and 52 mm/min.

3.3. Mechanical Properties of the As-Cast and the FSPed Al-5Si Alloy

3.3.1. Tensile Properties

To assess the tensile properties of the hypoeutectic Al-5Si alloy in both as-cast and FSPed conditions, tensile tests were performed. Figure 12 shows the stress–strain curves of the as-cast Al-5Si alloy samples before and after FSP at 1030 rpm and various travel speeds (26–52 mm/min). The resulting stress–strain curves, presented in Figure 12, provide crucial insights into the material's overall deformation behavior under tensile loading. Monitoring the stress–strain curves reveals a significant improvement in tensile properties, specifically ultimate tensile strength (UTS) and strain at fracture, for all of the FSPed specimens across the investigated travel speed range (26–52 mm/min) compared to the as-cast Al-5Si alloy. Furthermore, the yield strength (YS) also generally increased in the processed samples relative to the as-cast state. This overall enhancement in mechanical properties can be attributed to beneficial microstructural modifications induced by FSP, such as porosity reduction or elimination, grain refinement within the stir zone, and the disruption of the coarse as-cast dendritic structure [41,42,61]. In addition, analysis of these curves will focus on key mechanical properties such as UTS, YS, and strain at fracture, as given in Figure 13, which can indicate differences in the material's response to applied stress as a function of the FSP travel speed and the initial as-cast condition.

The relationship between UTS, YS, ductility, and grain size in the microstructure-modified Al-5Si alloy through FSP at various travel speeds reveals critical insights into the mechanical behavior of the material. In the as-cast condition, the alloy exhibits a relatively coarse grain size of 26.24 ± 4.79 μm, which correlates with the lowest observed UTS (108.7 MPa), YS (57 MPa), and strain at fracture (3.6 %) compared to the processed specimens. This reduced strength and ductility is primarily attributed to the dendritic microstructure and porosity, which lacks the ability to effectively resist applied loads during tensile testing. In the case of FSP, the alloy experiences significant grain refinement and porosity elimination which typically enhances its mechanical performance. For instance, at a travel speed of 26 mm/min, the grain size is refined to 12.26 ± 1.59 μm, accompanied by a notable increase in UTS (164.5 MPa), YS (73.5 MPa), and strain at fracture (10.2%). The sample FSPed at 33 mm/min shows slightly reduced, but still enhanced, mechanical properties compared to the 26 mm/min sample, with UTS at 157.4 MPa, YS at 65.3 MPa, and strain at fracture at 10.1%. This improvement in the lower travel speeds (26 and

33 mm/min) can be ascribed to multiple mechanisms associated with FSP, including the fragmentation and uniform redistribution of eutectic Si and the dendritic arms, which enhance the load-bearing capacity of the modified aluminum matrix. Grain refinement contributes not only to increased strength but also to improved ductility and toughness, as indicated by the strain at fracture (Figures 12 and 13), as it promotes uniform plastic deformation and mitigates casting-related defects. At the higher speeds of 42 and 52 mm/min the improvement in UTS and strain at fracture compared to the as-cast Al-5Si is still observed (these values were 143 MPa, 10.9%, and 131.2 MPa, 10%, respectively), but the improvement is lower than that obtained at the lower speeds of 26 and 33 mm/min. For YS, a noticeable improvement is observed at the lower processing speeds of 26 and 33 mm/min. This enhancement can be attributed to improved material flow and dynamic recrystallization during FSP, which strengthens the matrix and hinders dislocation motion.

However, no improvement was detected at the higher speeds of 42 and 52 mm/min compared to the as-cast material. While increased travel speeds in this range continue to refine grain size, the reduced heat input per unit length and insufficient plasticization at these higher speeds become the dominant factors compromising mechanical properties. This can lead to incomplete material consolidation, less-effective material stirring, and the formation of residual defects such as micro-voids, tunnels, or incomplete grain boundary bonding, which are detrimental to mechanical integrity. Consequently, these processing-induced defects override the strengthening effect of further grain refinement, leading to the observed decrease in strength at higher travel speeds. These results underscore the critical role of travel speed in influencing YS during FSP, where higher speeds, despite finer grains, may diminish mechanical performance due to this complex interplay of heat input, material flow, microstructural stability, and defect formation. In fact, the mechanical properties are highly dependent on processing conditions, highlighting the importance of optimizing FSP parameters to maximize mechanical performance while preserving structural integrity.

Figure 14 presents SEM fractographic images of the tensile fracture surfaces of the Al-5Si alloy in both the as-cast state and after FSP at various travel speeds (26–52 mm/min) at a fixed rotation speed of 1330 rpm. The tensile fracture surfaces, as observed in the SEM micrographs, reveal distinct features that correspond to the variations in mechanical properties and microstructure induced by different FSP travel speeds. For the as-cast alloy, the fracture surface typically displays a mainly brittle failure mode, characterized by flat and faceted regions with limited plastic deformation, as shown in Figure 14a,b. The presence of coarse dendritic structures and casting micro-cavities contributes to crack initiation and propagation along weak interfaces, resulting in a relatively smooth fracture surface with cleavage planes and limited ductile behaviors [61,67]. In contrast, the FSPed sample at 26 mm/min, which exhibited the highest UTS and significant grain refinement, shows a predominantly ductile fracture mode. The SEM micrograph, as shown in Figure 14c,d, reveals numerous equiaxed deep and shallow dimples, indicative of substantial plastic deformation before failure. The deep and uniformly distributed dimples suggest enhanced toughness and ductility. The results are in agreement with previous works [68,71], which reported that FSP entirely disrupted the eutectic structure of the Al-Si alloy, thereby eliminating the network of preferential crack paths inherent to the as-cast material. Consequently, fracture initiation during tensile loading predominantly originated from the brittle eutectic Si particles. Finally, at the highest travel speed of 52 mm/min, the fracture surface reveals mixed modes of fractured features. The SEM images in Figure 14e,f show larger, irregular cleavage planes interspersed with micro-cracks and small and large dimples. This fracture morphology corresponds well with the lowest Ys and UTSs observed among the FSPed specimens, emphasizing the detrimental impact of insufficient heat input and plastic flow at the highest travel speed.

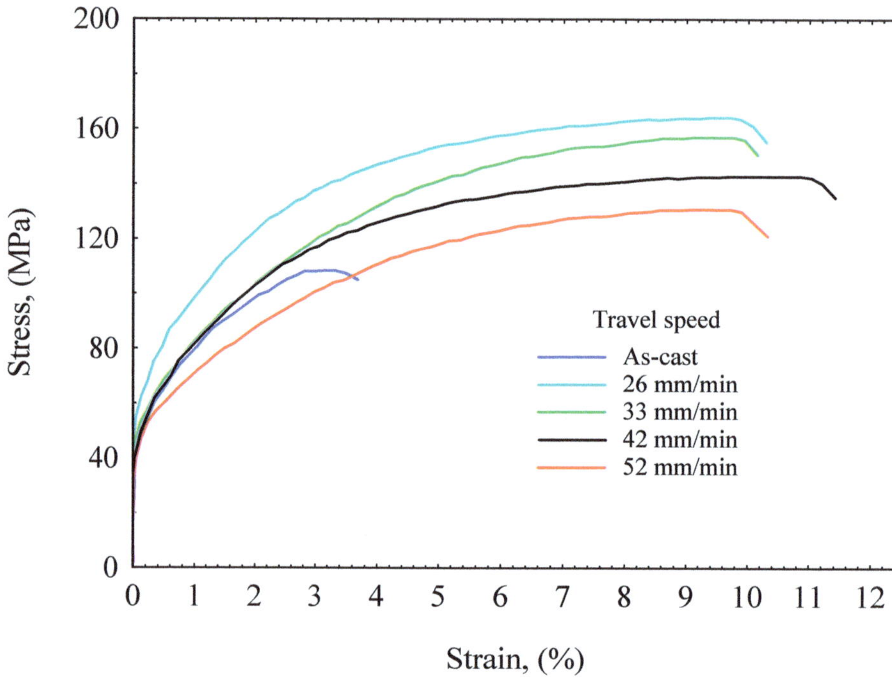

Figure 12. Stress–strain curves of the as-cast Al-5Si alloy samples before and after FSP at 1030 rpm and at various travel speeds (26–52 mm/min).

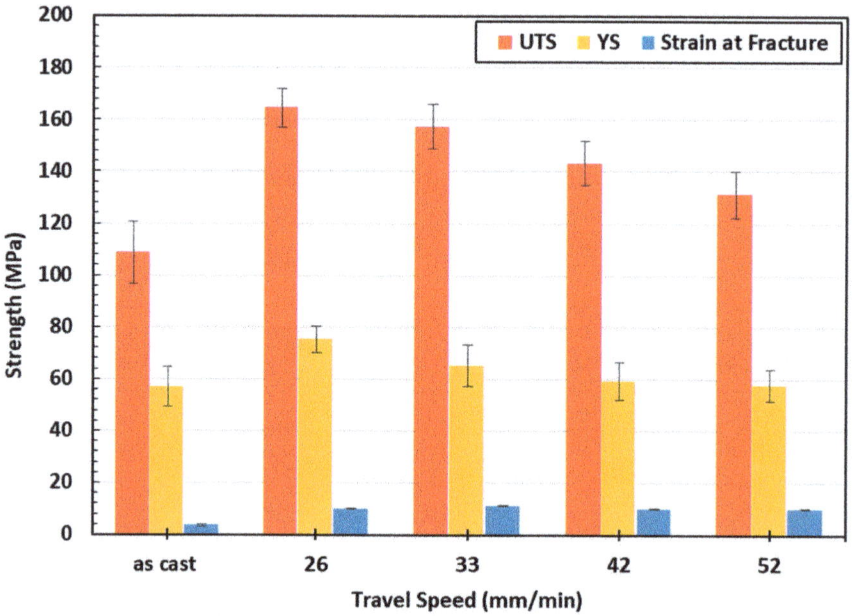

Figure 13. Ultimate tensile and yield strengths of the Al-5Si alloy before and after FSP at a constant rotation speed of 1030 rpm and various tool travel speeds (26–52 mm/min).

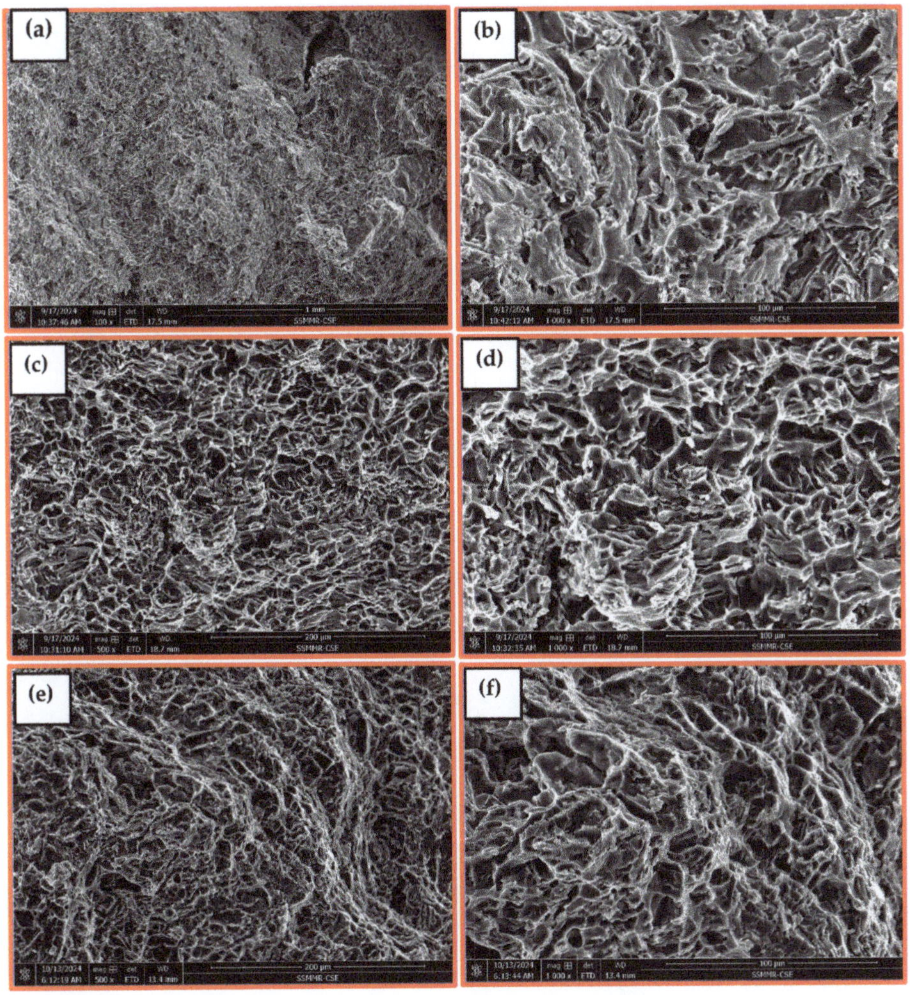

Figure 14. Fracture-surface SEM images of the as-cast Al-5Si alloy sample (**a**) at 100× and (**b**) at 1000×, and that of the FSPed specimens at 1330 rpm and various travel speeds of 26 mm/min (**c**) at 500× and (**d**) at 1000×, and 52 mm/min (**e**) at 500× and (**f**) at 1000×.

3.3.2. Microhardness

Hardness, which indicates a material's ability to resist localized plastic deformation, is a key characteristic of the Al-5Si alloy. When subjected to FSP with a fixed rotation speed of 1330 rpm but different traverse speeds (26–52 mm/min), the as-cast alloy's hardness is notably influenced by microstructure modification. Figure 15 represents the microhardness trend of the as-cast Al-5Si alloy and the FSP specimens at various travel speeds. The as-cast condition exhibits a microhardness of 45.5 ± 5.5 HV, which aligns with its coarse grain size, scattered eutectic phase, porosity formation, and relatively low tensile strength. Furthermore, it can be noted that the hardness values improved as a result of the FSP process at all applied travel speeds by an improvement rate of not less than 39% compared to the unprocessed material. Although there is also a slight improvement in the hardness values with increasing travel speed, the scattered measurements of hardness showed relatively large error bars for the as-cast material and the specimens processed at the higher travel speeds of 42 and 52 mm/min. The increase in microhardness may be attributed to the fact that the smaller the grain size, the higher the hardness, according to the Hall–Petch relationship. Conversely, the large error bars indicate a heterogeneous microstructure fracture which is due to insufficient heat input at high travel speeds [41,42,61].

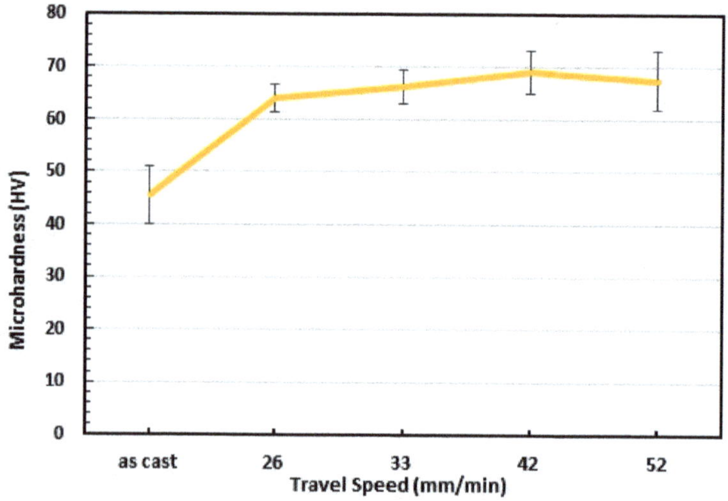

Figure 15. Microhardness of the as-cast Al-5Si alloy and the FSP specimens at a constant tool rotation speed of 1030 rpm and various travel speeds of 26, 33, 42, and 52 mm/min.

3.3.3. Impact

The impact energy results provide additional insights into the mechanical behavior of the Al-5Si alloy after FSP, particularly in relation to the observed tensile properties, microhardness, and grain size. Figure 16 shows the impact energy before and after FSP of the as-cast Al-5Si alloy specimens processed at 1330 rpm and at different travel speeds. All the processed Al-Si alloys show higher toughness behavior than the as-cast materials at all the applied travel speeds. The as-cast alloy, with its coarse grain structure, exhibits a low impact energy of 4.2 ± 1 J, indicating poor toughness due to the brittle nature of the as-cast microstructure [7,62]. The microstructure features show the presence of cracks, micro-voids, dendrite arm fractures, cleavage facets, and, in some areas, large and shallow dimples, indicating a predominantly brittle fracture mode (Figure 17a). The large and shallow dimples suggest limited localized ductility of the as-cast alloy. The FSP significantly refines the microstructure (Figure 11), leading to improved toughness (Figure 16). However, fracture features vary with travel speed, as shown in Figure 17a,b for the specimens processed at 26 and 52 mm/min, respectively. The impact energy reached its highest value of 12.5 ± 0.57 J at a travel speed of 26 mm/min. This increase is consistent with the grain refinement and improved material ductility resulting from FSP, which enhances the alloy's ability to absorb energy during impact. The fracture surface shows equiaxed deep dimples, indicating a mainly ductile fracture (Figure 17b). These observations are consistent with the grain refinement (Figure 11) and improved material ductility resulting from FSP (Figure 12), which enhances the alloy's ability to absorb energy during impact. However, as the travel speed increases, the impact energy slightly decreases to be around the same value of 11.2 J for the two speeds of 33 and 42 mm/min. At the highest travel speed of 52 mm/min it attains the lowest value of 10 ± 0.64 J and the fracture surface shows dimples of different sizes and fewer cleavage facets compared to the as-cast alloy, indicating a mixed mode of the ductile brittle mode.

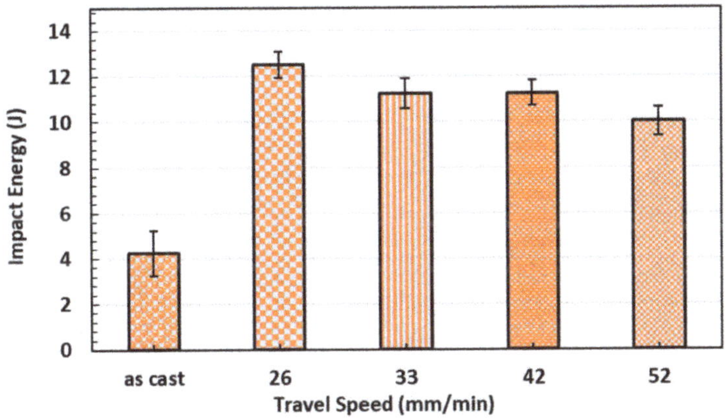

Figure 16. Impact energy of the as-cast Al-5Si alloy and the FSP materials at a constant tool rotation speed of 1030 rpm and at the various travel speeds of 26, 33, 42, and 52 mm/min.

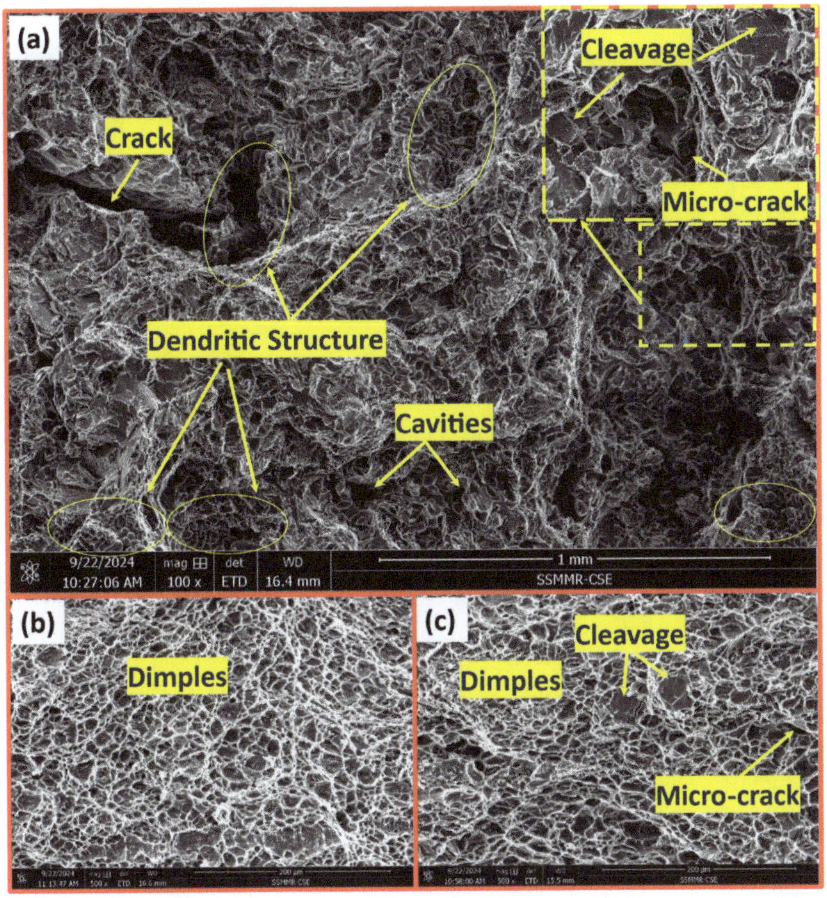

Figure 17. SEM fractography impact fracture surfaces of (**a**) the as-cast Al-5Si alloy at a magnification of 100×, and the FSPed specimens processed at a constant tool rotation speed of 1030 rpm and at the various travel speeds of (**b**) 26 mm/min at a magnification of 500× and of (**c**) 52 mm/min at a magnification of 500×.

3.3.4. Wear Behavior

The wear behavior of the as-cast Al-5Si alloy is significantly influenced by its microstructure, bulk hardness, and toughness, which are in turn affected by processing conditions such as the travel speed when the rotation speed is constant. The wear test was performed at a 10 N wear load for the as-cast Al-5Si and the FSPed alloys at the different travel speeds of 26, 33, 42, and 52 mm/min. The wear behavior is plotted as a weight loss measured value against travel speed (Figure 18). It can be noticed that the applied load was enough to cause surface damage on all the tested samples, and the FSPed samples lost lower weight than the as-cast samples, showing a lower wear rate, which is shown in the wear rate as a function of FSP travel speed in Figure 19. For the as-cast state, the Al-5Si alloy exhibits a relatively high wear rate of $9.23 \pm 0.84 \times 10^{-6}$ g/m, indicating a lower wear resistance compared to the processed specimens under dry sliding conditions. This elevated wear rate is primarily attributed to the coarse microstructure of the alloy, which is characterized by a large SDAS of 26.24 ± 4.79 µm, which represents a lower hardness and toughness. The presence of a non-uniformly distributed network of a hard eutectic silicon phase within the softer Al matrix leads to microstructural inhomogeneity, which in turn contributes to uneven wear and excessive material loss during sliding.

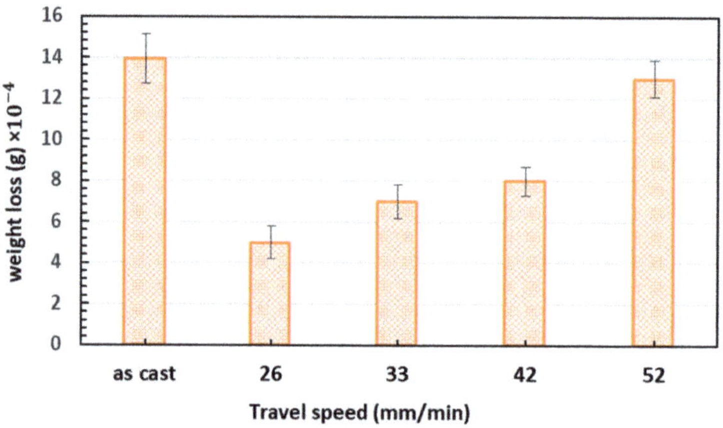

Figure 18. Weight loss as a function of tool travel speed of the as-cast Al-5Si alloy and the FSPed materials at a constant tool rotation speed of 1030 rpm and at the various travel speeds of 26, 33, 42, and 52 mm/min.

At a travel speed of 26 mm/min, the wear rate is markedly reduced to $3.31 \pm 0.8 \times 10^{-6}$ g/m, representing a substantial improvement over the as-cast condition. This enhancement is mainly a result of grain refinement achieved through FSP. More refined and homogeneously distributed microstructural constituents lead to improved hardness, toughness, and resistance against wear by promoting a uniform distribution of stress during sliding. Thus, it can be concluded that among all the Al-5Si samples subjected to the wear test, the sample processed at a speed of 26 showed the highest wear resistance. Finally, findings presented in Figure 19 highlight the substantial improvement in wear resistance following FSP, with the extent of enhancement being strongly influenced by the processing parameters, particularly the tool travel speed. Generally, increased hardness correlates with improved wear resistance, whereas higher toughness can reduce crack propagation under abrasive conditions. Thus, the weight loss and wear rate are expected to vary depending on the balance between these competing mechanical properties at different speeds. These findings underscore the necessity of optimizing FSP parameters to achieve the best possible wear resistance for specific applications, particularly under different sliding conditions. These results are consistent with those described in the previous studies found in [61,69].

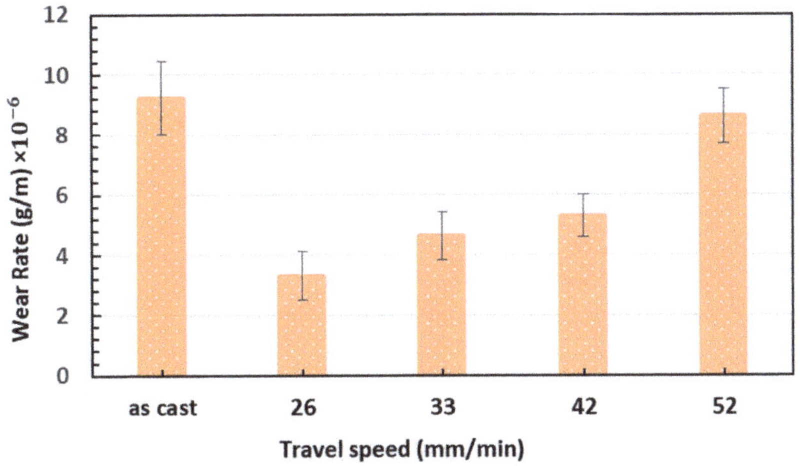

Figure 19. Wear rate against tool travel speed of the as-cast Al-5Si alloy and the FSPed specimens at a constant rotation speed of 1030 rpm and various travel speeds (26–52 mm/min).

Figure 20 illustrates the coefficient of friction (COF) for the as-cast and FSPed Al-Si alloy samples under a dry sliding condition, revealing a complex relationship between microstructural changes and frictional behavior. The as-cast material exhibits a relatively high COF of 1.1 ± 0.13, which directly reflects its coarse, inhomogeneous microstructure characterized by SDAS. This non-uniform distribution of a hard eutectic silicon phase within the softer aluminum matrix leads to irregular interactions with the counter surface, resulting in increased friction (Figure 20) and higher wear rates (Figure 19). When subjected to FSP the Al-5Si alloy generally shows notable improvements in COF under a dry sliding condition, with the most significant reduction observed at a travel speed of 26 mm/min where the COF decreases to 0.66 ± 0.01. This substantial reduction is primarily attributed to the significant grain refinement and increased microstructural homogeneity achieved through FSP, leading to a smoother and more consistent contact surface that effectively reduces friction. While the COF values for samples processed at 33 mm/min and 42 mm/min are still lower than the as-cast condition, they show a slight increase compared to 26 mm/min, despite further grain refinement (Figure 11). This suggests that at these higher speeds factors like increased grain boundary density or residual stresses may begin to offset the benefits of refinement. At the highest travel speed of 52 mm/min, the COF is 0.9 ± 0.13, which is lower than that of the as-cast state but higher than the optimal FSP conditions. This indicates a less favorable balance of microstructural factors, resulting in higher friction and correlating with the increased wear rates observed at these less optimal FSP travel speeds.

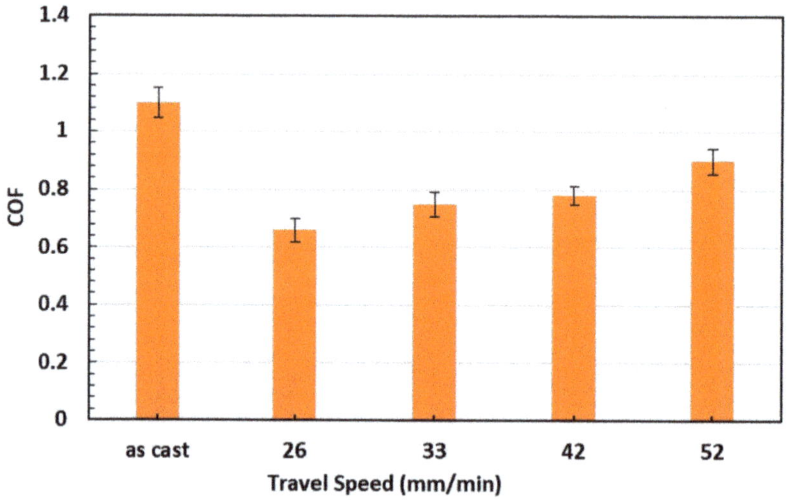

Figure 20. Coefficient of friction (COF) against tool travel speed of the as-cast Al-5Si alloy and the FSPed specimens at a constant rotation speed of 1030 rpm and at various travel speeds (26–52 mm/min).

4. Conclusions

The present study investigates the influence of FSP travel speed on the microstructure and mechanical performance of a hypoeutectic Al-5 wt.% Si alloy in its as-cast condition. FSP was conducted at a fixed rotational speed of 1330 rpm with varying travel speeds (26, 33, 42, and 52 mm/min). This work offers novel insights by systematically exploring the effect of varying travel speeds on this less-studied hypoeutectic Al-5Si alloy, filling a significant gap in the existing literature which is primarily focused on other Al-Si compositions or different processing parameters. The microstructure and mechanical properties (hardness, tensile strength, impact energy, and wear behavior) of the as-cast and processed materials were explored, and the following conclusions summarize the key outcomes of this study.

1. Friction stir processing (FSP) significantly modifies the microstructure and enhances the mechanical properties of as-cast Al-5Si alloy, with the most pronounced improvements generally observed at lower travel speeds. For instance, at a travel speed of 26 mm/min the hardness improved by up to 39%, ultimate tensile strength (UTS) by up to 51%, impact energy by up to 197%, and wear resistance by up to 64% compared to the as-cast state.
2. FSP results in a significant reduction in α-Al grain size. The as-cast Al-5Si alloy exhibited a coarse grain size of 26.24 ± 4.79 μm. Following FSP, the α-Al grain sizes were notably refined to 12.46 ± 1.59 μm, 9.55 ± 1.51 μm, 7.87 ± 1.38 μm, and 7.75 ± 1.34 μm for travel speeds of 26, 33, 42, and 52 mm/min, respectively.
3. Compared to the as-cast alloy, which exhibits a heterogeneous coarse structure and poor tensile properties (YS of 57 MPa, UTS of 108.7 MPa, and strain at fracture of 3.6%), FSP significantly enhances the UTS and strain at fracture of Al-5Si alloys, particularly at lower travel speeds (26 and 33 mm/min). The sample processed at 26 mm/min achieved the highest UTS (164.5 MPa) and strain at fracture (10.2%), primarily due to porosity elimination and the breakdown/dispersion of the dendrite arm structure and eutectic Si-phase. While higher travel speeds (42 and 52 mm/min) still showed improvement (e.g., UTS of 131.2 MPa and strain at fracture of 10% at 52 mm/min), the enhancement was less pronounced than at lower speeds. FPS improves the toughness

of Al-5Si alloys at all travel speeds, and the impact energy reached its highest value of 12.5 ± 0.57 J at a travel speed of 26 compared to 4.2 ± 1 J for the as-cast state, indicating a predominantly ductile fracture mode.

4. Analysis of tensile and impact fracture surfaces confirmed changes in failure mechanisms. The as-cast Al-5Si alloy primarily exhibited brittle fracture, characterized by flat faced regions and limited plastic deformation. In contrast, the FSPed alloy, especially at a travel speed of 26 mm/min, showed a predominantly ductile fracture mode with numerous deep, equiaxed dimples, consistent with improved toughness. At the highest processing travel speed of 52 mm/min, a mixed mode fracture (combining ductile dimples and brittle cleavage facets) was observed.
5. The wear resistance of the Al-5Si alloy was significantly enhanced by FSP. The wear rate of the as-cast Al-5Si alloy was $9.23 \pm 1.2 \times 10^{-6}$ g/m, which is substantially higher than that of all FSPed materials. The lowest wear rate, indicating the highest wear resistance, was achieved at a travel speed of 26 mm/min ($3.31 \pm 0.8 \times 10^{-6}$ g/m). While FSP generally improved wear resistance, the wear rate progressively increased with higher travel speeds, reaching $4.63 \pm 0.9 \times 10^{-6}$ g/m at 33 mm/min, $5.29 \pm 0.7 \times 10^{-6}$ g/m at 42 mm/min, and $8.6 \pm 0.9 \times 10^{-6}$ g/m at 52 mm/min. This trend highlights the optimal balance of microstructural refinement and heat input at lower travel speeds for superior wear performance.

Author Contributions: Conceptualization, A.E.-S., M.M.E.-S.S., A.E. and B.E.-E.; methodology, B.E.-E., M.M.E.-S.S. and S.A.; software, B.E.-E. and A.E.-S.; validation, B.E.-E., S.A., M.F.Y.S. and M.M.E.-S.S.; formal analysis, B.E.-E. and M.F.Y.S.; investigation, B.E.-E., M.M.E.-S.S. and S.A.; resources, A.E.-S., S.A. and A.E.; data curation, B.E.-E.; writing—original draft preparation, B.E.-E.; writing—review and editing, M.M.E.-S.S., S.A, M.F.Y.S. and A.E.-S.; visualization, B.E.-E., A.E. and M.F.Y.S.; supervision, A.E.-S., M.M.E.-S.S. and A.E.; project administration, M.F.Y.S. and S.A.; funding acquisition, S.A. and M.F.Y.S. All authors have read and agreed to the published version of the manuscript.

Funding: This work was supported and funded by the Deanship of Scientific Research at Imam Mohammad Ibn Saud Islamic University (IMSIU) (grant number IMSIU-DDRSP2503).

Data Availability Statement: The original contributions presented in this study are included in the article. Further inquiries can be directed to the corresponding author.

Conflicts of Interest: The authors declare no conflicts of interest.

References

1. Jeong, C.; Kang, C.-S.; Cho, J.-I.; Oh, I.-H.; Kim, Y.-C. Effect of microstructure on mechanical properties for A356 casting alloy. *Int. J. Cast Met. Res.* **2008**, *21*, 193–197. [CrossRef]
2. Kang, M.; Kwon, H.; Byon, E.; Lee, E. Anti-corrosion and wear performance of Al-Si coating deposited by wire arc spray for marine applications. *J. Adv. Mar. Eng. Technol.* **2023**, *47*, 175–186. [CrossRef]
3. Alzahrani, B.; El-Sayed Seleman, M.M.; Ahmed, M.M.Z.; Elfishawy, E.; Ahmed, A.M.; Touileb, K.; Jouini, N.; Habba, M.I. The applicability of die cast A356 alloy to additive friction stir deposition at various feeding speeds. *Materials* **2021**, *14*, 6018. [CrossRef]
4. Bohlooli, V.; Shabani Mahalli, M.; Boutorabi, S. Effect of ablation casting on microstructure and casting properties of A356 aluminium casting alloy. *Acta Metall. Sin.* **2013**, *26*, 85–91. [CrossRef]
5. Wang, Q. Microstructural effects on the tensile and fracture behavior of aluminum casting alloys A356/357. *Metall. Mater. Trans. A* **2003**, *34*, 2887–2899. [CrossRef]
6. Merlin, M.; Timelli, G.; Bonollo, F.; Garagnani, G.L. Impact behaviour of A356 alloy for low-pressure die casting automotive wheels. *J. Mater. Process. Technol.* **2009**, *209*, 1060–1073. [CrossRef]
7. Jiang, W.; Fan, Z.; Liu, D.; Liao, D.; Dong, X.; Zong, X. Correlation of microstructure with mechanical properties and fracture behavior of A356-T6 aluminum alloy fabricated by expendable pattern shell casting with vacuum and low-pressure, gravity casting and lost foam casting. *Mater. Sci. Eng. A* **2013**, *560*, 396–403. [CrossRef]

8. Nafisi, S.; Ghomashchi, R. Grain refining of conventional and semi-solid A356 Al–Si alloy. *J. Mater. Process. Technol.* **2006**, *174*, 371–383. [CrossRef]
9. Kolahdooz, A.; Dehkordi, S.A. Effects of important parameters in the production of Al-A356 alloy by semi-solid forming process. *J. Mater. Res. Technol.* **2019**, *8*, 189–198. [CrossRef]
10. Nafisi, S.; Ghomashchi, R. Effects of modification during conventional and semi-solid metal processing of A356 Al-Si alloy. *Mater. Sci. Eng. A* **2006**, *415*, 273–285. [CrossRef]
11. Jamaati, R.; Amirkhanlou, S.; Toroghinejad, M.R.; Niroumand, B. Significant improvement of semi-solid microstructure and mechanical properties of A356 alloy by ARB process. *Mater. Sci. Eng. A* **2011**, *528*, 2495–2501. [CrossRef]
12. Khosravi, H.; Eslami-Farsani, R.; Askari-Paykani, M. Modeling and optimization of cooling slope process parameters for semi-solid casting of A356 Al alloy. *Trans. Nonferrous Met. Soc. China* **2014**, *24*, 961–968. [CrossRef]
13. Chou, S.; Trask, M.; Danovitch, J.; Wang, X.; Choi, J.; Brochu, M. Pulsed laser powder bed fusion additive manufacturing of A356. *Mater. Charact.* **2018**, *143*, 27–33. [CrossRef]
14. Carneiro, V.H.; Rawson, S.; Puga, H.; Withers, P. Macro-, meso-and microstructural characterization of metallic lattice structures manufactured by additive manufacturing assisted investment casting. *Sci. Rep.* **2021**, *11*, 4974. [CrossRef]
15. Fan, H.; Witvrouw, A.; Wolf-Monheim, F.; Souschek, R.; Yang, S. Effects of substrate surface treatments on hybrid manufacturing of AlSi7Mg using die casting and selective laser melting. *J. Mater Sci. Technol.* **2023**, *156*, 142–156. [CrossRef]
16. Mehmood, A.; Shah, M.; Sheikh, N.A.; Qayyum, J.A.; Khushnood, S. Grain refinement of ASTM A356 aluminum alloy using sloping plate process through gravity die casting. *Alex. Eng. J.* **2016**, *55*, 2431–2438. [CrossRef]
17. Lin, J.-H.; Zhao, H.-D.; Huang, J.-M. Spatial interfacial heat transfer and surface characteristics during gravity casting of A356 alloy. *Trans. Nonferrous Met. Soc. China* **2019**, *29*, 43–50. [CrossRef]
18. Ceschini, L.; Morri, A.; Morri, A.; Pivetti, G. Predictive equations of the tensile properties based on alloy hardness and microstructure for an A356 gravity die cast cylinder head. *Mater. Des.* **2011**, *32*, 1367–1375. [CrossRef]
19. Huang, L.; Du, X.; Zhuang, Q.; Huang, C.; Li, J. Effect of alloying elements Mg and Cu on the modification of eutectic silicon in hypoeutectic Al–Si alloys. *Metals* **2023**, *13*, 1995. [CrossRef]
20. Gursoy, O.; Timelli, G. Lanthanides: A focused review of eutectic modification in hypoeutectic Al–Si alloys. *J. Mater. Res. Technol.* **2020**, *9*, 8652–8666. [CrossRef]
21. Callegari, B.; Lima, T.N.; Coelho, R.S. The influence of alloying elements on the microstructure and properties of Al-Si-based casting alloys: A review. *Metals* **2023**, *13*, 1174. [CrossRef]
22. Hegde, S.; Prabhu, K.N. Modification of eutectic silicon in Al–Si alloys. *J. Mater. Sci.* **2008**, *43*, 3009–3027. [CrossRef]
23. Fracchia, E.; Gobber, F.S.; Rosso, M. Effect of alloying elements on the Sr modification of Al-Si cast alloys. *Metals* **2021**, *11*, 342. [CrossRef]
24. Ding, W.; Gou, L.; Hu, L.; Zhang, H.; Zhao, W.; Ma, J.; Qiao, J.; Li, X. Modification of eutectic Si in hypoeutectic Al-Si alloy with novel Al-3Ti-4.35 La master alloy. *J. Alloys Compd.* **2022**, *929*, 167350. [CrossRef]
25. Jiang, D.; Yu, J. Simultaneous refinement and modification of the eutectic Si in hypoeutectic Al–Si alloys achieved via the addition of SiC nanoparticles. *J. Mater. Res. Technol.* **2019**, *8*, 2930–2943. [CrossRef]
26. Al-Helal, K.; Stone, I.; Fan, Z. Simultaneous primary Si refinement and eutectic modification in hypereutectic Al–Si alloys. *Trans. Indian Inst. Met.* **2012**, *65*, 663–667. [CrossRef]
27. Manani, S.; Pradhan, A. Effects of melt thermal treatment on cast Al-Si alloys: A review. *Mater. Today Proc.* **2022**, *62*, 6568–6572. [CrossRef]
28. Samuel, A.; Garza-Elizondo, G.; Doty, H.; Samuel, F. Role of modification and melt thermal treatment processes on the microstructure and tensile properties of Al–Si alloys. *Mater. Des.* **2015**, *80*, 99–108. [CrossRef]
29. Xu, C.; Yang, Y.; Wang, H.; Jiang, Q. Effects of modification and heat-treatment on the abrasive wear behavior of hypereutectic Al–Si alloys. *J. Mater. Sci.* **2007**, *42*, 6331–6338. [CrossRef]
30. Shabestari, S.; Shahri, F. Influence of modification, solidification conditions and heat treatment on the microstructure and mechanical properties of A356 aluminum alloy. *J. Mater. Sci.* **2004**, *39*, 2023–2032. [CrossRef]
31. Pan, T.-A.; Tzeng, Y.-C. Impact of Sr and La Modification and Post-Heat Treatment on Microstructural Evolution and Thermal Conductivity of Hypoeutectic Al-Si Alloys. *J. Alloys Compd.* **2025**, *1032*, 181114. [CrossRef]
32. Ma, X.; Pan, Y.; Chen, L.; Hua, H.; Hou, J.; Zhao, Y. Effects of Cu and Mg alloying and low-temperature aging on microstructure characteristics and mechanical properties of heat treatment-free cast Al-Si-Cu-Mg alloys. *J. Mater. Res. Technol.* **2025**, *36*, 7081–7093. [CrossRef]
33. Sahu, B.P.; Andani, M.T.; Ghosh, A.; Wang, J.; Misra, A. Crystallography and Interface Structures in As-Arc Melted and Laser Surface-Remelted Aluminum–Silicon Alloys with and without Strontium Addition. *Crystals* **2024**, *14*, 283. [CrossRef]
34. Bashir, N.; Iqbal, A. Characterizing Laser-Modified Microstructures and Electrical and Mechanical Properties of Al-15.3% Si (wt.%) Alloys. *J. Mater. Eng. Perform.* **2024**, *33*, 2196–2208. [CrossRef]

35. Babalou, R.; Azarbarmas, M.; Prashanth, K.G. Heat treatment and laser shock peening of AlSi10Mg alloy produced by selective laser melting: Microstructure, hardness and residual stress analysis. *Mater. Today Commun.* **2025**, *45*, 112408. [CrossRef]
36. Snopiński, P.; Yu, T.; Zhang, X.; Jensen, D.J. Effect of multi-pass shot peening on the microstructure of LPBF AlSi10Mg alloy. In Proceedings of the 44th Risø International Symposium on Materials Science (RISO 2024), Roskilde, Denmark, 2–6 September 2024; p. 012039. [CrossRef]
37. Asharsheikh, B.; Daneshifar, M.H.; Jabbareh, M.A. Effect of Friction Stir Processing on the Microstructure and Mechanical Properties of Iron-Bearing A356 Cast Aluminum Alloy. *Int. J. Met.* **2025**, 1–12. [CrossRef]
38. Wais, A.M.H.; Salman Ahmed, J.M.; Al-Roubaiy, O. Influence of friction stir processing on mechanical properties and the microstructure of aluminum-silicon cast alloys. *Jordan J. Mech. Ind. Eng.* **2025**, *19*, 203–214. [CrossRef]
39. Singh, K.; Hussain, M.I.; Pancholi, V.; Kashyap, B.P. Effect of friction stir processing on grain refinement and superplastic properties of binary Al-8 wt.% Si to Al-30 wt.% Si alloys. *Philos. Mag.* **2025**, *105*, 229–260. [CrossRef]
40. Traiano, D.; do Nascimento Rosa, S.; Lourençato, L.A.; Sánchez Roca, A.; Sánchez Orozco, M.C.; Carvajal Fals, H.D. Multiresponse optimization applied to friction-stir processing to enhance wear and corrosion performance in Al–Si alloys. *Discov. Appl. Sci.* **2024**, *6*, 643. [CrossRef]
41. Meenia, S.; Khan, F.; Babu, S.; Immanuel, R.; Panigrahi, S.; Ram, G.J. Particle refinement and fine-grain formation leading to enhanced mechanical behaviour in a hypo-eutectic Al–Si alloy subjected to multi-pass friction stir processing. *Mater. Charact.* **2016**, *113*, 134–143. [CrossRef]
42. Singh, S.K.; Immanuel, R.; Babu, S.; Panigrahi, S.; Ram, G.J. Influence of multi-pass friction stir processing on wear behaviour and machinability of an Al-Si hypoeutectic A356 alloy. *J. Mater. Process. Technol.* **2016**, *236*, 252–262. [CrossRef]
43. Charandabi, F.K.; Jafarian, H.R.; Mahdavi, S.; Javaheri, V.; Heidarzadeh, A. Modification of microstructure, hardness, and wear characteristics of an automotive-grade Al-Si alloy after friction stir processing. *J. Adhes. Sci. Technol.* **2021**, *35*, 2696–2709. [CrossRef]
44. Bates, W.P.; Patel, V.; Rana, H.; Andersson, J.; De Backer, J.; Igestrand, M.; Fratini, L. Properties augmentation of cast hypereutectic Al–Si alloy through friction stir processing. *Met. Mater. Int.* **2023**, *29*, 215–228. [CrossRef]
45. Cheng, W.; Liu, C.; Ge, Z. Optimizing the mechanical properties of Al–Si alloys through friction stir processing and rolling. *Mater. Sci. Eng. A* **2021**, *804*, 140786. [CrossRef]
46. Elfishawy, E.; Ahmed, M.M.Z.; El-Sayed Seleman, M.M. Additive Manufacturing of Aluminum Using Friction Stir Deposition. In Proceedings of the TMS 2020 149th Annual Meeting & Exhibition Supplemental Proceedings, San Diego, CA, USA, 23–27 February 2020; Minerals, Metals & Materials Society, Ed.; The Minerals, Metals & Materials Series. Springer: Cham, Switzerland, 2020.
47. El-Eraki, B.; El-Sissi, A.; Khafagi, S.; Nada, H. Process parameters optimization for producing AA6061/(Al_2O_3, Gr and Al_2O_3 + Gr) surface composites by friction stir processing. In Proceedings of the International Conference on Aerospace Sciences and Aviation Technology, Cairo, Egypt, 9–11 April 2019; pp. 1–13. [CrossRef]
48. El-Sayed Seleman, M.M.; Alateyah, A.I.; Mahmoud, E.A.E.; Elsoeudy, R.; Ahmed, M.M.Z.; Hafez, K.M.; El-Garaihy, W.H.; AS, A. Friction stir welding of 2507 super-duplex stainless steel: Feasibility of butt joint groove filling at different process parameters. *Adv. Mater. Process. Technol.* **2024**, 1–23. [CrossRef]
49. Ahmed, M.M.Z.; Jouini, N.; Alzahrani, B.; El-Sayed Seleman, M.M.; Jhaheen, M. Dissimilar friction stir welding of AA2024 and AISI 1018: Microstructure and mechanical properties. *Metals* **2021**, *11*, 330. [CrossRef]
50. Maqbool, A.; Lone, N.F.; Khan, N.Z.; Siddiquee, A.N.; Chen, D. Exceptional tensile strength-ductility synergy in friction stir processed Mg-Y-Nd-Zr alloy achieved through bimodal grain size distribution. *Mater. Sci. Eng. A* **2025**, *919*, 147521. [CrossRef]
51. Lone, N.F.; Bajaj, D.; Gangil, N.; Abidi, M.H.; Chen, D.; Arora, A.; Siddiquee, A.N. Implementing high entropy alloy synthesis via friction stir processing: Simultaneous solid-state alloying and ultra-refinement of copper. *Vacuum* **2024**, *222*, 113042. [CrossRef]
52. Lone, N.F.; Bajaj, D.; Gangil, N.; Khan, T.; Abidi, M.H.; Al-Ahmari, A.; Siddiquee, A.N. Multi principal element alloy particle reinforced metal matrix composites: Synthesis, microstructure, and mechanical aspects. *Manuf. Lett.* **2023**, *36*, 46–51. [CrossRef]
53. Maqbool, A.; Lone, N.F.; Ahmad, T.; Khan, N.Z.; Siddiquee, A.N. Effect of hybrid reinforcement and number of passes on microstructure, mechanical and corrosion behavior of WE43 Mg alloy based metal matrix composite. *J. Manuf. Process.* **2023**, *89*, 170–181. [CrossRef]
54. Besekar, A.; Kathiresan, M.; Jose Immanuel, R. Friction stir processing of recycled titanium reinforced A356 composite developed through stir casting. *Trans. Indian Inst. Met.* **2024**, *77*, 3037–3043. [CrossRef]
55. Cheremnov, A.; Chumaevskii, A.; Knyazhev, E.; Kolubaev, E. Effect of multipass friction stir processing on structure formation, mechanical and tribological properties of CuSn6 copper alloy. *Russ. Phys. J.* **2024**, *67*, 2049–2055. [CrossRef]
56. Ralls, A.M.; Kasar, A.K.; Menezes, P.L. Friction stir processing on the tribological, corrosion, and erosion properties of steel: A review. *J. Manuf. Mater. Process.* **2021**, *5*, 97. [CrossRef]
57. Ghasemi-Kahrizsangi, A.; Kashani-Bozorg, S.F. Microstructure and mechanical properties of steel/TiC nano-composite surface layer produced by friction stir processing. *Surf. Coat. Technol.* **2012**, *209*, 15–22. [CrossRef]

58. Aldajah, S.; Ajayi, O.; Fenske, G.; David, S. Effect of friction stir processing on the tribological performance of high carbon steel. *Wear* **2009**, *267*, 350–355. [CrossRef]
59. Ahmed, M.M.Z.; El-Sayed Seleman, M.M.; Eid, R.G.; Zawrah, M. Production of AA1050/silica fume composite by bobbin tool-friction stir processing: Microstructure, composition and mechanical properties. *CIRP J. Manuf. Sci. Technol.* **2022**, *38*, 801–812. [CrossRef]
60. Elshaghoul, Y.G.; El-Sayed Seleman, M.M.; Bakkar, A.; Elnekhaily, S.A.; Albaijan, I.; Ahmed, M.M.Z.; Abdel-Samad, A.; Reda, R. Additive friction stir deposition of AA7075-T6 alloy: Impact of process parameters on the microstructures and properties of the continuously deposited multilayered parts. *Appl. Sci.* **2023**, *13*, 10255. [CrossRef]
61. Ma, Z.; Sharma, S.; Mishra, R. Microstructural modification of as-cast Al-Si-Mg alloy by friction stir processing. *Metall. Mater. Trans. A* **2006**, *37*, 3323–3336. [CrossRef]
62. Jiang, H.; Liu, C.; Yang, Z.; Li, Y.; Huang, H.; Qin, F. Effect of friction stir processing on the microstructure, damping capacity, and mechanical properties of Al-Si alloy. *J. Mater. Eng. Perform.* **2019**, *28*, 1173–1179. [CrossRef]
63. Ahmed, M.M.Z.; El-Sayed Seleman, M.M.; Eid, R.G.; Albaijan, I.; Touileb, K. The influence of tool pin geometry and speed on the mechanical properties of the bobbin tool friction stir processed AA1050. *Materials* **2022**, *15*, 4684. [CrossRef]
64. Adi, S.S.; Malik, V.R. Friction stir processing of aluminum machining waste: Carbon nanostructure reinforcements for enhanced composite performance—A comprehensive review. *Mater. Manuf. Process.* **2025**, *40*, 285–334. [CrossRef]
65. Naumov, A.A.; Safi, S.V.; Safi, S.M. Advances in friction stir processing of aluminum 2024: A review of nanoparticle-reinforced surface composites. *Int. J. Adv. Manuf. Technol.* **2025**, *137*, 3141–3164. [CrossRef]
66. Guru, P.; Khan, F.; Panigrahi, S.; Ram, G.J. Enhancing strength, ductility and machinability of a Al–Si cast alloy by friction stir processing. *J. Manuf. Process.* **2015**, *18*, 67–74. [CrossRef]
67. Cui, G.; Ni, D.; Ma, Z.; Li, S. Effects of friction stir processing parameters and in situ passes on microstructure and tensile properties of Al-Si-Mg casting. *Metall. Mater. Trans. A* **2014**, *45*, 5318–5331. [CrossRef]
68. Alidokht, S.A.; Abdollah-Zadeh, A.; Soleymani, S.; Saeid, T.; Assadi, H. Evaluation of microstructure and wear behavior of friction stir processed cast aluminum alloy. *Mater. Charact.* **2012**, *63*, 90–97. [CrossRef]
69. Soleymanpour, M.; Aval, H.J.; Jamaati, R. Manufacturing of high-toughness Al–Si alloy by rolling and friction stir processing: Effect of traverse speed. *CIRP J. Manuf. Sci. Technol.* **2022**, *37*, 19–36. [CrossRef]
70. John Baruch, L.; Raju, R.; Balasubramanian, V.; Rao, A.; Dinaharan, I. Influence of multi-pass friction stir processing on microstructure and mechanical properties of die cast Al–7Si–3Cu aluminum alloy. *Acta Metall. Sin.* **2016**, *29*, 431–440. [CrossRef]
71. Kumar, H.; Prasad, R.; Kumar, P. Microstructural and mechanical Characterization of friction stir processed A356 alloy. *J. Phys. Conf. Ser.* **2024**, *2837*, 012023. [CrossRef]

crystals

MDPI

Review

Advances in Magnesium-Based Biomaterials: Strategies for Enhanced Corrosion Resistance, Mechanical Performance, and Biocompatibility

Yushun Liu [1], Jian Yin [1,2] and Guo-zhen Zhu [1,*]

1 Department of Mechanical Engineering and Manitoba Institute of Materials, University of Manitoba, Winnipeg, MB R3T 5V6, Canada; liuy11@myumanitoba.ca (Y.L.); yinjian@jgsu.edu.cn (J.Y.)
2 Key Laboratory of Jiangxi Province for Modern Agricultural Equipment, School of Mechanical and Electrical Engineering, Jinggangshan University, Ji'an 343009, China
* Correspondence: guozhen.zhu@umanitoba.ca

Abstract

Magnesium (Mg) and its alloys have emerged as promising biomaterials for orthopedic and cardiovascular applications, thanks to their good biodegradability, biocompatibility, and mechanical properties close to that of natural bone. However, the rapid degradation of Mg in physiological environments and limited mechanical performance tend to compromise the structural integrity of implants before healing is complete. These drawbacks have been heavily limiting the application of Mg and its alloys as biomaterials. In this paper, we review recent advancements in two common solutions to these problems: alloying and surface treatment, with a focus on controlling the corrosion resistance, mechanical performance, and biocompatibility of Mg-based biomaterials.

Keywords: magnesium; rare-earth; mechanical properties; biomaterials; corrosion resistance; biocompatibility; alloying; coating; surface severe plastic deformation

1. Introduction

Traditional biomaterials for clinical implants, as well as implantable sensors, have long been bioinert materials like stainless steel and titanium alloys [1]. The excellent corrosion resistance of these permanent biomaterials is, on the other hand, their biggest drawbacks in many cases, as additional removal surgery is required when the healing process is completed, leading to prolonged healing and/or increased risks of infection [2–4].

The lack of biodegradability of these permanent biomaterials has raised the need for biodegradable materials; such materials are intended to offer decreasing support and be completely absorbed once fulfilling their purpose, eliminating the necessity for extra surgery removal. The most interesting biodegradable materials include degradable polymers like PMMA [5] and metals like iron (Fe), zinc (Zn), and magnesium (Mg) [6]. Among these candidates, magnesium distinguishes itself with its good biodegradability, biocompatibility, and adequate mechanical properties [7]. Being one of the fundamentally required elements in the human body, the degradation product of Mg is generally considered unharmful or even beneficial to the bone healing process [8]. Another merit of Mg-based biomaterials lies in their similar mechanical performance to that of human bone. The density (1.74 g/cm^3) and Young's modulus (41 GPa to 45 GPa) of pure Mg [9] are very close to that of natural bone (8 GPa to 22.8 GPa for a compact bone [10] and 0.5 GPa to 1.5 GPa for a cancellous one [11]), making Mg-based biomaterials less susceptible to issues like stress shielding, which often troubles biomaterials with a high modulus like steel. Such a phenomenon occurs when an orthopedic implant (e.g., bone plates, screws, or joint replacements) absorbs more mechanical load than the surrounding bone, leading to reduced mechanical stress on the natural bone and possible bone resorption (bone loss) and weakening over time [12]. Consequently, increased interest in biomedical applications has been seen for Mg-based biomaterials in the past decades. Various fabrication techniques, including traditional methods like casting, wrought, powder metallurgy, etc., have been adopted for the manufacturing of Mg-based biomaterials. Additionally, additive manufacturing (AM) like laser additive manufacturing (LAM), powder bed fusion, friction stir AM, wire arc AM, etc., which has been showing great promise due to its high flexibility in manufacturing complex shaped structures, is also gaining increasing interest for Mg-based alloys for biomedical applications. These manufacturing techniques used for the manufacturing of Mg biomaterials have been well summarized in some recent review papers [13–15].

However, the good degradability of Mg, which is its most attractive merit as a biomaterial, is also its biggest problem. With the lowest standard electrode potential of −2.363 V among commonly used metallic biomaterials (Zn, −0.763 V; Fe, −0.44 V; Ti, −1.63 V) [16,17], the degradation rate of pure Mg is often too fast to fulfill its designed purpose. More importantly, unlike Al or Ti, where a dense surface oxide layer forms to prevent further corrosion, the passive oxide layer formed on the Mg surface is loose and easy to peel off, offering very limited protection for the underlying uncorroded materials. As illustrated in Figure 1a, ideally, the mechanical performance of a degradable implant should gradually decline due to degradation in a controlled manner as the healing proceeds, and eventually, the implant is completely absorbed when the healing is finished. The introduction of Mg ions released during the degradation of Mg implants can be beneficial to the healing process, such as accelerated bone regeneration, as illustrated in Figure 1b. However, due to the fast degradation rate of pure Mg in a physiological environment, the mechanical integrity of the implant can be severely compromised before the implant serves its purpose, which can be accompanied by the excessive release of hydrogen (H_2), causing problem with the healing process [18,19].

(a)

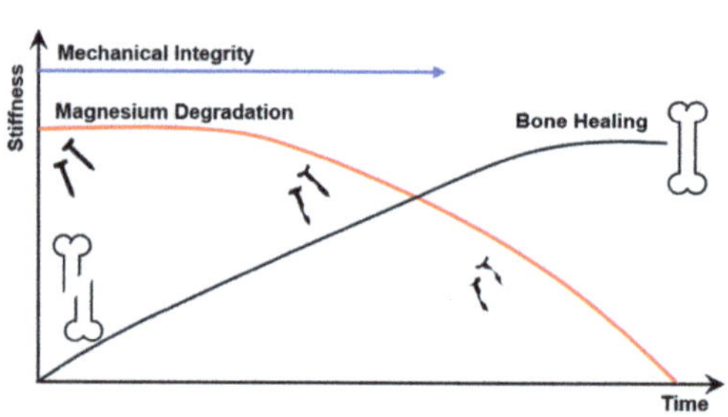

(b)

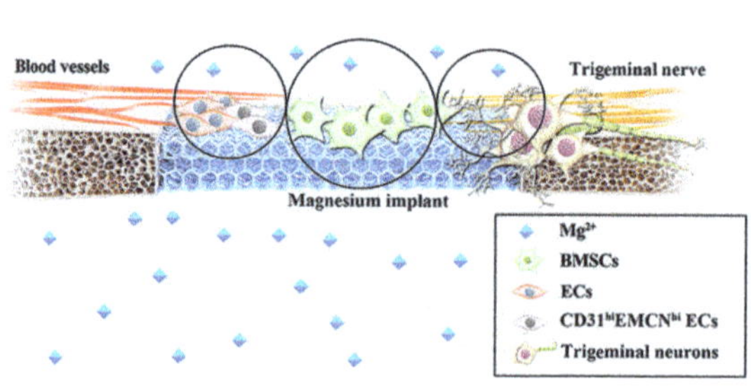

Figure 1. (**a**) Stiffness evolution with time for ideal biodegradable materials; (**b**) illustration of the degradation process of the Mg implant inside the human body [20]. The presence of Mg ions introduced by scaffolds promotes the osteogenic differentiation of bone marrow mesenchymal stem

cells (BMSCs). Enhanced osteogenesis has also been observed due to increased angiogenesis, including a higher percentage of CD31^{hi} Endomucin^{hi} (CD31^{hi}Emcn^{hi}) endothelial cells (ECs). Moreover, the scaffolds can produce the neuropeptide calcitonin gene-related peptide (CGRP) by triggering trigeminal neurons, leading to further improved angiogenesis and osteogenesis.

Aside from rapid corrosion, the relatively poor mechanical properties of pure Mg also hinder its application for high load-bearing Mg implants. With a hexagonal close-packed (hcp) crystal structure, the strength and ductility of Mg are generally inferior compared to metals with a cubic crystal structure like Al and Fe, along with a much higher mechanical anisotropy, making Mg implants more susceptible to premature failure. This is caused by limited independent slip systems and a dominating easy basal slip, which both originate from the reduced symmetry of the hexagonal lattice [21]. As can be seen in Figure 2, three dislocation slips and one twinning are commonly reported for Mg alloys deformed at room temperature (RT). However, the basal <a> slip operating on the (0001) plane often dominates the RT deformation of Mg alloys due to its much smaller critical resolved shear stress (CRSS) compared to other deformation modes, causing strong mechanical anisotropy during deformation. For example, the first principal calculation indicates that the CRSS of prismatic <a> and pyramidal <a+c> are ~46 and 82 times that of basal <a>, respectively [22]. As basal <a> can only introduce strain along the basal plane, more importantly, it offers only two independent slip systems, failing to meet the well-known von Mises–Taylor criterion, which states at least five independent slip systems are required for the homogeneous deformation of polycrystalline materials. Consequently, complementary deformation modes like twinning often occur to compensate for the lack of <c> axis strain and shortage of independent slip systems. The activation of twinning, however, tends to further increase the mechanical anisotropy of Mg alloys, as the twinning mode being activated is highly dependent on the loading direction, leading to a preferential twinning mode being activated [23]. Consequently, lower strength combined with high mechanical anisotropy has been reported for Mg alloys and is one of the key issues preventing the wider application of Mg alloys as both structural and biomedical materials.

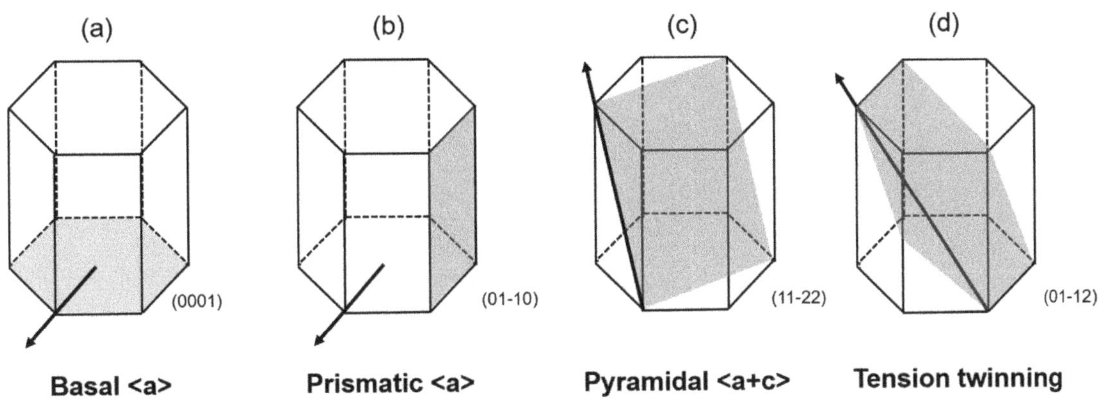

Figure 2. Major deformation modes at room temperature for Mg alloys [21].

Common solutions to these problems can be divided into three categories: alloying, which alters the alloy's overall mechanical/corrosion behavior; severe plastic deformation (SPD), which improves alloys' performance by introducing excessive defects like grain boundaries/dislocations; and surface treatment, which focuses on modifying the surface characteristics. In this paper, commonly used alloy elements and surface treatment techniques will be reviewed, focusing on their effects on corrosion resistance, mechanical performance, and biocompatibility of Mg biomaterials. Slip/twinning planes are visualized in gray color, slip/twinning directions are indicated by black arrow.

2. Alloying Elements and Their Effects on Mg Biomaterials

As a common practice of improving the performance of metals, alloying is also adopted for Mg biomaterials in the hope of achieving a better corrosion/mechanical response. The degradation rate of Mg alloys is highly dependent on the alloy composition [24], making it possible to achieve controlled degradation for different applications with adequate alloying. While the degradation of pure Mg, which is generally considered nontoxic and even beneficial to the bone healing process, the degradation product introduced by alloy elements is more complicated and can present serious health issues, even for essential trace elements like Fe, given enough amount [25]. Thus, it is worth noting that biocompatibility should be regarded as an important factor for consideration when choosing alloy elements for Mg biomaterials, as Mg-based biomaterials are often designed to be completely absorbed after implantation. The effects of common alloying elements used in Mg biomaterials on biocompatibility, mechanical performance, and corrosion resistance are summarized in Table 1.

Table 1. Effects of common alloying elements used in Mg biomaterials on biocompatibility, mechanical performance, and corrosion resistance.

Alloying Elements	Biocompatibility	Mechanical Performance	Corrosion Resistance	Refs.
Al	Neurotoxic, potentially causing Alzheimer's disease	Addition of Al can lead to improved strength, ductility, and formability	Depending on the chemical composition and manufacturing processes, the types and distribution of Al-containing intermetallic phases can have a distinct effect on alloys' corrosion resistance.	[26–35]
Y	High in vivo toxicity, strong affiliation to the liver and spleen. Potential health risks due to Y_2O_3 formed during casting. Due to the naturel presence of Y in the human body and the body's ability to slowly metabolize it, alloys with good biocompatibility can be obtained given controlled releasing rate of Y.	Addition of Y can lead to significant improvement on strength and ductility of Mg alloys via grain refining, texture randomization, solid solution strengthening, and precipitate strengthening etc.	Y can improve the corrosion resistance of Mg by forming Y-enriched network structure and modifying the passive protective layer with Y_2O_3. The solid dissolved Y in alloys with higher Y concentration can, however, lead to accelerated corrosion rate due to galvanic corrosion.	[36–41]
La, Ce	Highest toxicity among RE elements, strong hepatotoxicity. Should be avoided, if possible, for Mg alloys used for biomedical applications.	Addition of La/Ce can lead to excellent improvement on both strength and ductility.	Highly dependent on the nature of formed intermetallic phases. In Mg-Al alloys, La and Ce are reported to improve the corrosion resistance by modifying the Al-containing intermetallic phases. In Mg-Zn-Ca alloys, however, the presence of La and Ce leads to the formation of fine La/Ce-containing intermetallic phases, causing enhanced galvanic corrosion and deteriorated corrosion resistance.	[42–47]

Table 1. *Cont.*

Alloying Elements	Biocompatibility	Mechanical Performance	Corrosion Resistance	Refs.
Gd	Moderate compared to other RE elements. Potential tendency to accumulate in multiple organs. Lower tendency to form oxide which can present health concerns during casting compared to Y.	Due to its similar atomic radius to Y, the addition of Gd gives similar strengthening effects and is often used as a substitution to Y.	Addition of Gd up to 10 wt.% improves the corrosion resistance of Mg alloys, further increasing Gd, however, leads to accelerated corrosion due to formation of Mg_5Gd intermetallic.	[36,48–55]
Nd	Insufficiently studied. High toxicity, similar to La and Ce. Pronounced liver toxicity. Controversy between in vivo and in vitro results.	Small amount of Nd (<3 wt.%) has been reported to effectively improve the strength.	Enhancing the corrosion resistance by modifying the passive oxide layer.	[56–60]
Ca	Good biocompatibility due to its abundant presence in the human body, especially natural bone.	Small amount of Ca alloying can significantly improve the strength and ductility of Mg alloys via grain refining and precipitate strengthening.	Low concentration of Ca (<1 wt.%, the solubility of Ca in Mg) can enhance the corrosion resistance of Mg alloys by forming intermetallic phases at grain boundaries. Higher Ca concentration leads to an accelerated corrosion rate due to prominent micro-galvanic corrosion caused by increased number of precipitates.	[61–68]
Zn	Crucial to many biological functions, essential to the operation of more than 600 enzymes. Good biocompatibility in most cases, although excessive amount of Zn ions in the human body can be harmful to several vital organs.	Increasing Zn at low concentration (~4–5 wt.%) improves the mechanical performance of Mg-Zn based alloys, while further addition of Zn can be detrimental.	Similar to mechanical performance, Zn at low concentration (<5 wt.%) shows positive relationship with alloys' corrosion resistance. Excessive amount of Zn, however, promotes galvanic corrosion.	[69–74]
Sr	Essential component of natural bone, exhibits good biocompatibility and is capable of promoting bone formation	Increasing Sr concentration up to 2 wt.% leads to improved ultimate tensile strength and elongation, with slightly reduced yield strength.	Small concentration of Sr appears to be beneficial to the corrosion resistance of Mg, although controversy exists regarding the exact threshold beyond which Sr presents detrimental effects on corrosion resistance. Reports show that 0.2 wt.% Sr seems to have the optimum combination of mechanical performance and corrosion resistance.	[75–79]

Table 1. *Cont.*

Alloying Elements	Biocompatibility	Mechanical Performance	Corrosion Resistance	Refs.
Li	Widely used as an effective treatment for mood disorders like bipolar disorder. Potential risks regarding the possibility in developing nervous system disorders and kidney damage.	Addition of Li alters the crystal structure of Mg alloys from hcp to bcc, leading to improved ductility but usually at the cost of strength due to the low hardness of Li. Appropriate heat treatment is reported capable of compensating the loss in strength.	Single-phase Mg-Li (Li < 5.7 wt.%, Li > 10.3 wt.%), especially single β-phase Mg-Li alloys with high concentration of Li, show improved corrosion resistance due to homogenous microstructure and modified passive layer with Li-containing oxide. Dual phase Mg-Li alloys with intermediate Li concentration (Li~5.7–10.3 wt.%) show accelerated corrosion due to galvanic corrosion introduced by the multi-phase microstructures.	[80–88]
Mn	Essential nutrient for intracellular activities and plays vital roles in many critical processes like digestion, reproduction, antioxidant defense, energy production, immune response, and neuronal activities. Overexposure can lead to neurological disorder.	Alloying of Mn at low concentration leads to improved strength and hardness, tendency for brittle fracture has been noticed for higher Mn concentration.	Due to its affinity to impurity Fe, Mn at low concentration can effectively eliminate Fe by forming Fe-Mn binary compounds, leading to improved corrosion resistance. At higher concentration, however, shows detrimental effects on corrosion resistance due to precipitation of α-Mn particles.	[89–95]

2.1. *Aluminum (Al)*

Al is one of the most commonly used alloying elements for Mg as structural materials, leading to the development of widely used commercial alloy systems such as AZ (Mg-Al-Zn), AS (Mg-Al-Si), AM (Mg-Al-Mn) series, etc. [26–29]. Mg-Al alloys are known for their good casting properties, thermal conductivity, and mechanical properties. Both the mechanical and corrosion resistance of Mg can be improved with Al and are closely related to the type and distribution of formed intermetallic phases. The addition of Al into pure Mg leads to the formation of a $Mg_{17}Al_{12}$ intermetallic phase that can improve mechanical properties [30]. In alloy systems with more elements, various intermetallic phases like Al_2Ca, Mg_2Ca, Al_8Mn_5, $(Mg, Al)_2Ca$, etc. [31,32] can be formed, causing complex influences on the alloy performance. Despite its wide use in both structural and biomedical applications, the impact of Al on the biocompatibility of Mg remains controversial, as it has been shown that the introduction of Al inside the human body can cause several symptoms like muscle damage, reduced activities of osteoclasts, and potentially Alzheimer's disease [33–35]. Thus, Al is not considered an ideal alloying element for Mg used in biomedical applications. Although some in vitro studies show little negative influence of Mg-Al based alloys in terms of biocompatibility [96,97], the long-term effects of Al introduced by Mg-Al alloys remain unclear, and care should be taken when introducing Al to Mg used for biomedical applications [98].

2.2. *Rare-Earth (RE) Elements*

Alternatively, rare-earth (RE) elements like yttrium (Y), lanthanum (La), cerium (Ce), gadolinium (Gd), neodymium (Nd), etc. have also been proven efficient in improving the strength, ductility, and corrosion resistance of Mg alloys [56,99], thus attracting increasing interest in the past decades. While not fully understood, RE ions released during alloy degradation are generally considered to have negative effects inside the human body, causing potential health risks, including cytotoxicity [42], inflammatory response [100], reactive oxygen species (ROS) production, DNA damage, etc. [101]. As a result, the releasing rate of RE elements in Mg biomaterials should be carefully controlled to prevent excessive absorption of RE elements into the human body.

RE elements are generally believed to be beneficial to the mechanical performance and corrosion resistance of Mg alloys by most research [102,103]. Doubled corrosion resistance can be achieved by the addition of RE elements in a Mg-Zn-Zr alloy, as reported by Chen et al. [56]. RE elements have been reported capable of stabilizing the rather fragile passive oxide/hydroxide protective layers naturally formed in pure Mg when subjected to an aqueous environment, leading to increased corrosion resistance [57,99]. The improved corrosion resistance of Mg-RE alloys is also believed to be associated with microstructural optimizations like reduced micro-galvanic corrosion benefitting from grain refinement and continuous distribution of secondary phases acting as corrosion barriers [104–106]. However, it is important to note that the beneficial effects of RE elements on corrosion resistance and degradability are highly microstructural and concentration-dependent, and factors like detrimental secondary phases (e.g., β-$Mg_{17}Al_{12}$ phase [107]), inhomogeneous distribution, and excessive RE-containing secondary phases can lead to compromised corrosion behaviors [108].

There has yet to be a systematic study regarding the effects of individual RE elements on the biocompatibility of Mg biomaterials. Research has been largely limited to RE elements commonly used in engineering materials like Y, La, Ce, Gd, Nd, etc., and large discrepancies between in vivo and in vitro results have been noticed [36,42].

2.2.1. Yttrium (Y)

Despite showing high in vivo toxicity in animal experiments [36], Y is one of the most heavily studied RE elements in both engineering and biomedical Mg alloys due to its excellent effects on improving mechanical and corrosion performance [109]. Y-containing Mg alloys like WE43 (W represents yttrium, and E represents rare-earth) used as cardiovascular stents have received considerable success during in vivo applications of Mg implants [42].

With relatively high solubility (maximum 3.75%) in Mg [110], Y exhibits a strong solid solution-strengthening effect. Gao et al. [37] studied the solid solution strengthening of Y using Mg-Y binary alloys with varying Y concentrations. Their results suggest that the hardness and yield strength increase linearly with the increasing Y concentration (0.2–1.88 at.%) at room temperature. A significantly higher solid solution strengthening rate has been observed for Y compared to that of Al and Zn, which is attributed to the large difference in both atomic size and shear modulus between Mg and Y. Additionally, Y alloying has also been shown to be effective in grain refining, texture randomization, and precipitate strengthening, collectively contributing to the high mechanical performance of Y-containing Mg alloys [38,39].

In terms of corrosion resistance, the addition of Y can modify the porous and loose passive layer naturally formed on the Mg surface. Yu et al. [40] reported that a compact protective layer enriched in Y_2O_3 can be formed in Mg-Y binary alloys, effectively shielding the underlying materials from further corrosion. Consequently, the alloy with the highest Y concentration shows the best corrosion resistance in NaCl solution. Solid dissolved Y, however, seems to deteriorate the corrosion resistance of Mg alloys, especially in relatively low concentrations. Liu et al. [41] also studied the effect of Y on the corrosion resistance of Mg. Their results, however, indicated an increased corrosion rate with increasing Y at low concentrations (below 1.16 wt.%). These solid dissolved Y, along with a Y-rich zone, accelerate galvanic corrosion, and while a modified protective layer with a higher corrosion resistance does form, it is not strong enough to prevent further corrosion when the Y concentration is low. This changes when more Y is introduced, as increased corrosion resistance is observed due to the formation of a Y-rich network structure, which effectively impedes the corrosion reaction. This finding highlights the importance of the microstructure to the corrosion resistance of Mg. Similarly, Yu et al. [111] reported that changing the microstructure by pre-homogenization can effectively compensate the compromised corrosion resistance due to a higher Y concentration.

The biocompatibility of Y is still under debate. A strong affiliation to the liver and spleen has been reported for Y ions, along with significant calcium depositions into said organs [112]. Additionally, yttrium oxide (Y_2O_3), which often forms during the manufacturing process [48], also presents potential oxidative stress risks, as revealed by in vitro experiments [113]. Due to its natural presence in the human body, although very limited, a small amount of Y intake can be slowly metabolized, and in vivo tests on mice estimated the elimination half-time to be 144 days at a dose of 1 mg Y/rat [112]. It is also important to note that the extent of the toxicity is influenced by factors such as concentration, exposure duration, and the specific biological environment [114]. Thus, it is possible to minimize the adverse effects of Y on the biocompatibility of Mg alloy by carefully controlling the Y releasing rate in Y-containing Mg alloys.

2.2.2. Lanthanum (La) and Cerium (Ce)

La and Ce are believed to be more toxic compared to other RE elements [42,43], as both exhibit strong hepatotoxicity. Thus, despite their good strengthening effect and wide adoption in Mg alloys used for structural materials, more consideration should be taken for their application in the biomedical field. Tekumalla et al. [44] reported excellent improvement in both the strength (by 182%) and elongation (by 93%) of Mg under tensile tests by the addition of 0.4 wt.% Ce. While an accelerated corrosion resistance is observed, heat treatment effectively removes the detrimental effects, restoring the corrosion resistance comparable to pure Mg. In more complex alloy systems, the corrosion resistance of La-/Ce-containing alloys is highly dependent on the nature of the formed intermetallic phases. In Al-containing Mg alloys, Ce and La are reported to improve the corrosion resistance by altering the type/distribution of intermetallic phases and forming a modified protective layer [45,46]. The research from Tong et al. [47] suggested that the addition of La and Ce leads to the formation of new La-/Ce-containing intermetallic phases with refined sizes. These fine intermetallic phases, while effective in enhancing the mechanical performance, can serve as the cathodic site and accelerate the galvanic corrosion rate.

The biocompatibility is the biggest issue preventing the addition of La and Ce in biomedical Mg alloys. The injection of acetate salt of La and Ce is reported to cause many acute symptoms, including fever, muscle pains, abdominal cramps, hemoglobinemia, and hemoglobinuria [58,115]. Elmar et al. [116] reported that a low amount of La and Ce does not present significant cytotoxicity. However, the study considered only very short-term effects, and a suppressed ability to stimulate bone healing near the implant was noticed, which might suggest potential long-term adverse effects. Thus, La and Ce are generally considered less suitable for Mg alloys used in biomedical application, and extra caution should be taken if unavoidable [42].

2.2.3. Gadolinium (Gd)

The atomic radius of Gd is very close to that of Y, and its strengthening effect also closely resembles Y, making these two RE elements easily interchangeable in alloy composition while maintaining comparable performance. Liu et al. [49] studied the aging response of Mg-Gd binary alloy. Strong precipitate strengthening was observed, leading to significant improvement in hardness and the ultimate tensile strength. Due to its lower tendency to form oxide inclusions, Gd is also used to substitute Y to suppress the formation of Y_2O_3, which is detrimental to the mechanical performance. Luo et al. [48] replaced Y with Gd to address the oxide inclusion problem often seen in Mg-Y alloys. A 50% reduction of Y replaced by Gd has been achieved while maintaining almost the same mechanical performance, effectively reducing Y-containing oxides, which can also raise health concerns [48].

The addition of Gd is believed to be beneficial to the corrosion resistance of Mg before a certain threshold. Hort et al. reported [52] improved corrosion resistance with increasing the Gd concentration up to 10 wt.%, agreeing with the studies by Rokhlin and Zidane et al. [53,54]. A drastic decrease in corrosion resistance has also been noticed for Gd 15 wt.%, despite its finer grain size, possibly related to the Mg_5Gd intermetallic phases formed at grain boundaries. A similar reduction on corrosion resistance due to the precipitation of Mg_5Gd has been reported by Lu et al. [55], although at much lower Gd concentrations.

The toxicity of Gd is generally believed moderate among RE elements. Although some in vitro results suggest a lower toxicity of Gd ions compared to other RE elements [42], the in vivo results showed no significant differences [36]. Another alarming sign is the constant accumulation of Gd in multiple organs, including the spleen, lung, liver, and kidney, in in vivo experiments even 36 weeks after implantation [50]. Immune system impairing [36] and nephrogenic systemic fibrosis (NSF) [51] are also reported to be positively related to the exposure of Gd.

2.2.4. Neodymium (Nd)

Nd is another commonly used alloying element for Mg. The addition of Nd can effectively improve the corrosion resistance and strength of Mg. Lilach et al. [59] reported optimal mechanical and corrosion performance can be achieved with 2% Nd addition into a Mg-Y-Zn-Zr alloy, and the improvement was attributed to the formation of secondary phases. Another study conducted by Wu et al. [60] examined the effect of Nd on the mechanical behavior of Mg-Zn-Zr alloys. Their finding revealed that a 3% Nd addition significantly enhanced the yield strength of the alloy across various temperatures, suggesting that Nd contributes to improved mechanical stability. Nd is also reported to enhance the corrosion resistance of the passive layer formed on the Mg surface. Jin et al. [57] ion-implanted Nd in a WE43 Mg alloy, resulting in a smooth and hydrophobic layer that significantly slows down the corrosion process.

With insufficient research, the biocompatibility of Nd remains debatable. Similar to La and Ce, several symptoms and pronounced liver toxicity have been noticed for Nd [58,115], with toxicity sitting between La and Ce, according to long-term in vivo results by Cheng et al. [58], although some in vitro studies have suggested good biocompatibility of Nd-containing Mg alloys [56,57]. The solubility of Nd (3.6 wt.%) is limited inside the Mg matrix [116], giving its alloys a tendency to form secondary phases, which can also cause health concerns.

While all RE elements are considered somewhat toxic, the toxicity is heavily related to the ion releasing rate; thus, it is still possible to avoid serious health issues while improving the corrosion and mechanical performance of Mg-RE alloys given a carefully controlled degradation rate is achieved.

2.3. Biocompatible Alloy Elements

Compared to Al and RE elements, which biocompatibility remains questionable, elements that naturally exist in and are essential to the human body, like calcium (Ca), zinc (Zn), strontium (Sr), lithium (Li), manganese (Mn), etc. [6], would be more preferable if a comparable corrosion/mechanical performance can be achieved.

2.3.1. Calcium (Ca)

As an essential element and crucial bone component, Ca has been considered a promising alloy element for Mg in bone implant application [62] and has been receiving increasing interest in the past decade. Due to its wide existence inside the human body, the addition of Ca generally does not pose an obvious threat to the biocompatibility of Mg alloys [61,62]. The addition of Ca can effectively improve the mechanical performance of Mg. A recent study conducted by Akhmetshina et al. [63] showed that, with a small amount of Ca alloying, an UTS ~420 MPa or a ductility ~36% can be achieved in Mg-Ca binary alloys, comparable to that of many high-performance Mg-RE alloys with heavy RE alloying [64]. The alloys' superior mechanical performance is attributed to the formation of Mg_2Ca intermetallic phases and increased probability of pyramidal/basal slip transition caused by Ca addition, achieving high strength while maintaining excellent ductility by reducing the mechanical anisotropy. The strengthening effect of Ca is believed to be dependent on the alloying amount. At low concentrations, Yin et al. [61] reported that increasing the Ca addition results in a slight improved yield strength, while the tensile strength tends to increase first before decreasing, indicating an optimal concentration of Ca for the mechanical performance, while other studies have shown a constant increase in strength and decrease in ductility as the Ca concentration gets higher [65,66]. Grain refining and fine Ca-containing intermetallic phases are considered crucial to the improved mechanical properties of Mg-Ca alloys. The impact of Ca on the corrosion rate of Mg alloys is more complex, while a low amount of Ca (<1 wt.%, the solubility of Ca in Mg) is reported to enhance the corrosion resistance of Mg alloys by forming intermetallic phases at grain boundaries that prevent further corrosion [65,67,68], and excessive Ca alloying has been shown to be substantially detrimental to the alloy's corrosion resistance [65]. The accelerated corrosion rate found in Mg alloys with a high Ca concentration is believed to be caused by more prominent micro-galvanic corrosion caused by an increased number of precipitates acting as cathode sides [61]. In summary, calcium at a low concentration (usually below 1 wt.%) can effectively enhance Mg alloys' mechanical properties while maintaining good biocompatibility. However, excessive Ca content can lead to an accelerated corrosion rate and cause serious mechanical integrity problems. Therefore, careful control of the calcium concentration is essential to develop Mg-Ca-based biomaterials with balanced properties.

2.3.2. Zinc (Zn)

Zn, similar to Ca, is an essential element that is crucial to many biological functions, and approximately 600 enzymes require the presence of Zn^{2+} to function properly [69]. Therefore, Mg-Zn alloys generally possess good biocompatibility, although an excessive amount of Zn ions in the human body can be harmful to several vital organs [70]. Zn alloying can enhance the mechanical properties of Mg alloys through grain refining, solid solution strengthening, and precipitate hardening [117–119]. Zn alloying is often accompanied by Ca, leading to the development of Mg-Zn-Ca alloys, which show great promise in the biomedical field due to their good biocompatibility, adequate corrosion rate, and improved mechanical properties [117,118]. Increased Zn at low concentrations (~4–5 wt.%) can improve the mechanical properties of both Mg-Zn binary and Mg-Zn-Ca alloys; the excessive addition of Zn, on the other hand, tends to be detrimental [71–73]. The standard electrode potential of Zn (−0.763 V) is higher than that of Mg (−2.363 V), giving Zn a slower corrosion rate compared to Mg [120]. Thus, it is possible to reduce the corrosion rate of Mg by adding Zn to increase the overall corrosion potential [74]. However, this works only when the Zn concentration is low. Cai et al. reported that Zn alloying below 5 wt.% can enhance the corrosion resistance of Mg-Zn binary alloys, but further increasing the Zn content leads to an increased corrosion rate [72]. The compromised corrosion resistance at high Zn concentrations is likely associated with the precipitate of the network intermetallic phase, which accelerates micro-galvanic corrosion by acting as the cathode. Similar micro-galvanic corrosion has been also observed for Mg-Zn-Ca alloys, and the type and distribution of the intermetallic phases, which alter the corrosion behaviors heavily, are found to be closely related to the Zn/Ca ratio [61,73,121].

2.3.3. Strontium (Sr)

Sr, as an essential component of natural bone, exhibits good biocompatibility and has been shown to promote bone formation [75]. Fan et al. [76] studied the effect of Sr concentrations on the mechanical and corrosion performance of AZ91D Mg alloys. Their results suggest that increasing the Sr concentration (up to 2 wt.%) leads to improved ultimate tensile strength and elongation, with slightly reduced yield strength. Improved corrosion resistance has also been observed with the slowest corrosion rate-associating alloy with the highest Sr (2 wt.%) concentration, showing a ~1/3 corrosion rate in NaCl solution compared to Sr-free base materials. The continuous β phase is believed to prevent corrosion progression, resulting in improved corrosion resistance. Lai et al. [77] also studied the effect of Sr concentrations on Mg-2Zn alloys. Their results, however, suggest that 0.2 wt.% of Sr leads to the optimum combination of the best mechanical and corrosion resistance. Similar results were obtained by Wen et al. [78] in Mg-1Zn-1Sr ternary alloys, where they found that excessive Sr alloying over 0.2 wt.% leads to an accelerated corrosion rate. Tang et al. [79] reported that the mechanical and corrosion performance of Mg-Sr binary alloy can be further improved by adding extra RE elements (La, Nd, and Gd), achieving 0.5–2 times ductility and 45–66% reduction of the corrosion rate. Despite the involvement of the RE elements, good biocompatibility has been found for the prepared Mg-Sr-RE alloys, with improved endothelial cell attachment, spreading, and proliferation on the surface. The uncompromised biocompatibility is likely due to the limited alloying concentration (0.5 wt.%), which ensures a relatively slow releasing rate of RE ions.

2.3.4. Lithium (Li)

Li has been widely used for a long time as an effective treatment for mood disorders like bipolar disorder [80]. Although the clinical usage of Li is generally considered within the safety range [122], some concerns have been raised regarding the possibility of developing nervous system disorders and potential kidney damage [81,82]. The addition of Li into Mg tends to change the hcp crystal structure of Mg to body-centric cubic (bcc). Three kinds of microstructures can be found for Mg-Li binary alloys with different Li concentrations: (i) single α-phase (hcp) when Li is 0–5.7 wt.%; (ii) dual-phase structure comprising both a hcp α-phase and bcc-phase when Li is 5.7–10.3 wt.%; and (iii) a single β-phase when Li is over 10.3 wt.%. As mentioned earlier, the inferior mechanical performance of Mg largely originates from the limited available slip systems of hcp crystal structures. Therefore, the bcc crystal structure transition caused by the addition of Li can effectively improve the mechanical properties of Mg by offering more homogenous deformation. However, due to the extremely low hardness of Li, a higher concentration of Li alloying, while capable of effectively promoting ductility, usually leads to a decrease in strength. Liu et al. [83] reported significant improvement (~two times) in the yield strength and elongation of Mg-Li-Zn ternary alloys compared to that of pure Mg. The yield strength and ultimate tensile strength, however, appear to decrease with the increasing Li concentration. Heat treatment is often performed in order to overcome such drawbacks of Mg-Li alloys. Xu et al. [84] reported a heat-treated β-Mg-Li-Al-based alloy with a remarkable combination of good strength, ductility, and corrosion resistance. The high strength was later attributed to the spinodal decomposition of the β-phase caused by the heat treatment [85]. Despite the potential benefits, Mg-Li alloys can exhibit rapid corrosion, especially in multi-phase compositions. The lack of excellent corrosion resistance is a key issue limiting their broader application. For single-phase Mg-Li alloys with either low or high Li concentrations, improved corrosion resistance has been reported due to the homogenous single-phase microstructure and more stable passive layers with Li-containing oxides [84,86,87]. For Mg-Li alloys with intermediate Li concentrations, however, rapid corrosion has been reported [87,88]. Galvanic coupling due to the dual-phase microstructures is believed to be responsible for the accelerated corrosion rate.

2.3.5. Manganese (Mn))

As an essential nutrient for intracellular activities, Mn plays vital roles in many critical processes in the human body, like digestion, reproduction, antioxidant defense, energy production, immune response, and neuronal activities [89]. Overexposure to Mn, however, can lead to neurological disorders [90,91]. Due to its affinity to Fe, a common impurity in Mg alloy manufacturing that has detrimental effects on corrosion resistance, Mn is often added into Mg alloys to reduce the impurity level of Fe by forming Fe-Mn binary compounds with higher density [92], leading to improved corrosion resistance. A small amount of Mn alloying is reported to effectively improve both the corrosion and mechanical performance of Mg alloys. Hu et al. [93] manufactured Mg-Mn-Zn alloys with varying Mn concentrations for biomedical applications via powder metallurgy. Their results suggest a positive relationship between alloys' strength/hardness/corrosion resistance and Mn concentration, although a tendency toward brittle fractures has been observed for alloys with higher Mn concentrations. Similarly, Yao et al. [94] reported that the addition of Mn into Mg-3Al alloys significantly improved the alloys' corrosion resistance up to 0.45 wt.% Mn. The presence of modified Al-containing intermetallic phases like Al_8Mn_5 is believed to cause improved corrosion resistance. An excessive amount of Mn alloying, however, can lead to deteriorated corrosion resistance. Li et al. [95] studied the effect of Mn concentrations on Mg-Zn-Gd-Y; alloys with 1 wt.% Mn, despite higher mechanical properties, showed lower corrosion resistance compared to alloys with 0.5 wt.% Mn. The precipitation of α-Mn particles in alloys with higher Mn concentrations is believed to be responsible, which falls off and causes cavities that sustain further corrosion.

3. Severe Plastic Deformation (SPD)

Apart from alloying, plastic deformation is another commonly used strengthening strategy for metallic materials like Mg. A large number of defects like grain boundaries and dislocations can be introduced by plastic deformation, effectively causing grain refining and strengthening. However, conventional plastic deformation methods like rolling are incapable of introducing a large plastic strain without cracking, limiting the strengthening capacity. To address this problem, a number of SPD techniques have been proposed, including equal-channel angular pressing (ECAP), high-pressure torsion (HPT), accumulative roll bonding (ARB), multi-directional forging (MDF), twist–extrusion (TE), etc. [123]. Despite their different approaches, these techniques can all introduce extremely high strain without cracking, leading to significant grain refining and microstructure homogenization and, consequently, more superior alloy performance in both mechanical and corrosion resistance.

Among them, ECAP and HPT are the most commonly used ones for Mg alloys and, thus, will be reviewed in this section, focusing on their impact on Mg-based biomaterials.

3.1. Equal-Channel Angular Pressing (ECAP)

The principle of ECAP is illustrated in Figure 3a, where a metal billet is forced through a die with two intersecting channels of equal cross-sections. Unlike conventional forging or rolling, the cross-sectional dimensions of the billet remain constant, allowing multiple passes without altering the overall shape. This repeated deformation is capable of introducing high strain without cracking, leading to significant grain refinement and microstructural evolution. The key parameter controlling the strain introduced by ECAP is the intersecting channels angle (φ, usually at 90° in conventional ECAP), as a smaller φ offers more intensive plastic deformation but also makes the deformation process harder and more susceptible to cracking [124]. The outer arc of the curvature angle (ψ), on the other hand, shows relatively small influence. Due to the limited slip systems of Mg available at RM, early attempts at ECAP on Mg were often conducted on elevated temperatures to reduce the possibility of cracking [125]. This, however, leads to dynamic crystallization that hinders further grain refining, making it difficult to produce true nanoscale grains [126]. A lot of research has been devoted to addressing this problem, and it has been found that a higher φ and slower pressing speed can effectively reduce the tendency to crack [127,128], allowing for a more refined microstructure compared to ECAP conducted at high temperatures. Additionally, strategies like applying back pressure and performing pre-treatment prior to ECAP are also proven viable to improve the workability of Mg processed by ECAP at RT [129–131].

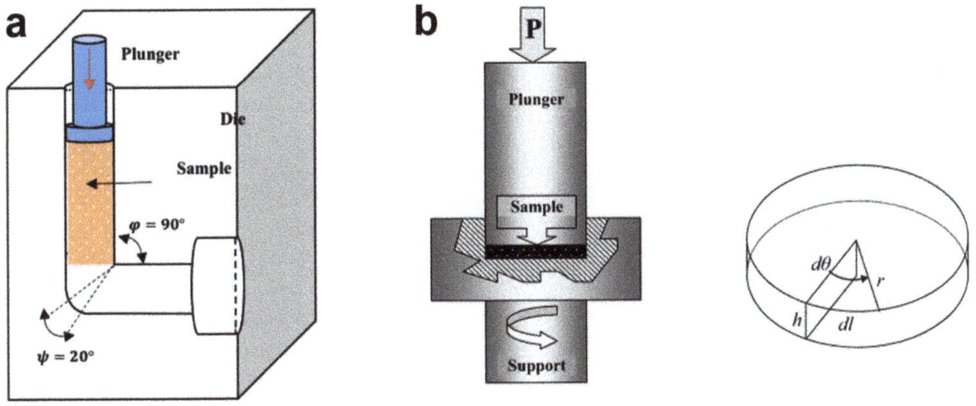

Figure 3. Schematic illustration of (**a**) ECAP [132] and (**b**) HPT [133].

According to the well-known Hall–Petch effect, a smaller grain size leads to stronger materials. Thus, most studies have suggested a significant improvement in strength and hardness for Mg alloys subjected to ECAP due to their refined grain size. Through a combination of solid solution treatment, ECAP, and rolling at RT, Klu et al. [131] fabricated a Mg-9Li alloy with UTS ~205 MPa, almost double that of its solution-treated counterpart, while maintaining good elongation to a failure of ~21%. ECAP is also reported to effectively modify the texture distribution, causing strong mechanical anisotropy. Xu et al. [134] reported simultaneous improvement in both strength and ductility along three orthotropic directions, indicating high mechanical anisotropy. A strong texture with most of the c-axis tilted ~30° away from the ED and the basal plane parallel to the TD has been observed, which evolves continuously upon further ECAP passes.

Grain refining, apart from the strengthening effect, is also believed to be beneficial to the corrosion resistance of polycrystalline materials [135–137]. The excessive number of grain boundaries in fine-grained Mg alloys subjected to ECAP promotes corrosion resistance in two different mechanisms. Firstly, a large number of grain boundaries helps the diffusion of many alloying elements into the naturally formed passive oxide layer on the Mg surface, modifying it to be more stable and corrosion-resistant [45,46,57,99,135]. Secondly, SPD tends to create a more homogenous microstructure; this, alongside the smaller grain size, leads to a more homogenous instead of local corrosion [138]. It is worth noting that, although many believe that grain refining introduced by ECAP is beneficial to corrosion resistance, there have also been reports regarding the unaffected or even deteriorated corrosion resistance of ECAP Mg alloys [139]. The compromised corrosion resistance is believed to be surface defects introduced by the ECAP process [140] and heterogeneous microstructure due to less symmetric deformation modes [141].

As ECAP does not introduce extra alloying elements or fundamentally change the surface characteristics, its effects on the biocompatibility of Mg alloys appear to be minimal. As shown in a recent review [139], most alloys processed by ECAP exhibit good biocompatibilities both in vivo and in vitro.

3.2. *High-Pressure Torsion (HPT)*

As shown in Figure 3b, HPT introduces SPD by placing thin disks for processing between two anvils and rotating one anvil relative to the other under high pressure. Compared to ECAP, where only pure shear deformation is applied, HPT simultaneously introduces both high hydrostatic pressure and torsional shear deformation. Due to the high hydrostatic pressure and continuous shear stress, HPT is capable of causing higher strain on Mg alloys at RT without cracking, leading to ultra-fine or even nanocrystalline materials, which is difficult to achieve using other SPD techniques like ECAP. The disadvantages of HPT compared to ECAP are the rather ununiform deformation and limited scalability [142]. Due to its rotational nature, samples located at different radii are inherently subjected to varying amounts of strain, causing severe plastic deformation at the edge of the thin plate, and consequently, a heterogeneous microstructure is obtained. This, however, can be compensated for by increasing the HPT turns. Sabbaghianrad and Langdon [143] reported that the center of 7075 Al alloys, which initially had a heterogenous structure compared to the edge part due to a smaller strain imposed, can be gradually transformed to a rather homogenous structure after 20 HPT turns. The maximum size for the thin plate used in HPT is also very limited, usually a few mm in diameter, making it difficult to process bulk materials. Therefore, HPT is more suitable for applications which have small dimensions and require high mechanical strength like screws or stents. The most important processing parameter of HPT is the number of HTP turns, which directly determines the total amount of plastic strain being introduced into the thin plate. Gu et al. [144] reported that the grain size of HPT Mg decreased continuously as more HPT turns were applied, causing the hardness of Mg to increase up to ~530 MPa before decreasing to a steady-state level at a large strain. They believed the dynamic recovery or recrystallization caused by high temperatures during continuous HPT was responsible for the decreased hardness. Other parameters like processing temperature, rotation speed, and applied pressure were also found to be influencing the microstructure of HPT Mg alloys [145].

The effects of HPT on the mechanical performance and corrosion resistance of Mg alloys are similar to that of ECAP, as both techniques effectively produce fine grains, which are the main origin of the improved performance. Generally, due to its higher capability in grain refining, samples subjected to HPT generally have finer grain and, consequently, better mechanical performance and corrosion resistance compared to their

ECAP counterparts. A recent review paper [139] comprehensively compared the different effects of ECAP and HPT on both the flow stress and corrosion rate of Mg alloys. It can be seen from Figure 4 that an overall trend for a better combination of flow stress and corrosion rate can be seen for the HPT samples. A more homogenous and refined microstructure produced by HPT is believed to be accountable for the improved overall performance.

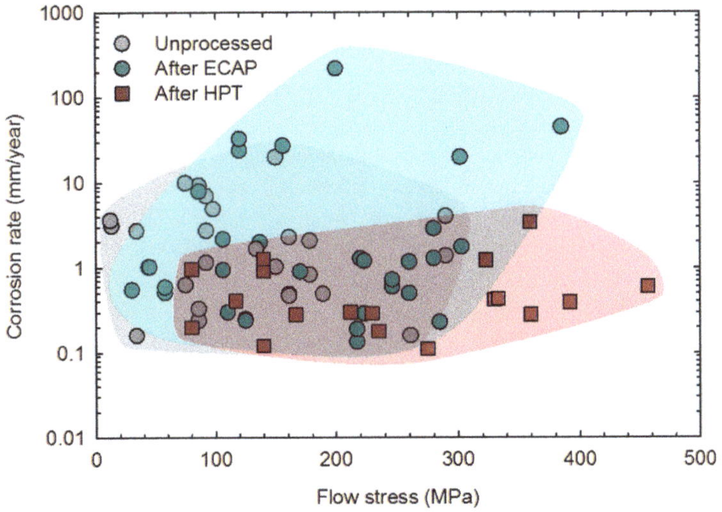

Figure 4. Comparison of the corrosion rate and flow stress of Mg alloys subjected to ECAP and HPT from Ref. [139].

4. Surface Treatment

In ideal cases, the degradation of biodegradable materials should occur slowly, maintaining its mechanical integrity for a given period to allow the healing process to finish. Many implants require the materials to maintain their mechanical integrity for quite a long time initially, for example, a Mg coronary stent is expected to last for 6–12 months before it loses enough support due to degradation [146]. However, the degradation rate of most Mg alloys is way too quick to meet such a requirement. Alloying elements, while capable of significantly improving the mechanical performance of Mg, often cause nonhomogeneous corrosion, causing the premature failure of the implants. Additionally, the degradation of both Mg (mostly H_2 gas pocket) and alloy elements can raise health concerns given an inadequate releasing rate [25,42,100,101,147,148]. Alternatively, surface modification can effectively slow down the degradation of Mg without compromising the overall mechanical performance [149]. Not only can the modified surface improve the mechanical and corrosion performance of Mg, being the interface between tissue and implant, the surface characteristics of Mg can also significantly influence the biological response of the human body to implants [150–152], making room for different clinical applications. Currently, surface modifications of Mg can be categorized into three kinds: surface coating, which adds an additional layer of materials onto the alloy surface; surface treatment, which modifies the microstructure of the material surface by methods like surface severe plastic deformation (SSPD); and a combination of both.

4.1. Surface Coating

A passive oxide layer forms when Mg and its alloys are subjected to a physiological environment. However, such a layer is porous and loose, unable to shield the underneath

Mg from further corrosion. H_2 generated during Mg corrosion is also reported to further weaken the mechanical integrity of such a layer [147], contributing to the fast corrosion rate of Mg inside the body. Therefore, in order to improve the corrosion resistance of Mg and its alloys, it is imperative to either strengthen the existing weak oxide layers or have additional layers to achieve a dense, protective surface. Techniques like chemical conversion coating, anodization, plasma electrolytic oxidation (PEO), and biomimetic deposition are among the most commonly used ones [153–156].

4.1.1. Chemical Conversion Coating

The idea of chemical conversion coating is to modify the oxide layers of Mg alloys (mainly comprised of MgO and $Mg(OH)_2$) by chemically treating the Mg surface with various elements and substances like fluoride, phosphates, RE, carbonate, etc. [157–160], making it strong enough to protect the underneath materials from corrosion. In fact, many alloying elements have been found to cause modification of the passive layer on the Mg surface and consequently alter the corrosion resistance [45,46,57,99]. Due to the chemical bonding between the modified layers and Mg matrix, the adhesion is generally strong, making it an easy yet effective way to produce a protective coating for Mg alloys.

Among the substrates used for chemical conversion coating, fluoride is probably one of the most commonly used chemical conversion coating methods for Mg due to its simplicity, good biocompatibility, and excellent corrosion resistance. As an element that naturally exists in human bone and teeth, fluorine exhibits good biocompatibility. Fluoride treatment of Mg alloys can not only effectively enhance their corrosion resistance by replacing the natural oxide layer with a dense and homogenous fluoride coating but also shows a positive effect on the bone healing process similar to that of Mg ions [153]. The formation of a MgF_2 and/or $Mg(OH)_{2-x}F_x$ layer is believed to be accountable for the improved corrosion resistance.

Immersion fluoride (HF) and ultrasonic immersion fluoride (UHF) are two major chemical conversion approaches for fluoride coatings. The preparation of immersion fluoride (HF) is straightforward, simply submerging Mg for coating into hydrofluoric acid (HF) solutions, and coatings made of MgF_2, $Mg(OH)_2$, and other substances can form on the surface due to the chemical reaction between the Mg surface and HF. The simplicity and efficiency have made HF a popular fluoride coating method. Alloying elements, HF concentration, and immersion time are considered critical to HF treatment [161–164], as they can significantly influence the composition, quality, and corrosion resistance of the final coating. Liu et al. [165] showed that a protective MgF_2 coating can be formed by a simple two-step immersion treatment, resulting in improvement on both the corrosion resistance and biocompatibility. Barajas et al. reported that the $Mg(OH)_{2-x}F_x$ layers formed after conversion treatment decreased the corrosion rate of the treated samples by about three orders of magnitude and can be further improved by increasing the fluoride concentration of the conversion layers [161]. It is worth noting that the coating layer formed from HF treatment is usually porous and tends to peel off severely during dynamic loading in the simulated body fluid, possibly due to H_2 gas generated by the chemical reaction [166]. This deteriorates the corrosion resistance of HF-treated Mg alloys, limiting their application in clinical treatments [167].

To address this problem, a modified approach called ultrasonic immersion fluoride (UHF) is proposed, leading to improved coating quality. The UHF enhances the traditional HF approach by introducing ultrasonics into the preparation, generating denser, nonporous, and smoother coating layers, which result in improved corrosion resistance of the coated Mg alloys [167]. The ultrasonic energy induces cavitation—rapid formation and collapse of microbubbles in the solution—which enhances mass transport and promotes a more uniform reaction between the Mg surface and fluoride ions [168,169]. The uniform and defect-free coating resulting from the ultrasonic treatment is believed to be accountable for the enhanced corrosion resistance. Li et al. [170] reported a pore-free fluoride coating induced by ultrasonic treatment; a significant reduction in both H_2 generation and pH value waws observed, indicating improved corrosion resistance.

Other elements like chromate, phosphate, stannate, rare-earth elements, silicate, molybdate, vanadate, etc. are also effective elements that are widely used for chemical conversion coating and were reviewed in detail in Refs. [159,160]

4.1.2. Anodization

Anodization is an electrolytic process that can be used to effectively tune the degradation rate of Mg alloys, as it is capable of producing coatings with varying thickness, composition, and structures by changing the electrolyte and voltage [155,171,172], making it more versatile for different applications. The anodized alloys usually exhibit better corrosion resistance and mechanical properties [173,174]. A study conducted by Hiromoto and Yamamoto [155] demonstrated the possibility of a tunable degradation rate by using different voltages, and a clear relationship between the coating thickness/surface morphology and the anodizing voltage has been found. While most studies have indicated a positive relationship between the applied voltage and coating thickness and, subsequently, better corrosion resistance [155,175], excessive voltage resulting in a compromised coating stability has been reported in some cases [176].

Pores are often found in the coating formed by anodization and are considered harmful to the corrosion resistance of Mg alloys [177,178]. These pores often exhibit through-thickness and are prone to generating microcracks upon cyclic loading, which compromises the integrity of the coating, exposing the new surface to sustaining further corrosion [179]. Consequently, Mg alloys subjected to anodization processes are often followed by additional sealing treatment to further improve the corrosion resistance [178,180,181]. Fitriana and Anawati [181] reported that a sealing treatment using beeswax-colophony resin on a anodized AZ31 Mg alloy resulted in a corrosion rate five times lower than that of unsealed ones.

4.1.3. Plasma Electrolytic Oxidation (PEO)

Plasma electrolytic oxidation (PEO), also known as microarc oxidation (MAO), is another promising electrolytic preparation technique for Mg biomaterial coating. While often considered as a superior alternative to conventional anodization, significant differences can be found for PEO in terms of electrolytes, current density, voltage ranges, and coating formation mechanisms [182]. The high voltage range used in PEO (~250–750 V [183]) leads to a local dielectric breakdown of the formed coatings [184], causing localized plasma discharge, which promotes the formation of thick, ceramic-like oxidation layers with varying porosities [185–187]. The characteristics of the formed coatings are also fundamentally different for PEO and traditional anodization. Due to the low electrical conductivity of the oxide layer formed from the traditional anodization process, it is difficult to further thicken the layer, since it effectively isolates the Mg surface [188]. Consequently, an oxidation layer with a thickness of only a few tens of microns is achievable for traditional anodization [189,190]. In contrast, in the PEO treatment, such an oxidation layer can be effectively broken down by the intense sparkling caused by high applied voltage [188], making it possible to produce thick oxidation layers over 100 microns [185,191]. High temperatures and pressures created by the sparkling promote the formation of a thick, compact, hard ceramic-like layer, substantially enhancing the corrosion resistance of the coated materials. Shao et al. [154] reported that the corrosion time of PEO coating in the neutral salt spray test (NSST) was ~20 times longer compared to the anodization coating given the same coating thickness, showing good protection of the substrate even after 1500 h of long-term testing. Although the density of the pores inside the PEO coating is relatively high in many cases compared to anodized coating, they generally present a less detrimental effect on the corrosion resistance [183]. Because of the localized high temperature caused by the plasma discharge, melting of the oxides occurs, which automatically seals the formed pores, reducing the chance of through-thickness pores that are believed to deteriorate corrosion resistance, as in the case of anodized coating [178,180,181]. It should be noted that, while pores are generally considered detrimental to the corrosion of the coating, the porous nature of the PEO surface can prove beneficial to biocompatibility. Research shows that a porous surface can lead to improved tissue adhesion and bone generation [192–194]. Therefore, not only can PEO coating substantially improve the corrosion resistance of Mg biomaterials, it also offers promising potential to change the biocompatibility by tuning the surface characteristics of Mg.

4.1.4. Biomimetic Deposition

Biomimetic deposition is favored due to its preparation simplicity and ease in tuning the coating characteristics like composition, crystallinity, and surface conditions. It utilizes the biomineralization process [156] that naturally occurs in living organisms to produce a protective coating comprised of bioactive minerals like calcium phosphate (CaP), hydroxyapatite (HA), and graphene oxide (GO) on the substrate surface. Several active molecules have been found to promote biomineralization, including arginine–glycine–aspartic-acid–cysteine (RGDC) peptide [195], polydopamine [196], dipped GO [197], etc. Aside from the enhancing corrosion resistance, due to the bioactive nature of the deposited substance, biomimetic deposition can also promote better integration with the surrounding tissues, facilitating cell adhesion and proliferation [198]. CaP as a coating material for Mg has been shown to possess good biocompatibility and osteoconductivity due to its similar structure with bone tissue [199]. By introducing graphene oxide into the deposition solution, Wang et al. [156] manufactured a dense and more adhesive CaP composite coating. Enhancement in both in vitro biocompatibility and corrosion resistance have been noted for the coated samples. Aghamohammadi et al. [200] studied the influence of the microstructure and crystalline texture of the substrate Mg on the deposition biomimetic CaP coating. Their results suggest a strong dependence of the coating morphology and, consequently, corrosion resistance on the nature of the Mg substrate. Hernández et al. [194] studied the effect of the pre-treatment and soaking time on the corrosion resistance of the Mg HA coating prepared by a modified biomimetic method, and their results offer the possibility of tuning the corrosion rate of a biomimetic deposited HA coating. Chu et al. [201] fabricated hydride GO-based biomimetic coatings on Mg, and a significant improvement in the corrosion resistance was observed for bistriethoxysilylethane (BTSE)-bridged GO coating compared to both pure Mg and non-BTSE-bridged Go coating. The introduction of such a polymer coupling agent to the GO coating not only enhances the integrity and continuity of the coating [202,203] but also reduces the galvanic corrosion that often troubles GO coatings subjected to long-term corrosion due to its excellent electrical conductivity [204].

4.2. *Surface Severe Plastic Deformation (SSPD)*

SSPD is a surface modification technique that can be used to enhance the properties of many metallic materials. SSPD introduces severe plastic deformation (SPD) on the sample surface, leading to surface nanocrystallization (SNC) [205] or surface amorphization [206]. Compared to conventional SPD, SSPD can introduce a much higher strain near the sample surface, creating a gradient structure with unique mechanical properties. Additionally, being a surface modification technique, SSPD allows fine tuning of the surface characteristics that are crucial to the body response to the implant, offering more flexibility to the base implant alloy without changing much of its bulk mechanical properties. Some popular techniques for SSPD include ultrasonic nanocrystal surface modification (UNSM), surface mechanical attrition treatment (SMAT), shot peening (SP), etc. According to the well-known Hall–Petch phenomenon, the strength of the crystalline materials increases as the grain size is reduced. Thus, a surface layer with high strength and hardness can be obtained via the SNC process. Additionally, due to the gradient nature of the applied strain gradient (severe at places closer to the surface), many sample surfaces treated by SNC naturally exhibit a gradient microstructure, and the grain size increases as it moves away from the heavily deformed surface. Such a gradient structure has been shown to possess an excellent combination of high strength and good ductility [205], which are usually two contradictory properties for most strengthening methods. Grain refinement introduced by SSPD is believed to be beneficial to corrosion resistance by stabilizing the corrosion product layer [207]. However, compromised corrosion resistance has also been reported for SSPD-treated samples, mainly due to high roughness and/or surface contamination caused by the treatment [208–210]. Similar to conventional coating techniques, tunable biocompatibility has also been reported for materials subjected to SNC due to the modified surface nature [211,212].

4.2.1. Ultrasonic Nanocrystal Surface Modification (UNSM)

UNSM is a widely used surface modification technique for biomedical devices. As illustrated in Figure 5a, ultrasonic waves are imposed on the sample surface through a tip with high-frequency vibration (10–30 kHz) moving across the sample surface, causing the SPD and SNC layers near the tip [206,213]. Ye et al. [206] prepared an amorphous layer with UNSM on Ti-Ni alloys; the complete surface amorphization led to a remarkable improvement in the wear resistance while maintaining good biocompatibility. Their results suggest that the hardness of the top surface was almost doubled due to the amorphization compared to the untreated substrate. Most researchers believe the effect of UNSM on corrosion resistance to be beneficial. Many studies have suggested [214–217] an improved corrosion resistance of the UNSM-treated samples due to the formation of a denser and more stable passive oxidation layer on the sample surface caused by grain refining [215], along with the possible chemical modification of the layer introduced by the alloy elements [216], while others have reported no obvious [218] or even detrimental effect [208] of UNSM on corrosion resistance, possibly due to the high surface roughness and defect density caused by the excessive loading. Additionally, UNSM is reported to be capable of repairing a corroded surface and enhancing its corrosion resistance. A study done by Huang et al. [219] showed that UNSM can effectively repair the defects in the corroded surface of a Al alloy subjected to NaCl solution, improving its resistance to stress corrosion cracks (SCCs) and pitting corrosion. They believe the improved corrosion resistance is caused by the compressive residual stress and grain refinement introduced by UNSM. Sun et al. combined UNSM with high-density electric pulse, successfully creating SNC layers with thicknesses three times that of conventional UNSM [220]. However, unlike conventional SNC layers, which have the maximum hardness at the outermost surface, a softer surface has been noticed for the pulse-assisted UNSM, likely due to surface recrystallization caused by the electric pulse [221,222].

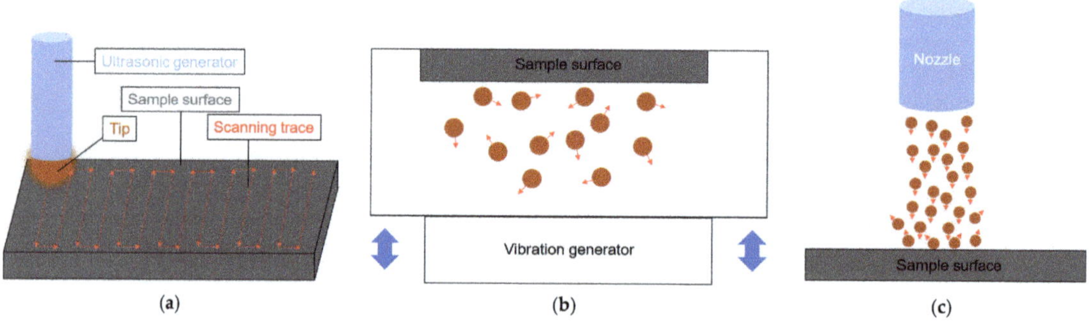

Figure 5. Schematic diagram showing the principles of different SSPDs: (**a**) UNSM, (**b**) SMAT, and (**c**) SP.

4.2.2. Surface Mechanical Attrition Treatment (SMAT)

As shown in Figure 5b, SMAT achieves SSPD by subjecting the sample surface to a closed chamber where high-velocity attrition media with random orientation are accelerated by a vibration generator to intensively bombard the sample surface [223]. Various attrition media like zirconia (ZrO_2), alumina (Al_2O_3), steel balls, etc. can be used for SMAT. The corrosion response of the SMAT-treated surface is highly dependent on the preparation parameters. Parameters like treatment time, attrition media size, attrition media type, and attrition media quantity, which influence the severity of the introduced plastic deformation, are considered to be crucial for an optimized treated surface. Due to the grain refinement introduced by SMAT, improved corrosion resistance has been reported for both pure Mg and Mg alloys subjected to SMAT treatment [224,225]. Kumbhar et al. [212] explored the effect of SMAT on the corrosion resistance and biocompatibility of Mg alloys. A significant improvement in the surface hardness, along with a reduced corrosion rate, has been observed for the SMAT-treated samples. Additionally, the improved corrosion resistance leads to reduced cytotoxicity due to a lower Mg^{2+}-releasing rate of the SMAT-treated samples. An enhanced protein absorption has also been observed for the SMAT-treated surface, demonstrating its good biocompatibility. Similarly, Singh and Keshiri [226] found that SMAT leads to a significant increase in corrosion potential and decrease in corrosion current density. The improved corrosion resistance is attributed to grain refinement and smaller surface roughness. There are also reports regarding possible adverse effects of SMAT on corrosion resistance, perhaps due to the chemical contamination and excessive defects introduced by the attrition media. Fabijanic et al. [209] studied the effect of three different attrition media on the corrosion resistance of SMAT-treated pure Mg. Surface contamination introduced by applied attrition media was found for all the treated surfaces, leading to a slightly different corrosion resistance between samples treated with different media and the overall deteriorated corrosion resistance compared to pure Mg. Similarly, Wen et al. [210] also noted the effect of different attrition media on the corrosion resistance of Al alloys. They found that SMAT using ceramic balls improves the corrosion resistance while iron deteriorates it. An iron-contaminated layer has also been noted for the samples treated with iron balls, leading to enhanced galvanic corrosion and a defective passive film. Laleh and Kargar [224] studied the attrition media size on the microstructure and corrosion resistance of Mg alloys. A similar and significant improvement on hardness was observed for all media sizes due to their similar grain size at the top surface. Increasing the media size effectively thickened the deformed layer, but the surface treated with a larger media size showed poorer corrosion resistance, which was attributed to the increased surface roughness and defect density associated with a larger media size.

4.2.3. Shot Peening (SP)

Similar to SMAT, the SP treatment bombards the sample surface with small metallic or ceramic balls (shots) to introduce SSPD. The working principle of SP is shown in Figure 5c; a nozzle fixed at a given angle to the sample surface instead of a vibration generator, as in the case of SMAT, is used to accelerate the shots. Most shots of SP arrive on the sample surface at a nearly constant angle compared to the random bombarding of SMAT. Additionally, smaller shots with much higher velocity are often used in SP compared to SMAT [227], making the surface deformation more intensive, and a modified layer with higher roughness and defect density is often obtained for the SP-treated samples [228,229]. Peral et al. [230] studied the effect of the grain size and surface roughness of SP-treated surfaces on the corrosion resistance of Mg alloys. Despite significant grain refinement introduced by the SP treatment, which is generally believed to promote corrosion resistance [207], a much higher corrosion rate has been found for all SP-treated samples compared to their untreated counterparts. However, a significant improvement in corrosion resistance of the SP-treated samples, which, in some cases, surpasses the untreated ones, has been observed after applying an extra electropolishing process to reduce the surface roughness. Similarly, Bagherifard et al. [231] reported high surface roughness caused by the SP treatment on Mg alloys, leading to accelerated corrosion and decreased osteoblast viability at the beginning. However, they suggest that, by reducing the surface roughness via grinding, the corrosion resistance can be effectively improved and no long-term cytotoxicity has been observed, indicating good biocompatibility of the SP-treated samples. Maliwemu et al. [232,233] studied the effects of the shot distance and shot angle on the surface condition and corrosion resistance of 316L stainless steel used as biomaterial, and it was suggested that a smaller shot distance and larger shot angle led to optimized surface conditions with the lowest corrosion rate. A study by Khany et al. [234] showed that the mechanical performance was also dependent on the shot angle, and 45° was found to be the optimum shot angle in this case. These results highlight the critical role of the surface condition and, in turn, processing parameters on the corrosion resistance of the SP-treated samples.

5. Conclusions

Mg and its alloys have been attracting increasing interest in the past decades for biomedical applications due to their excellent biodegradability and biocompatibility combined with adequate mechanical properties. The smaller discrepancy of Young's modulus from natural bone, accompanied by its small density, makes Mg and its alloys ideal candidates for load-bearing bone implants, reducing the risks of stress shielding, which predominantly occurs in conventional biomaterials with a high elastic modulus like steel and titanium. However, achieving a controlled degradation rate while maintaining mechanical integrity remains challenging due to the fast corrosion rate of Mg in a physiological environment. Alloying strategies, while demonstrated to be effective in promoting the corrosion resistance and mechanical properties of Mg, raises concerns over the biocompatibility of certain alloying elements, particularly RE elements, which are the essential ingredient of many high-performance Mg alloys. Diverse results and large discrepancies between in vivo and in vitro studies on the biocompatibility of alloying elements emphasize the importance of future in-depth investigation. To further optimize Mg-based biomaterials, other potential alloying elements like silver, tin (Sn), germanium (Ge), etc. [235,236] may need to be considered in the future alongside the commonly used ones reviewed in this paper. Apart from alloying, SPP has also proven to be an effective way of improving the bulk mechanical and corrosion resistance of the Mg biomaterials without compromising the biocompatibility. Alternatively, surface modification techniques like chemical conversion coatings, anodization, PEO, biomimetic deposition, SSPD, etc. are also promising in terms of effectively enhancing the corrosion resistance and bioactivity of Mg biomaterials. A much more controllable degradation rate and highly tunable bioactive surface can be achieved compared to alloying. The treated surface is, however, highly sensitive to the preparation parameters, and further investigation is required to refine the surface treatment processes. These solutions show great promise for the biomedical application of Mg, and a combination of multiple techniques is likely necessary for specific clinical applications in the future, offering Mg with a tunable corrosion/mechanical performance.

Author Contributions: Writing—original draft preparation, Y.L.; writing—review and editing, Y.L., J.Y. and G.-z.Z.; supervision, G.-z.Z. All authors have read and agreed to the published version of the manuscript.

Funding: This research was funded by the Natural Sciences and Engineering Research Council of Canada (NSERC) Discovery Grant RGPIN-2019-05882 (G.-z.Z.) and Canada Research Chairs Program CRC-2021-00512 (G.-z.Z.).

Conflicts of Interest: The authors declare no conflicts of interest.

References

1. Prasadh, S.; Ratheesh, V.; Manakari, V.; Parande, G.; Gupta, M.; Wong, R. The Potential of Magnesium Based Materials in Mandibular Reconstruction. *Metals* **2019**, *9*, 302. [CrossRef]
2. Seitz, J.M.; Lucas, A.; Kirschner, M. Magnesium-Based Compression Screws: A Novelty in the Clinical Use of Implants. *JOM* **2016**, *68*, 1177–1182. [CrossRef]
3. Luthringer, B.J.C.; Feyerabend, F.; Römer, R.W. Magnesium-Based Implants: A Mini-Review. *Magnes. Res.* **2014**, *27*, 142–154. [CrossRef]
4. Chakraborty Banerjee, P.; Al-Saadi, S.; Choudhary, L.; Harandi, S.E.; Singh, R. Magnesium Implants: Prospects and Challenges. *Materials* **2019**, *12*, 136. [CrossRef]
5. Zegre, M.; Poljańska, E.; Caetano, L.A.; Gonçalves, L.; Bettencourt, A. Research progress on biodegradable polymeric platforms for targeting antibiotics to the bone. *Int. J. Pharm.* **2023**, *648*, 123584. [CrossRef]
6. Zheng, Y.F.; Gu, X.N.; Witte, F. Biodegradable metals. *Mater. Sci. Eng. R Rep.* **2014**, *77*, 1–34. [CrossRef]
7. Thomas, K.K.; Zafar, M.N.; Pitt, W.G.; Husseini, G.A. Biodegradable Magnesium Alloys for Biomedical Implants: Properties, Challenges, and Surface Modifications with a Focus on Orthopedic Fixation Repair. *Appl. Sci.* **2024**, *14*, 10. [CrossRef]
8. Zhang, Y.; Xu, J.; Ruan, Y.C.; Yu, M.K.; O'Laughlin, M.; Wise, H.; Chen, D.; Tian, L.; Shi, D.; Wang, J.; et al. Implant-derived magnesium induces local neuronal production of CGRP to improve bone-fracture healing in rats. *Nat. Med.* **2016**, *22*, 1160–1169. [CrossRef] [PubMed]
9. Zhang, T.; Wang, W.; Liu, J.; Wang, L.; Tang, Y.; Wang, K. A review on magnesium alloys for biomedical applications. *Front. Bioeng. Biotechnol.* **2022**, *10*, 953344. [CrossRef]
10. Bosisio, M.R.; Talmant, M.; Skalli, W.; Laugier, P.; Mitton, D. Apparent Young's modulus of human radius using inverse finite-element method. *J. Biomech.* **2007**, *40*, 2022–2028. [CrossRef]
11. Sylvester, A.D.; Kramer, P.A. Young's Modulus and Load Complexity: Modeling Their Effects on Proximal Femur Strain. *Anat. Rec.* **2018**, *301*, 1189–1202. [CrossRef] [PubMed]
12. Chmielewska, A.; Dean, D. The role of stiffness-matching in avoiding stress shielding-induced bone loss and stress concentration-induced skeletal reconstruction device failure. *Acta Biomater.* **2024**, *173*, 51–65. [CrossRef] [PubMed]
13. Akbarzadeh, F.Z.; Sarraf, M.; Ghomi, E.R.; Kumar, V.V.; Salehi, M.; Ramakrishna, S.; Bae, S. A state-of-the-art review on recent advances in the fabrication and characteristics of magnesium-based alloys in biomedical applications. *J. Magnes. Alloys* **2024**, *12*, 2569–2594. [CrossRef]
14. Nasr Azadani, M.; Zahedi, A.; Bowoto, O.K.; Oladapo, B.I. A review of current challenges and prospects of magnesium and its alloy for bone implant applications. *Prog. Biomater.* **2022**, *11*, 1–26. [CrossRef]
15. Kaushik, V.; Kumar, N.; Vignesh, M. Magnesium role in additive manufacturing of biomedical implants—Challenges and opportunities. *Addit. Manuf.* **2022**, *55*, 102802. [CrossRef]
16. Callister, W.D., Jr.; Rethwisch, D.G. *Materials Science and Engineering: An Introduction*; John Wiley & Sons: Hoboken, NJ, USA, 2020.
17. Bard, A. *Standard Potentials in Aqueous Solution*; Routledge: London, UK, 2017.
18. Agarwal, S.; Curtin, J.; Duffy, B.; Jaiswal, S. Biodegradable magnesium alloys for orthopaedic applications: A review on corrosion, biocompatibility and surface modifications. *Mater. Sci. Eng. C* **2016**, *68*, 948–963. [CrossRef]

19. Li, X.; Liu, X.; Wu, S.; Yeung, K.W.K.; Zheng, Y.; Chu, P.K. Design of magnesium alloys with controllable degradation for biomedical implants: From bulk to surface. *Acta Biomater.* **2016**, *45*, 2–30. [CrossRef]
20. Zhang, D.; Ni, N.; Su, Y.; Miao, H.; Tang, Z.; Ji, Y.; Wang, Y.; Gao, H.; Ju, Y.; Sun, N.; et al. Targeting Local Osteogenic and Ancillary Cells by Mechanobiologically Optimized Magnesium Scaffolds for Orbital Bone Reconstruction in Canines. *ACS Appl. Mater. Interfaces* **2020**, *12*, 27889–27904. [CrossRef]
21. Shi, B.; Yang, C.; Peng, Y.; Zhang, F.; Pan, F. Anisotropy of wrought magnesium alloys: A focused overview. *J. Magnes. Alloys* **2022**, *10*, 1476–1510. [CrossRef]
22. Tian, L.; Zhou, Y.; Jing, X.; Zheng, T.; Zhou, S.; Zhang, J.; Peng, P.; Tang, A. Calculation of critical shear stress for binary magnesium alloys: A first-principles study. *Mech. Mater.* **2023**, *184*, 104711. [CrossRef]
23. Christian, J.W.; Mahajan, S. Deformation twinning. *Prog. Mater. Sci.* **1995**, *39*, 1–157. [CrossRef]
24. Liu, R.L.; Scully, J.R.; Williams, G.; Birbilis, N. Reducing the corrosion rate of magnesium via microalloying additions of group 14 and 15 elements. *Electrochim. Acta* **2018**, *260*, 184–195. [CrossRef]
25. Chen, Q.; Thouas, G.A. Metallic implant biomaterials. *Mater. Sci. Eng. R Rep.* **2015**, *87*, 1–57. [CrossRef]
26. Huang, X.; Suzuki, K.; Chino, Y.; Mabuchi, M. Texture and stretch formability of AZ61 and AM60 magnesium alloy sheets processed by high-temperature rolling. *J. Alloys Compd.* **2015**, *632*, 94–102. [CrossRef]
27. Kang, D.H.; Kim, D.W.; Kim, S.; Bae, G.T.; Kim, K.H.; Kim, N.J. Relationship between stretch formability and work-hardening capacity of twin-roll cast Mg alloys at room temperature. *Scr. Mater.* **2009**, *61*, 768–771. [CrossRef]
28. Thanabal, N.; Silambarasan, R.; Seenuvasaperumal, P.; Basha, D.A.; Elayaperumal, A. Microstructure and mechanical behavior of AXM Mg alloy systems—A review. *J. Magnes. Alloys* **2024**, *12*, 2624–2646. [CrossRef]
29. Bamberger, M. Structural refinement of cast magnesium alloys. *Mater. Sci. Technol.* **2001**, *17*, 15–24. [CrossRef]
30. Cao, H.; Wessén, M. Effect of microstructure on mechanical properties of as-cast Mg-Al alloys. *Metall. Mater. Trans. A* **2004**, *35*, 309–319. [CrossRef]
31. Samimi, G.; Mirdamadi, S.; Razavi, H.; Boutorabi, M.-A. Improvement in impression creep property of as-cast AC515 Mg alloy by Mn addition. *Mater. Sci. Eng. A* **2013**, *587*, 213–220. [CrossRef]
32. Li, Z.T.; Qiao, X.G.; Xu, C.; Kamado, S.; Zheng, M.Y.; Luo, A.A. Ultrahigh strength Mg-Al-Ca-Mn extrusion alloys with various aluminum contents. *J. Alloys Compd.* **2019**, *792*, 130–141. [CrossRef]
33. Okazaki, Y.; Rao, S.; Ito, Y.; Tateishi, T. Corrosion resistance, mechanical properties, corrosion fatigue strength and cytocompatibility of new Ti alloys without Al and V. *Biomaterials* **1998**, *19*, 1197–1215. [CrossRef] [PubMed]
34. Shingde, M.; Hughes, J.; Boadle, R.; Wills, E.J.; Pamphlett, R. Macrophagic myofasciitis associated with vaccine-derived aluminium. *Med. J. Aust.* **2005**, *183*, 145–146. [CrossRef]
35. Mirza, A.; King, A.; Troakes, C.; Exley, C. Aluminium in brain tissue in familial Alzheimer's disease. *J. Trace Elem. Med. Biol.* **2017**, *40*, 30–36. [CrossRef] [PubMed]
36. Evans, C.H. Toxicology and Pharmacology of the Lanthanides. In *Biochemistry of the Lanthanides*; Evans, C.H., Ed.; Springer: Boston, MA, USA, 1990; pp. 339–389. [CrossRef]
37. Gao, L.; Chen, R.S.; Han, E.H. Solid solution strengthening behaviors in binary Mg–Y single phase alloys. *J. Alloys Compd.* **2009**, *472*, 234–240. [CrossRef]
38. Zemková, M.; Minárik, P.; Jablonská, E.; Veselý, J.; Bohlen, J.; Kubásek, J.; Lipov, J.; Ruml, T.; Havlas, V.; Král, R. Concurrence of High Corrosion Resistance and Strength with Excellent Ductility in Ultrafine-Grained Mg-3Y Alloy. *Materials* **2022**, *15*, 7571. [CrossRef]
39. Minárik, P.; Veselý, J.; Čížek, J.; Zemková, M.; Vlasák, T.; Krajňák, T.; Kubásek, J.; Král, R.; Hofman, D.; Stráská, J. Effect of secondary phase particles on thermal stability of ultra-fine grained Mg-4Y-3RE alloy prepared by equal channel angular pressing. *Mater. Charact.* **2018**, *140*, 207–216. [CrossRef]
40. Yu, X.; Jiang, B.; He, J.; Liu, B.; Pan, F. Oxidation resistance of Mg-Y alloys at elevated temperatures and the protection performance of the oxide films. *J. Alloys Compd.* **2018**, *749*, 1054–1062. [CrossRef]
41. Liu, X.; Shan, D.; Song, Y.; Han, E.-h. Influence of yttrium element on the corrosion behaviors of Mg–Y binary magnesium alloy. *J. Magnes. Alloys* **2017**, *5*, 26–34. [CrossRef]
42. Feyerabend, F.; Fischer, J.; Holtz, J.; Witte, F.; Willumeit, R.; Drücker, H.; Vogt, C.; Hort, N. Evaluation of short-term effects of rare earth and other elements used in magnesium alloys on primary cells and cell lines. *Acta Biomater.* **2010**, *6*, 1834–1842. [CrossRef]
43. Evans, C.H. The Occurrence and Metabolism of Lanthanides. In *Biochemistry of the Lanthanides*; Evans, C.H., Ed.; Springer: Boston, MA, USA, 1990; pp. 285–337. [CrossRef]
44. Tekumalla, S.; Seetharaman, S.; Quy Bau, N.; Leong Eugene Wong, W.; Sim Goh, C.; Shabadi, R.; Gupta, M. Influence of Cerium on the Deformation and Corrosion of Magnesium. *J. Eng. Mater. Technol.* **2016**, *138*, 031011. [CrossRef]
45. Song, Y.L.; Liu, Y.H.; Wang, S.H.; Yu, S.R.; Zhu, X.Y. Effect of cerium addition on microstructure and corrosion resistance of die cast AZ91 magnesium alloy. *Mater. Corros.* **2007**, *58*, 189–192. [CrossRef]

46. Chenghao, L.; Shusen, W.; Naibao, H.; Zhihong, Z.; Shuchun, Z.; Jing, R. Effects of Lanthanum and Cerium Mixed Rare Earth Metal on Abrasion and Corrosion Resistance of AM60 Magnesium Alloy. *Rare Met. Mater. Eng.* **2015**, *44*, 521–526. [CrossRef]
47. Tong, L.B.; Zhang, Q.X.; Jiang, Z.H.; Zhang, J.B.; Meng, J.; Cheng, L.R.; Zhang, H.J. Microstructures, mechanical properties and corrosion resistances of extruded Mg-Zn-Ca-xCe/La alloys. *J. Mech. Behav. Biomed. Mater.* **2016**, *62*, 57–70. [CrossRef] [PubMed]
48. Luo, K.; Zhang, L.; Wu, G.; Liu, W.; Ding, W. Effect of Y and Gd content on the microstructure and mechanical properties of Mg–Y–RE alloys. *J. Magnes. Alloys* **2019**, *7*, 345–354. [CrossRef]
49. Liu, Y.; Song, Y.; Li, N.; Sha, X.; Xu, M.; Chen, B.; Gao, B.; Xiao, L.; Zhou, H. Mechanical Properties and Microstructure Evolution of Mg-Gd Alloy during Aging Treatment. *Metals* **2021**, *12*, 39. [CrossRef]
50. Myrissa, A.; Braeuer, S.; Martinelli, E.; Willumeit-Römer, R.; Goessler, W.; Weinberg, A.M. Gadolinium accumulation in organs of Sprague–Dawley® rats after implantation of a biodegradable magnesium-gadolinium alloy. *Acta Biomater.* **2017**, *48*, 521–529. [CrossRef]
51. Bernstein, E.J.; Schmidt-Lauber, C.; Kay, J. Nephrogenic systemic fibrosis: A systemic fibrosing disease resulting from gadolinium exposure. *Best Pract. Res. Clin. Rheumatol.* **2012**, *26*, 489–503. [CrossRef]
52. Hort, N.; Huang, Y.; Fechner, D.; Störmer, M.; Blawert, C.; Witte, F.; Vogt, C.; Drücker, H.; Willumeit, R.; Kainer, K.U.; et al. Magnesium alloys as implant materials—Principles of property design for Mg–RE alloys. *Acta Biomater.* **2010**, *6*, 1714–1725. [CrossRef]
53. Rokhlin, L.L. *Magnesium Alloys Containing Rare Earth Metals: Structure and Properties*; CRC Press: Boca Raton, FL, USA, 2003.
54. Zidane, N.; Albrimi, Y.A.; Addi, A.A.; Akbour, R.A.; Douch, J.; Nahle, A.; Hamdani, M. Effect of gadolinium content on the corrosion behavior of magnesium alloys in 1 wt.% NaCl solution. *Port. Electrochim. Acta* **2015**, *33*, 289–304. [CrossRef]
55. Lu, Y.; Huang, Y.; Bode, J.; Vogt, C.; Willumeit-Römer, R.; Kainer, K.U.; Hort, N. Effects of Minor Gadolinium Addition and T4 Heat Treatment on Microstructure and Properties of Magnesium. *Adv. Eng. Mater.* **2022**, *24*, 2200966. [CrossRef]
56. Chen, J.; Zhan, J.; Kolawole, S.K.; Tan, L.; Yang, K.; Wang, J.; Su, X. Effects of Different Rare Earth Elements on the Degradation and Mechanical Properties of the ECAP Extruded Mg Alloys. *Materials* **2022**, *15*, 627. [CrossRef] [PubMed]
57. Jin, W.; Wu, G.; Feng, H.; Wang, W.; Zhang, X.; Chu, P.K. Improvement of corrosion resistance and biocompatibility of rare-earth WE43 magnesium alloy by neodymium self-ion implantation. *Corros. Sci.* **2015**, *94*, 142–155. [CrossRef]
58. Cheng, J.; Li, N.; Cai, J.; Cheng, Z.; Hu, R.; Zhang, Q.; Wan, F.; Sun, Q.; Gui, S.; Sang, X.; et al. Organ Histopathological Changes and its Function Damage in Mice Following Long-term Exposure to Lanthanides Chloride. *Biol. Trace Elem. Res.* **2012**, *145*, 361–368. [CrossRef] [PubMed]
59. Elkaiam, L.; Hakimi, O.; Goldman, J.; Aghion, E. The Effect of Nd on Mechanical Properties and Corrosion Performance of Biodegradable Mg-5%Zn Alloy. *Metals* **2018**, *8*, 438. [CrossRef]
60. Wu, W.; Wang, Y.; Zeng, X.; Chen, L.; Liu, Z. Effect of neodymium on mechanical behavior of Mg-Zn-Zr magnesium alloy. *J. Mater. Sci. Lett.* **2003**, *22*, 445–447. [CrossRef]
61. Yin, P.; Li, N.F.; Lei, T.; Liu, L.; Ouyang, C. Effects of Ca on microstructure, mechanical and corrosion properties and biocompatibility of Mg–Zn–Ca alloys. *J. Mater. Sci. Mater. Med.* **2013**, *24*, 1365–1373. [CrossRef]
62. Li, Z.; Gu, X.; Lou, S.; Zheng, Y. The development of binary Mg–Ca alloys for use as biodegradable materials within bone. *Biomaterials* **2008**, *29*, 1329–1344. [CrossRef]
63. Akhmetshina, T.; Berger, L.; Basu, I.; Montibeller, S.; Rubin, W.; Rich, A.M.; Schäublin, R.E.; Löffler, J.F. High-performance ultra-lean biodegradable Mg–Ca alloys and guidelines for their processing. *Acta Mater.* **2024**, *278*, 120247. [CrossRef]
64. Jin, X.; Xu, W.; Shan, D.; Guo, B.; Jin, B.C. Mechanism of high-strength and ductility of Mg-RE alloy fabricated by low-temperature extrusion and aging treatment. *Mater. Des.* **2021**, *199*, 109384. [CrossRef]
65. Li, Y.; Hodgson, P.D.; Wen, C. The effects of calcium and yttrium additions on the microstructure, mechanical properties and biocompatibility of biodegradable magnesium alloys. *J. Mater. Sci.* **2011**, *46*, 365–371. [CrossRef]
66. Kim, J.H.; Kang, N.E.; Yim, C.D.; Kim, B.K. Effect of calcium content on the microstructural evolution and mechanical properties of wrought Mg–3Al–1Zn alloy. *Mater. Sci. Eng. A* **2009**, *525*, 18–29. [CrossRef]
67. Deng, M.; Wang, L.; Höche, D.; Lamaka, S.V.; Wang, C.; Snihirova, D.; Jin, Y.; Zhang, Y.; Zheludkevich, M.L. Approaching "stainless magnesium" by Ca micro-alloying. *Mater. Horiz.* **2021**, *8*, 589–596. [CrossRef] [PubMed]
68. Seong, J.W.; Kim, W.J. Mg-Ca binary alloy sheets with Ca contents of ≤1wt.% with high corrosion resistance and high toughness. *Corros. Sci.* **2015**, *98*, 372–381. [CrossRef]
69. Khan, A.R.; Grewal, N.S.; Zhou, C.; Yuan, K.; Zhang, H.-J.; Jun, Z. Recent advances in biodegradable metals for implant applications: Exploring in vivo and in vitro responses. *Results Eng.* **2023**, *20*, 101526. [CrossRef]
70. Nriagu, J. Zinc Toxicity in Humans. In *Encyclopedia of Environmental Health, Volume 1–5*; Elsevier: Amsterdam, The Netherlands, 2011; Volume 5, pp. 1–7.
71. Li, H.; Liu, D.; Zhao, Y.; Jin, F.; Chen, M. The Influence of Zn Content on the Corrosion and Wear Performance of Mg-Zn-Ca Alloy in Simulated Body Fluid. *J. Mater. Eng. Perform.* **2016**, *25*, 3890–3895. [CrossRef]

72. Cai, S.; Lei, T.; Li, N.; Feng, F. Effects of Zn on microstructure, mechanical properties and corrosion behavior of Mg–Zn alloys. *Mater. Sci. Eng. C* **2012**, *32*, 2570–2577. [CrossRef]
73. Zhang, B.; Hou, Y.; Wang, X.; Wang, Y.; Geng, L. Mechanical properties, degradation performance and cytotoxicity of Mg–Zn–Ca biomedical alloys with different compositions. *Mater. Sci. Eng. C* **2011**, *31*, 1667–1673. [CrossRef]
74. Shi, Z.; Song, G.; Atrens, A. Corrosion resistance of anodised single-phase Mg alloys. *Surf. Coat. Technol.* **2006**, *201*, 492–503. [CrossRef]
75. Marie, P.J. Strontium ranelate: A novel mode of action optimizing bone formation and resorption. *Osteoporos. Int.* **2005**, *16*, S7–S10. [CrossRef]
76. Fan, Y.; Wu, G.H.; Zhai, C.Q. Effect of strontium on mechanical properties and corrosion resistance of AZ91D. *Mater. Sci. Forum.* **2007**, *546–549*, 567–570.
77. Lai, H.; Li, J.; Li, J.; Zhang, Y.; Xu, Y. Effects of Sr on the microstructure, mechanical properties and corrosion behavior of Mg-2Zn-xSr alloys. *J. Mater. Sci. Mater. Med.* **2018**, *29*, 87. [CrossRef]
78. Wen, Y.; Liu, Q.; Wang, J.; Yang, Q.; Zhao, W.; Qiao, B.; Li, Y.; Jiang, D. Improving in vitro and in vivo corrosion resistance and biocompatibility of Mg–1Zn–1Sn alloys by microalloying with Sr. *Bioact. Mater.* **2021**, *6*, 4654–4669. [CrossRef] [PubMed]
79. Tang, H.; Wang, F.; Li, D.; Gu, X.; Fan, Y. Mechanical properties, degradation behaviors and biocompatibility of micro-alloyed Mg-Sr-RE alloys for stent applications. *Mater. Lett.* **2020**, *264*, 127285. [CrossRef]
80. Machado-Vieira, R.; Manji, H.K.; Zarate, C.A., Jr. The role of lithium in the treatment of bipolar disorder: Convergent evidence for neurotrophic effects as a unifying hypothesis. *Bipolar Disord.* **2009**, *11*, 92–109. [CrossRef] [PubMed]
81. Liaugaudaite, V.; Mickuviene, N.; Raskauskiene, N.; Naginiene, R.; Sher, L. Lithium levels in the public drinking water supply and risk of suicide: A pilot study. *J. Trace Elem. Med. Biol.* **2017**, *43*, 197–201. [CrossRef]
82. Aurell, M.; Svalander, C.; Wallin, L.; Alling, C. Renal function and biopsy findings in patients on long-term lithium treatment. *Kidney Int.* **1981**, *20*, 663–670. [CrossRef] [PubMed]
83. Liu, Y.; Wu, Y.; Bian, D.; Gao, S.; Leeflang, S.; Guo, H.; Zheng, Y.; Zhou, J. Study on the Mg-Li-Zn ternary alloy system with improved mechanical properties, good degradation performance and different responses to cells. *Acta Biomater.* **2017**, *62*, 418–433. [CrossRef]
84. Xu, W.; Birbilis, N.; Sha, G.; Wang, Y.; Daniels, J.E.; Xiao, Y.; Ferry, M. A high-specific-strength and corrosion-resistant magnesium alloy. *Nat. Mater.* **2015**, *14*, 1229–1235. [CrossRef]
85. Xin, T.; Zhao, Y.; Mahjoub, R.; Jiang, J.; Yadav, A.; Nomoto, K.; Niu, R.; Tang, S.; Ji, F.; Quadir, Z.; et al. Ultrahigh specific strength in a magnesium alloy strengthened by spinodal decomposition. *Sci. Adv.* **2021**, *7*, eabf3039. [CrossRef]
86. Cain, T.W.; Labukas, J.P. The development of β phase Mg–Li alloys for ultralight corrosion resistant applications. *npj Mater. Degrad.* **2020**, *4*, 17. [CrossRef]
87. Dobkowska, A.; Adamczyk-Cieslak, B.; Mizera, J.; Kubásek, J.; Vojtěch, D. Corrosion Behaviour of Magnesium Lithium Alloys in NaCl Solution. *Solid State Phenom.* **2015**, *227*, 87–90. [CrossRef]
88. Li, C.Q.; Xu, D.K.; Chen, X.B.; Wang, B.J.; Wu, R.Z.; Han, E.H.; Birbilis, N. Composition and microstructure dependent corrosion behaviour of Mg-Li alloys. *Electrochim. Acta* **2018**, *260*, 55–64. [CrossRef]
89. Chen, P.; Bornhorst, J.; Aschner, M.A. Manganese metabolism in humans. In *Postprints der Universität Potsdam Mathematisch-Naturwissenschaftliche Reihe*; Universität Potsdam: Potsdam, Germany, 2019.
90. Miah, M.R.; Ijomone, O.M.; Okoh, C.O.A.; Ijomone, O.K.; Akingbade, G.T.; Ke, T.; Krum, B.; da Cunha Martins, A.; Akinyemi, A.; Aranoff, N.; et al. The effects of manganese overexposure on brain health. *Neurochem. Int.* **2020**, *135*, 104688. [CrossRef]
91. Martins, A.C.; Morcillo, P.; Ijomone, O.M.; Venkataramani, V.; Harrison, F.E.; Lee, E.; Bowman, A.B.; Aschner, M. New Insights on the Role of Manganese in Alzheimer's Disease and Parkinson's Disease. *Int. J. Environ. Res. Public Health* **2019**, *16*, 3546. [CrossRef]
92. Gu, D.-D.; Peng, J.; Wang, J.-W.; Liu, Z.-T.; Pan, F.-S. Effect of Mn Modification on the Corrosion Susceptibility of Mg–Mn Alloys by Magnesium Scrap. *Acta Metall. Sin. (Engl. Lett.)* **2021**, *34*, 1–11. [CrossRef]
93. Hu, Y.; Dong, D.; Wang, X.; Chen, H.; Qiao, Y. Synthesis and Properties of Mg-Mn-Zn Alloys for Medical Applications. *Materials* **2021**, *14*, 1855. [CrossRef]
94. Yao, S.; Liu, S.; Zeng, G.; Li, X.; Lei, T.; Li, Y.; Du, Y. Effect of Manganese on Microstructure and Corrosion Behavior of the Mg-3Al Alloys. *Metals* **2019**, *9*, 460. [CrossRef]
95. Li, J.; Liu, C.; Wang, Z.; Cai, Y.; Li, J. Effect of Mn Content on Microstructure, Mechanical Properties, and Corrosion Behavior of Mg-Zn-Gd-Y Alloy. *J. Mater. Eng. Perform.* **2024**, *33*, 10574–10588. [CrossRef]
96. Liu, C.; He, P.; Wan, P.; Li, M.; Wang, K.; Tan, L.; Zhang, Y.; Yang, K. The in vitro biocompatibility and macrophage phagocytosis of $Mg_{17}Al_{12}$ phase in Mg–Al–Zn alloys. *J. Biomed. Mater. Res. Part A* **2015**, *103*, 2405–2415. [CrossRef]
97. Graddala, S.B.; Chittaranjan Das, V. Role of surface grain refinement in AZ31 Mg alloy by shot peening on surface energy, biomineralization and degradation behavior. *Eng. Res. Express* **2024**, *6*, 045545. [CrossRef]

98. Pogorielov, M.; Husak, E.; Solodivnik, A.; Zhdanov, S. Magnesium-based biodegradable alloys: Degradation, application, and alloying elements. *Interv. Med. Appl. Sci.* **2017**, *9*, 27–38. [CrossRef]
99. Calado, L.M.; Carmezim, M.J.; Montemor, M.F. Rare Earth Based Magnesium Alloys—A Review on WE Series. *Front. Mater.* **2022**, *8*, 804906. [CrossRef]
100. Drynda, A.; Deinet, N.; Braun, N.; Peuster, M. Rare earth metals used in biodegradable magnesium-based stents do not interfere with proliferation of smooth muscle cells but do induce the upregulation of inflammatory genes. *J. Biomed. Mater. Res. Part A* **2009**, *91*, 360–369. [CrossRef] [PubMed]
101. Brouziotis, A.A.; Giarra, A.; Libralato, G.; Pagano, G.; Guida, M.; Trifuoggi, M. Toxicity of rare earth elements: An overview on human health impact. *Front. Environ. Sci.* **2022**, *10*, 948041. [CrossRef]
102. Zhang, J.; Niu, X.; Qiu, X.; Liu, K.; Nan, C.; Tang, D.; Meng, J. Effect of yttrium-rich misch metal on the microstructures, mechanical properties and corrosion behavior of die cast AZ91 alloy. *J. Alloys Compd.* **2009**, *471*, 322–330. [CrossRef]
103. Weng, W.; Biesiekierski, A.; Li, Y.; Dargusch, M.; Wen, C. A review of the physiological impact of rare earth elements and their uses in biomedical Mg alloys. *Acta Biomater.* **2021**, *130*, 80–97. [CrossRef] [PubMed]
104. Pereira, G.S.; Koga, G.Y.; Avila, J.A.; Bittencourt, I.M.; Fernandez, F.; Miyazaki, M.H.; Botta, W.J.; Bose Filho, W.W. Corrosion resistance of WE43 Mg alloy in sodium chloride solution. *Mater. Chem. Phys.* **2021**, *272*, 124930. [CrossRef]
105. Xie, J.; Zhang, J.; You, Z.; Liu, S.; Guan, K.; Wu, R.; Wang, J.; Feng, J. Towards developing Mg alloys with simultaneously improved strength and corrosion resistance via RE alloying. *J. Magnes. Alloys* **2021**, *9*, 41–56. [CrossRef]
106. Song, G.L. 1—Corrosion behavior and prevention strategies for magnesium (Mg) alloys. In *Corrosion Prevention of Magnesium Alloys*; Song, G.-L., Ed.; Woodhead Publishing: Cambridge, UK, 2013; pp. 3–37. [CrossRef]
107. Song, G.; Atrens, A.; Wu, X.; Zhang, B. Corrosion behaviour of AZ21, AZ501 and AZ91 in sodium chloride. *Corros. Sci.* **1998**, *40*, 1769–1791. [CrossRef]
108. Eivani, A.R.; Mehdizade, M.; Chabok, S.; Zhou, J. Applying multi-pass friction stir processing to refine the microstructure and enhance the strength, ductility and corrosion resistance of WE43 magnesium alloy. *J. Mater. Res. Technol.* **2021**, *12*, 1946–1957. [CrossRef]
109. Polmear, I.J. Magnesium alloys and applications. *Mater. Sci. Technol.* **1994**, *10*, 1–16. [CrossRef]
110. Massalski, T.B. Binary Alloy Phase Diagrams. In *Alloy Phase Diagrams*; Okamoto, H., Schlesinger, M.E., Mueller, E.M., Eds.; ASM International: Almere, The Netherlands, 2016; Volume 3.
111. Yu, H.; Dong, X.R.; Kang, S.M.; Yu, W.; Wang, Z.F.; Mu, J.Y.; Cui, X.W.; Li, J.H.; Yin, F.X.; Shin, K.S. Effect of the pre-homogenization on the precipitation behaviors, mechanical and corrosion properties of as-extruded MgY binary alloys. *Mater. Charact.* **2021**, *178*, 111307. [CrossRef]
112. Hirano, S.; Kodama, N.; Shibata, K.; Suzuki, K.T. Metabolism and Toxicity of Intravenously Injected Yttrium Chloride in Rats. *Toxicol. Appl. Pharmacol.* **1993**, *121*, 224–232. [CrossRef]
113. Moriyama, A.; Takahashi, U.; Mizuno, Y.; Takahashi, J.; Horie, M.; Iwahashi, H. The truth of toxicity caused by yttrium oxide nanoparticles to yeast cells. *J. Nanosci. Nanotechnol.* **2019**, *19*, 5418–5425. [CrossRef] [PubMed]
114. He, X. Yttrium, Biological Effects. In *Encyclopedia of Metalloproteins*; Kretsinger, R.H., Uversky, V.N., Permyakov, E.A., Eds.; Springer: New York, NY, USA, 2013; pp. 2338–2340. [CrossRef]
115. Beaser, S.B.; Segel, A.; Vandam, L. The anticoagulant effects in rabbits and man of the intravenous injection of salts of the rare earths. *J. Clin. Investig.* **1942**, *21*, 447–454. [CrossRef]
116. Willbold, E.; Gu, X.; Albert, D.; Kalla, K.; Bobe, K.; Brauneis, M.; Janning, C.; Nellesen, J.; Czayka, W.; Tillmann, W.; et al. Effect of the addition of low rare earth elements (lanthanum, neodymium, cerium) on the biodegradation and biocompatibility of magnesium. *Acta Biomater.* **2015**, *11*, 554–562. [CrossRef]
117. Hofstetter, J.; Becker, M.; Martinelli, E.; Weinberg, A.M.; Mingler, B.; Kilian, H.; Pogatscher, S.; Uggowitzer, P.J.; Löffler, J.F. High-Strength Low-Alloy (HSLA) Mg–Zn–Ca Alloys with Excellent Biodegradation Performance. *JOM* **2014**, *66*, 566–572. [CrossRef]
118. Holweg, P.; Berger, L.; Cihova, M.; Donohue, N.; Clement, B.; Schwarze, U.; Sommer, N.G.; Hohenberger, G.; van den Beucken, J.J.J.P.; Seibert, F.; et al. A lean magnesium–zinc–calcium alloy ZX00 used for bone fracture stabilization in a large growing-animal model. *Acta Biomater.* **2020**, *113*, 646–659. [CrossRef]
119. Boehlert, C.J.; Knittel, K. The microstructure, tensile properties, and creep behavior of Mg–Zn alloys containing 0–4.4 wt.% Zn. *Mater. Sci. Eng. A* **2006**, *417*, 315–321. [CrossRef]
120. Amukarimi, S.; Mozafari, M. Biodegradable magnesium-based biomaterials: An overview of challenges and opportunities. *MedComm* **2021**, *2*, 123–144. [CrossRef]
121. Zander, D.; Zumdick, N.A. Influence of Ca and Zn on the microstructure and corrosion of biodegradable Mg–Ca–Zn alloys. *Corros. Sci.* **2015**, *93*, 222–233. [CrossRef]
122. McKnight, R.F.; Adida, M.; Budge, K.; Stockton, S.; Goodwin, G.M.; Geddes, J.R. Lithium toxicity profile: A systematic review and meta-analysis. *Lancet* **2012**, *379*, 721–728. [CrossRef]

123. Edalati, K.; Bachmaier, A.; Beloshenko, V.A.; Beygelzimer, Y.; Blank, V.D.; Botta, W.J.; Bryła, K.; Čížek, J.; Divinski, S.; Enikeev, N.A.; et al. Nanomaterials by severe plastic deformation: Review of historical developments and recent advances. *Mater. Res. Lett.* **2022**, *10*, 163–256. [CrossRef]
124. Valiev, R.Z.; Langdon, T.G. Principles of equal-channel angular pressing as a processing tool for grain refinement. *Prog. Mater. Sci.* **2006**, *51*, 881–981. [CrossRef]
125. Langdon, T.G. Twenty-five years of ultrafine-grained materials: Achieving exceptional properties through grain refinement. *Acta Mater.* **2013**, *61*, 7035–7059. [CrossRef]
126. Yamashita, A.; Horita, Z.; Langdon, T.G. Improving the mechanical properties of magnesium and a magnesium alloy through severe plastic deformation. *Mater. Sci. Eng. A* **2001**, *300*, 142–147. [CrossRef]
127. Figueiredo, R.B.; Cetlin, P.R.; Langdon, T.G. The processing of difficult-to-work alloys by ECAP with an emphasis on magnesium alloys. *Acta Mater.* **2007**, *55*, 4769–4779. [CrossRef]
128. Kang, F.; Wang, J.T.; Peng, Y. Deformation and fracture during equal channel angular pressing of AZ31 magnesium alloy. *Mater. Sci. Eng. A* **2008**, *487*, 68–73. [CrossRef]
129. Xia, K.; Wang, J.T.; Wu, X.; Chen, G.; Gurvan, M. Equal channel angular pressing of magnesium alloy AZ31. *Mater. Sci. Eng. A* **2005**, *410–411*, 324–327. [CrossRef]
130. Poggiali, F.S.J.; Silva, C.L.P.; Pereira, P.H.R.; Figueiredo, R.B.; Cetlin, P.R. Determination of mechanical anisotropy of magnesium processed by ECAP. *J. Mater. Res. Technol.* **2014**, *3*, 331–337. [CrossRef]
131. Klu, E.E.; Song, D.; Li, C.; Wang, G.; Gao, B.; Ma, A.; Jiang, J. Achieving ultra-fine grains and high strength of Mg–9Li alloy via room-temperature ECAP and post rolling. *Mater. Sci. Eng. A* **2022**, *833*, 142371. [CrossRef]
132. Khani, S.; Aboutalebi, M.R.; Salehi, M.T.; Samim, H.R.; Palkowski, H. Microstructural development during equal channel angular pressing of as-cast AZ91 alloy. *Mater. Sci. Eng. A* **2016**, *678*, 44–56. [CrossRef]
133. Zhilyaev, A.P.; Nurislamova, G.V.; Kim, B.K.; Baró, M.D.; Szpunar, J.A.; Langdon, T.G. Experimental parameters influencing grain refinement and microstructural evolution during high-pressure torsion. *Acta Mater.* **2003**, *51*, 753–765. [CrossRef]
134. Xu, B.; Sun, J.; Yang, Z.; Xiao, L.; Zhou, H.; Han, J.; Liu, H.; Wu, Y.; Yuan, Y.; Zhuo, X.; et al. Microstructure and anisotropic mechanical behavior of the high-strength and ductility AZ91 Mg alloy processed by hot extrusion and multi-pass RD-ECAP. *Mater. Sci. Eng. A* **2020**, *780*, 139191. [CrossRef]
135. Miyamoto, H. Revealing What Enhance the Corrosion Resistance beside Grain Size in Ultrafine Grained Materials by Severe Plastic Deformation: Stainless Steels Case. *Mater. Trans.* **2023**, *64*, 1419–1428. [CrossRef]
136. Wang, P.J.; Ma, L.W.; Cheng, X.Q.; Li, X.G. Influence of grain refinement on the corrosion behavior of metallic materials: A review. *Int. J. Miner. Metall. Mater.* **2021**, *28*, 1112–1126. [CrossRef]
137. Wu, B.L.; Chen, B.; Wang, C.W.; Jiao, J.Y.; Shen, Q.C.; Zhou, T.T. Corrosion behavior of a novel Mg–13Li–X alloy with different grain sizes by rapid solidification rate. *Rare Met.* **2022**, *41*, 3197–3204. [CrossRef]
138. Miyamoto, H. Corrosion of Ultrafine Grained Materials by Severe Plastic Deformation, an Overview. *Mater. Trans.* **2016**, *57*, 559–572. [CrossRef]
139. Medeiros, M.P.; Lopes, D.R.; Kawasaki, M.; Langdon, T.G.; Figueiredo, R.B. An Overview on the Effect of Severe Plastic Deformation on the Performance of Magnesium for Biomedical Applications. *Materials* **2023**, *16*, 2401. [CrossRef]
140. Ramesh, S.; Anne, G.; Kumar, G.; Jagadeesh, C.; Nayaka, H.S. Influence of Ball Burnishing Process on Equal Channel Angular Pressed Mg-Zn-Si Alloy on the Evolution of Microstructure and Corrosion Properties. *Silicon* **2021**, *13*, 1549–1560. [CrossRef]
141. Figueiredo, R.B.; Langdon, T.G. Grain refinement and mechanical behavior of a magnesium alloy processed by ECAP. *J. Mater. Sci.* **2010**, *45*, 4827–4836. [CrossRef]
142. Bryła, K.; Horky, J. Magnesium Alloys Processed by Severe Plastic Deformation (SPD) for Biomedical Applications: An Overview. *Mater. Trans.* **2023**, *64*, 1709–1723. [CrossRef]
143. Sabbaghianrad, S.; Langdon, T.G. A critical evaluation of the processing of an aluminum 7075 alloy using a combination of ECAP and HPT. *Mater. Sci. Eng. A* **2014**, *596*, 52–58. [CrossRef]
144. Ji, G.; Xiaohui, Y.; Song, N.; Min, S. Structural and Hardness Evolution of Pure Magnesium Subjected to High Pressure Torsion. *Rare Met. Mater. Eng.* **2018**, *47*, 1347–1351. [CrossRef]
145. Yan, Z.; Zhu, J.; Zhang, Z.; Wang, Q.; Xue, Y. The microstructural, textural, and mechanical effects of high-pressure torsion processing on Mg alloys: A review. *Front. Mater.* **2022**, *9*, 964992. [CrossRef]
146. Hermawan, H.; Dubé, D.; Mantovani, D. Developments in metallic biodegradable stents. *Acta Biomater.* **2010**, *6*, 1693–1697. [CrossRef]
147. Witte, F.; Kaese, V.; Haferkamp, H.; Switzer, E.; Meyer-Lindenberg, A.; Wirth, C.J.; Windhagen, H. In vivo corrosion of four magnesium alloys and the associated bone response. *Biomaterials* **2005**, *26*, 3557–3563. [CrossRef] [PubMed]
148. Kuhlmann, J.; Bartsch, I.; Willbold, E.; Schuchardt, S.; Holz, O.; Hort, N.; Höche, D.; Heineman, W.R.; Witte, F. Fast escape of hydrogen from gas cavities around corroding magnesium implants. *Acta Biomater.* **2013**, *9*, 8714–8721. [CrossRef]

149. Yang, J.; Cui, F.; Lee, I.S. Surface Modifications of Magnesium Alloys for Biomedical Applications. *Ann. Biomed. Eng.* **2011**, *39*, 1857–1871. [CrossRef]
150. Ong, J.L.; Lucas, L.C. Auger electron spectroscopy and its use for the characterization of titanium and hydroxyapatite surfaces. *Biomaterials* **1998**, *19*, 455–464. [CrossRef]
151. Zhao, Q.; Zhai, G.-J.; Ng, D.H.L.; Zhang, X.-Z.; Chen, Z.-Q. Surface modification of Al_2O_3 bioceramic by NH^{+2} ion implantation. *Biomaterials* **1999**, *20*, 595–599. [CrossRef]
152. Mani, G.; Feldman, M.D.; Patel, D.; Agrawal, C.M. Coronary stents: A materials perspective. *Biomaterials* **2007**, *28*, 1689–1710. [CrossRef]
153. Jiang, H.; Wang, J.; Chen, M.; Liu, D. Biological activity evaluation of magnesium fluoride coated Mg-Zn-Zr alloy in vivo. *Mater. Sci. Eng. C* **2017**, *75*, 1068–1074. [CrossRef] [PubMed]
154. Shao, L.; Li, H.; Jiang, B.; Liu, C.; Gu, X.; Chen, D. A Comparative Study of Corrosion Behavior of Hard Anodized and Micro-Arc Oxidation Coatings on 7050 Aluminum Alloy. *Metals* **2018**, *8*, 165. [CrossRef]
155. Hiromoto, S.; Yamamoto, A. Control of degradation rate of bioabsorbable magnesium by anodization and steam treatment. *Mater. Sci. Eng. C* **2010**, *30*, 1085–1093. [CrossRef]
156. Wang, T.; Dong, Y.; Xu, Y.; Li, G.; Guo, Y.; Lian, J.; Zhang, Z.; Ren, L. A calcium phosphate coating improving corrosion resistance of the biodegradable magnesium alloy with graphene oxide modifying the deposition. *Ceram. Int.* **2023**, *49*, 11926–11935. [CrossRef]
157. Wandelt, K. *Encyclopedia of Interfacial Chemistry: Surface Science and Electrochemistry*; Elsevier: Amsterdam, The Netherlands, 2018.
158. Rahim, S.A.; Joseph, M.A.; Sampath Kumar, T.S.; Hanas, T. Recent Progress in Surface Modification of Mg Alloys for Biodegradable Orthopedic Applications. *Front. Mater.* **2022**, *9*, 848980. [CrossRef]
159. Chen, X.B.; Birbilis, N.; Abbott, T.B. Review of Corrosion-Resistant Conversion Coatings for Magnesium and Its Alloys. *Corrosion* **2011**, *67*, 035005-1–035005-16. [CrossRef]
160. Pesode, P.; Barve, S. Surface Modification of Biodegradable Zinc Alloy for Biomedical Applications. *BioNanoScience* **2023**, *13*, 1381–1398. [CrossRef]
161. Barajas, J.D.; Joya, J.C.; Durán, K.S.; Hernández-Barrios, C.A.; Coy, A.E.; Viejo, F. Relationship between microstructure and formation-biodegradation mechanism of fluoride conversion coatings synthesised on the AZ31 magnesium alloy. *Surf. Coat. Technol.* **2019**, *374*, 424–436. [CrossRef]
162. Bita, T.; Antoniac, A.; Ciuca, I.; Miculescu, M.; Cotrut, C.M.; Paltanea, G.; Dura, H.; Corneschi, I.; Antoniac, I.; Carstoc, I.D.; et al. Effect of Fluoride Coatings on the Corrosion Behavior of Mg–Zn–Ca–Mn Alloys for Medical Application. *Materials* **2023**, *16*, 4508. [CrossRef]
163. Lin, B.; Xu, Y.; Li, E. Effect of phosphating additives on corrosion resistance of phosphate coatings on AZ91D magnesium alloy. *Adv. Mater. Res.* **2011**, *337*, 112–115. [CrossRef]
164. Zhang, Z.; Yu, G.; Ouyang, Y.; He, X.; Hu, B.; Zhang, J.; Wu, Z. Studies on influence of zinc immersion and fluoride on nickel electroplating on magnesium alloy AZ91D. *Appl. Surf. Sci.* **2009**, *255*, 7773–7779. [CrossRef]
165. Liu, X.; Zhen, Z.; Liu, J.; Xi, T.; Zheng, Y.; Guan, S.; Zheng, Y.; Cheng, Y. Multifunctional MgF_2/Polydopamine Coating on Mg Alloy for Vascular Stent Application. *J. Mater. Sci. Technol.* **2015**, *31*, 733–743. [CrossRef]
166. Chen, Y.; Song, Y.; Zhang, S.; Li, J.; Wang, H.; Zhao, C.; Zhang, X. Effect of fluoride coating on in vitro dynamic degradation of Mg–Zn alloy. *Mater. Lett.* **2011**, *65*, 2568–2571. [CrossRef]
167. Sun, J.; Jin, S.; Zhao, B.C.; Cai, H.; Sun, L.; Xuan, X.; Oh, J.-R.; Jiang, H.B. Enhanced corrosion resistance of biodegradable Mg alloys via ultrasonically treated fluoride coating. *Surf. Topogr. Metrol. Prop.* **2019**, *7*, 025009. [CrossRef]
168. Zhan, Y.; Wu, J.; Xia, H.; Yan, N.; Fei, G.; Yuan, G. Dispersion and Exfoliation of Graphene in Rubber by an Ultrasonically-Assisted Latex Mixing and In situ Reduction Process. *Macromol. Mater. Eng.* **2011**, *296*, 590–602. [CrossRef]
169. Ma, F.-f.; Zhang, D.; Huang, T.; Zhang, N.; Wang, Y. Ultrasonication-assisted deposition of graphene oxide on electrospun poly(vinylidene fluoride) membrane and the adsorption behavior. *Chem. Eng. J.* **2019**, *358*, 1065–1073. [CrossRef]
170. Li, S.; Yi, L.; Zhu, X.; Liu, T. Ultrasonic Treatment Induced Fluoride Conversion Coating without Pores for High Corrosion Resistance of Mg Alloy. *Coatings* **2020**, *10*, 996. [CrossRef]
171. Makar, G.L.; Kruger, J. Corrosion of magnesium. *Int. Mater. Rev.* **1993**, *38*, 138–153. [CrossRef]
172. Blawert, C.; Dietzel, W.; Ghali, E.; Song, G. Anodizing Treatments for Magnesium Alloys and Their Effect on Corrosion Resistance in Various Environments. *Adv. Eng. Mater.* **2006**, *8*, 511–533. [CrossRef]
173. Mahmud, S.; Rahman, M.M.; Khatun, H.; Ali, M.O.; Ali, M.R.; Rahaman, M.; Islam, S.; Ali, Y. Ceramic coating on Mg alloy for enhanced degradation resistance as implant material. *Surf. Coat. Technol.* **2024**, *479*, 130544. [CrossRef]
174. Lin, E.; Li, X.; Kure-Chu, S.-Z.; Li, X.; Xiao, X. Effect of Electrical Parameters on the Microstructure and Corrosion Resistance of Anodized Film of Mg-1Zn-1Gd Alloy Based on Orthogonal Experiment Method. *J. Mater. Eng. Perform.* **2024**, *33*, 5049–5060. [CrossRef]

175. Hussain, F.; Manzoor, M.U.; Kamran, M.; Ahmad, T.; Riaz, F.; Mukhtar, S.; Tariq, H.M.R.; Ishtiaq, M. Optimizing Biocompatibility of Mg-AZ31B Alloy through Varied Surface Roughness and Anodization Time. *Iran. J. Mater. Sci. Eng.* **2024**, *21*. [CrossRef]
176. Dai, C.Y.; Gao, X.; Zhai, C.; Jia, Q.; Zhao, B.C.; Shi, H.; Gao, Q.; Cai, H.; Lee, E.-S.; Jiang, H.B. Corrosion Evaluation of Pure Mg Coated by Fluorination in 0.1 M Fluoride Electrolyte. *Scanning* **2021**, *2021*, 5574946. [CrossRef]
177. Si, Y.; Xiong, Z.; Zheng, X.; Li, M.; Yang, Q. Improving the Anti-Corrosion Ability of Anodization Film of AZ31B Magnesium Alloy by Addition of NH_4VO_3 in the Electrolyte. *Int. J. Electrochem. Sci.* **2016**, *11*, 3261–3268. [CrossRef]
178. Setiawan, A.R.; Rachman, M.D. An investigation of pulse anodization duty ratio and sealing treatment on the corrosion behavior of the anodic coating layer in magnesium AZ31B. *Corros. Sci. Technol.* **2021**, *20*, 45–51. [CrossRef]
179. Wang, J.; Tang, J.; Zhang, P.; Li, Y.; Wang, J.; Lai, Y.; Qin, L. Surface modification of magnesium alloys developed for bioabsorbable orthopedic implants: A general review. *J. Biomed. Mater. Res. Part B Appl. Biomater.* **2012**, *100B*, 1691–1701. [CrossRef] [PubMed]
180. Lkhagvaa, T.; Rehman, Z.U.; Choi, D. Post-anodization methods for improved anticorrosion properties: A review. *J. Coat. Technol. Res.* **2021**, *18*, 1–17. [CrossRef]
181. Fitriana, M.F.; Anawati, A. Remarkable improvement in corrosion resistance of anodized AZ31 alloy by sealing with beeswax-colophony resin. *J. Phys. Conf. Ser.* **2019**, *1191*, 012032. [CrossRef]
182. Krishna, L.R.; Purnima, A.S.; Sundararajan, G. A comparative study of tribological behavior of microarc oxidation and hard-anodized coatings. *Wear* **2006**, *261*, 1095–1101. [CrossRef]
183. Clyne, T.W.; Troughton, S.C. A review of recent work on discharge characteristics during plasma electrolytic oxidation of various metals. *Int. Mater. Rev.* **2019**, *64*, 127–162. [CrossRef]
184. Xue, W.; Deng, Z.; Lai, Y.; Chen, R. Analysis of Phase Distribution for Ceramic Coatings Formed by Microarc Oxidation on Aluminum Alloy. *J. Am. Ceram. Soc.* **1998**, *81*, 1365–1368. [CrossRef]
185. Yerokhin, A.L.; Nie, X.; Leyland, A.; Matthews, A.; Dowey, S.J. Plasma electrolysis for surface engineering. *Surf. Coat. Technol.* **1999**, *122*, 73–93. [CrossRef]
186. Hussein, R.O.; Zhang, P.; Nie, X.; Xia, Y.; Northwood, D.O. The effect of current mode and discharge type on the corrosion resistance of plasma electrolytic oxidation (PEO) coated magnesium alloy AJ62. *Surf. Coat. Technol.* **2011**, *206*, 1990–1997. [CrossRef]
187. Hussein, R.; Northwood, D.; Nie, X. Processing-microstructure relationships in the plasma electrolytic oxidation (PEO) coating of a magnesium alloy. *Mater. Sci. Appl.* **2014**, *5*, 124–139. [CrossRef]
188. Lugovskoy, A.; Zinigrad, M. Plasma electrolytic oxidation of valve metals. *Mater. Sci. Adv. Top.* **2013**, *1*, 85–102.
189. Franco, M.; Hari Krishna, T.; Pillai, A.M.; Rajendra, A.; Sharma, A.K. A comparative study on the corrosion behaviour of hard anodic coatings on AA 6061 obtained using DC and pulsed DC power sources. *Acta Metall. Sin. (Engl. Lett.)* **2013**, *26*, 647–656. [CrossRef]
190. Zuo, Y.; Zhao, P.-H.; Zhao, J.-M. The influences of sealing methods on corrosion behavior of anodized aluminum alloys in NaCl solutions. *Surf. Coat. Technol.* **2003**, *166*, 237–242. [CrossRef]
191. Shen, X.; Nie, X.; Hu, H.; Tjong, J. Effects of coating thickness on thermal conductivities of alumina coatings and alumina/aluminum hybrid materials prepared using plasma electrolytic oxidation. *Surf. Coat. Technol.* **2012**, *207*, 96–101. [CrossRef]
192. Dicu, M.; Matei, A.; Abrudeanu, M.; Ducu, C. Synthesis and properties of the porous litania coatings formed on titanium by plasma electrolytic oxidation for biomedical application. *J. Optoelectron. Adv. Mater.* **2011**, *13*, 318.
193. Whiteside, P.; Matykina, E.; Gough, J.E.; Skeldon, P.; Thompson, G.E. In vitro evaluation of cell proliferation and collagen synthesis on titanium following plasma electrolytic oxidation. *J. Biomed. Mater. Res. Part A* **2010**, *94A*, 38–46. [CrossRef]
194. Hernández, L.; González, J.E.; Barranco, V.; Veranes-Pantoja, Y.; Galván, J.C.; Gattorno, G.R. Biomimetic hydroxyapatite (HAp) coatings on pure Mg and their physiological corrosion behavior. *Ceram. Int.* **2022**, *48*, 1208–1222. [CrossRef]
195. Cao, L.; Wang, L.; Fan, L.; Xiao, W.; Lin, B.; Xu, Y.; Liang, J.; Cao, B. RGDC Peptide-Induced Biomimetic Calcium Phosphate Coating Formed on AZ31 Magnesium Alloy. *Materials* **2017**, *10*, 358. [CrossRef]
196. Lin, B.; Zhong, M.; Zheng, C.; Cao, L.; Wang, D.; Wang, L.; Liang, J.; Cao, B. Preparation and characterization of dopamine-induced biomimetic hydroxyapatite coatings on the AZ31 magnesium alloy. *Surf. Coat. Technol.* **2015**, *281*, 82–88. [CrossRef]
197. Gao, F.; Xu, C.; Hu, H.; Wang, Q.; Gao, Y.; Chen, H.; Guo, Q.; Chen, D.; Eder, D. Biomimetic synthesis and characterization of hydroxyapatite/graphene oxide hybrid coating on Mg alloy with enhanced corrosion resistance. *Mater. Lett.* **2015**, *138*, 25–28. [CrossRef]
198. Kreller, T.; Sahm, F.; Bader, R.; Boccaccini, A.R.; Jonitz-Heincke, A.; Detsch, R. Biomimetic calcium phosphate coatings for bioactivation of titanium implant surfaces: Methodological approach and in vitro evaluation of biocompatibility. *Materials* **2021**, *14*, 3516. [CrossRef]
199. Surmenev, R.A.; Surmeneva, M.A.; Ivanova, A.A. Significance of calcium phosphate coatings for the enhancement of new bone osteogenesis—A review. *Acta Biomater.* **2014**, *10*, 557–579. [CrossRef]

200. Aghamohammadi, H.; Rabiee, S.M.; Hosseinipour, S.J.; Jamaati, R. Effect of microstructure and texture of AZ31 magnesium alloy substrate on nucleation and growth of biomimetic calcium phosphate coating. *J. Mater. Res. Technol.* **2023**, *27*, 5154–5164. [CrossRef]
201. Chu, J.H.; Tong, L.B.; Wang, W.; Jiang, Z.H.; Sun, G.X.; Zou, D.N.; Wang, K.S.; Zhang, H.J. Sequentially bridged biomimetic graphene-based coating via covalent bonding with an effective anti-corrosion/wear protection for Mg alloy. *Colloids Surf. A: Physicochem. Eng. Asp.* **2021**, *610*, 125707. [CrossRef]
202. Cheng, Y.; Peng, J.; Xu, H.; Cheng, Q. Glycera-Inspired Synergistic Interfacial Interactions for Constructing Ultrastrong Graphene-Based Nanocomposites. *Adv. Funct. Mater.* **2018**, *28*, 1800924. [CrossRef]
203. Ahmadi, A.; Ramezanzadeh, B.; Mahdavian, M. Hybrid silane coating reinforced with silanized graphene oxide nanosheets with improved corrosion protective performance. *RSC Adv.* **2016**, *6*, 54102–54112. [CrossRef]
204. Schriver, M.; Regan, W.; Gannett, W.J.; Zaniewski, A.M.; Crommie, M.F.; Zettl, A. Graphene as a Long-Term Metal Oxidation Barrier: Worse Than Nothing. *ACS Nano* **2013**, *7*, 5763–5768. [CrossRef]
205. Fang, T.; Li, W.; Tao, N.; Lu, K. Revealing extraordinary intrinsic tensile plasticity in gradient nano-grained copper. *Science* **2011**, *331*, 1587–1590. [CrossRef] [PubMed]
206. Ye, C.; Zhou, X.; Telang, A.; Gao, H.; Ren, Z.; Qin, H.; Suslov, S.; Gill, A.S.; Mannava, S.R.; Qian, D.; et al. Surface amorphization of NiTi alloy induced by Ultrasonic Nanocrystal Surface Modification for improved mechanical properties. *J. Mech. Behav. Biomed. Mater.* **2016**, *53*, 455–462. [CrossRef] [PubMed]
207. Gao, M.; Etim, I.P.; Yang, K.; Tan, L.; Ma, Z. Enhancing mechanical property and corrosion resistance of Mg–Zn-Nd alloy wire by a combination of SPD techniques, extrusion and hot drawing. *Mater. Sci. Eng. A* **2022**, *829*, 142058. [CrossRef]
208. Kajánek, D.; Hadzima, B.; Pastorek, F.; Neslušan Jacková, M. Corrosion performance of AZ31 magnesium alloy treated by ultrasonic impact peening (UIP). *Mater. Today Proc.* **2018**, *5*, 26687–26692. [CrossRef]
209. Fabijanic, D.; Taylor, A.; Ralston, K.D.; Zhang, M.X.; Birbilis, N. Influence of Surface Mechanical Attrition Treatment Attrition Media on the Surface Contamination and Corrosion of Magnesium. *Corrosion* **2012**, *69*, 527–535. [CrossRef]
210. Wen, L.; Wang, Y.; Jin, Y.; Ren, X. Comparison of corrosion behaviour of nanocrystalline 2024-T4 Al alloy processed by surface mechanical attrition treatment with two different mediums. *Corros. Eng. Sci. Technol.* **2015**, *50*, 425–432. [CrossRef]
211. Kheradmandfard, M.; Kashani-Bozorg, S.F.; Lee, J.S.; Kim, C.-L.; Hanzaki, A.Z.; Pyun, Y.-S.; Cho, S.-W.; Amanov, A.; Kim, D.-E. Significant improvement in cell adhesion and wear resistance of biomedical β-type titanium alloy through ultrasonic nanocrystal surface modification. *J. Alloys Compd.* **2018**, *762*, 941–949. [CrossRef]
212. Kumbhar, N.K.; Yamamoto, A.; Wadhonkar, K.; Baig, M.S.; Hosmani, S.S. The effect of severe surface deformation on microstructure refinement, corrosion, and biocompatibility of Mg5Zn0.2Ca alloy. *J. Alloys Compd.* **2025**, *1010*, 178259. [CrossRef]
213. Biffi, C.A.; Bassani, P.; Nematollahi, M.; Shayesteh Moghaddam, N.; Amerinatanzi, A.; Mahtabi, M.J.; Elahinia, M.; Tuissi, A. Effect of Ultrasonic Nanocrystal Surface Modification on the Microstructure and Martensitic Transformation of Selective Laser Melted Nitinol. *Materials* **2019**, *12*, 3068. [CrossRef] [PubMed]
214. Wang, Q.; Li, Y.; Lu, Z.; Zhang, Y.; Zou, Y. Effects of Ultrasonic Nanocrystal Surface Modification on Mechanical and Corrosion Behavior of LZ91 Mg–Li Alloy. *Mater. Trans.* **2020**, *61*, 1258–1264. [CrossRef]
215. Zheng, Z.J.; Gao, Y.; Gui, Y.; Zhu, M. Corrosion behaviour of nanocrystalline 304 stainless steel prepared by equal channel angular pressing. *Corros. Sci.* **2012**, *54*, 60–67. [CrossRef]
216. Olumi, S.; Sadrnezhaad, S.K.; Atai, M. Correction to: The Influence of Surface Nanocrystallization Induced by Shot Peening on Corrosion Behavior of NiTi Alloy. *J. Mater. Eng. Perform.* **2023**, *32*, 3368. [CrossRef]
217. Baek, S.M.; Choi, I.Y.; Moon, J.H.; Rakhmatjon, U.; Kim, J.H.; Kim, N.J.; Kim, J.K.; Amanov, A.; Kim, H.S. Control of magnesium in vitro degradation based on ultrafine-grained surface gradient structure using ultrasonic nanocrystalline surface modification. *Materialia* **2020**, *12*, 100821. [CrossRef]
218. Hou, X.; Qin, H.; Gao, H.; Mankoci, S.; Zhang, R.; Zhou, X.; Ren, Z.; Doll, G.L.; Martini, A.; Sahai, N.; et al. A systematic study of mechanical properties, corrosion behavior and biocompatibility of AZ31B Mg alloy after ultrasonic nanocrystal surface modification. *Mater. Sci. Eng. C* **2017**, *78*, 1061–1071. [CrossRef]
219. Huang, B.; Ye, Y.; Wang, K.; Zhang, L.; Dong, Y.; Ye, C. Corrosion damage repair of 7075-T6 aluminum alloy by ultrasonic nanocrystal surface modification. *Surf. Coat. Technol.* **2023**, *474*, 130085. [CrossRef]
220. Sun, P.F.; Qu, C.X.; Duan, C.F.; Li, F.L.; Li, X.Q.; Qu, S.G. Explore the nature of gradient deformation layer constructed by high-density electric pulse assisted ultrasonic nanocrystalline surface modification for Ti6Al4V alloy. *J. Alloys Compd.* **2025**, *1010*, 177144. [CrossRef]
221. Zeng, W.; Shen, Y.; Zhang, N.; Huang, X.; Wang, J.; Tang, G.; Shan, A. Rapid hardening induced by electric pulse annealing in nanostructured pure aluminum. *Scr. Mater.* **2012**, *66*, 147–150. [CrossRef]
222. Grimm, T.J.; Mears, L.M. Skin effects in electrically assisted manufacturing. *Manuf. Lett.* **2022**, *34*, 67–70. [CrossRef]
223. Olugbade, T.O.; Olutomilola, E.O.; Olorunfemi, B.J. Review of passivity and electrochemical properties of nanostructured stainless steels obtained by surface mechanical attrition treatment (SMAT): Trend and progress. *Corros. Rev.* **2022**, *40*, 189–203. [CrossRef]

224. Laleh, M.; Kargar, F. Effect of surface nanocrystallization on the microstructural and corrosion characteristics of AZ91D magnesium alloy. *J. Alloys Compd.* **2011**, *509*, 9150–9156. [CrossRef]
225. Skowron, K.; Dryzek, E.; Wróbel, M.; Nowak, P.; Marciszko-Wiąckowska, M.; Le Joncour, L.; François, M.; Panicaud, B.; Baczmański, A. Gradient Microstructure Induced by Surface Mechanical Attrition Treatment (SMAT) in Magnesium Studied Using Positron Annihilation Spectroscopy and Complementary Methods. *Materials* **2020**, *13*, 4002. [CrossRef]
226. Singh, S.; Keshri, A.K. Improving Corrosion Resistance of Low Carbon Steel through Surface Mechanical Attrition Treatment Duration. *J. Mater. Eng. Perform.* **2024**. [CrossRef]
227. Azadmanjiri, J.; Berndt, C.C.; Kapoor, A.; Wen, C. Development of Surface Nano-Crystallization in Alloys by Surface Mechanical Attrition Treatment (SMAT). *Crit. Rev. Solid State Mater. Sci.* **2015**, *40*, 164–181. [CrossRef]
228. Gallitelli, D.; Retraint, D.; Rouhaud, E. Comparison between conventional shot peening (SP) and surface mechanical attrition treatment (SMAT) on a titanium alloy. *Adv. Mater. Res.* **2014**, *996*, 964–968. [CrossRef]
229. Zhou, J.; Retraint, D.; Sun, Z.; Kanouté, P. Comparative study of the effects of surface mechanical attrition treatment and conventional shot peening on low cycle fatigue of a 316L stainless steel. *Surf. Coat. Technol.* **2018**, *349*, 556–566. [CrossRef]
230. Peral, L.B.; Zafra, A.; Bagherifard, S.; Guagliano, M.; Fernández-Pariente, I. Effect of warm shot peening treatments on surface properties and corrosion behavior of AZ31 magnesium alloy. *Surf. Coat. Technol.* **2020**, *401*, 126285. [CrossRef]
231. Bagherifard, S.; Hickey, D.J.; Fintová, S.; Pastorek, F.; Fernandez-Pariente, I.; Bandini, M.; Webster, T.J.; Guagliano, M. Effects of nanofeatures induced by severe shot peening (SSP) on mechanical, corrosion and cytocompatibility properties of magnesium alloy AZ31. *Acta Biomater.* **2018**, *66*, 93–108. [CrossRef]
232. Kondi Maliwemu, E.U.; Malau, V.; Iswanto, P.T. Effect of Shot Peening in Different Shot Distance and Shot Angle on Surface Morphology, Surface Roughness and Surface Hardness of 316L Biomaterial. *IOP Conf. Ser. Mater. Sci. Eng.* **2018**, *299*, 012051. [CrossRef]
233. Kondi Maliwemu, E.U.; Malau, V.; Tri Iswanto, P. Corrosion Resistance of 316L Biomaterial in Simulated Body Fluid by Modification of Shot Distance and Shot Angle of Shot Peening. *IOP Conf. Ser. Mater. Sci. Eng.* **2019**, *553*, 012053. [CrossRef]
234. Khany, S.E.; Moyeed, M.A.; Siddiqui, M.S.; Ahmed, G.M.S.; Baig, M.M.A. An Experimental Study of the Effect of Shot Peening on the Low Carbon Steel and Identification of Optimal Process Parameters. *Mater. Today Proc.* **2015**, *2*, 3363–3370. [CrossRef]
235. Wei, X.; Ma, S.; Meng, J.; Qing, H.; Zhao, Q. In vitro evaluation of a novel Mg–Sn–Ge ternary alloy for orthopedic applications. *J. Alloys Compd.* **2023**, *953*, 169813. [CrossRef]
236. Renuga Devi, K.; Dondapati, S. A novel multicomponent micro-alloyed magnesium for orthopaedic applications: A microstructural, mechanical, degradation and bioactivity study. *Mater. Technol.* **2024**, *39*, 2428902. [CrossRef]

confirm that they exhibit good quality. Jinkeum [21] et al. studied fatigue properties and microstructure characteristics of TC4 joints welded by electron beam and found that pores exist at the weld seam of the joint. Stress concentration in the pores causes the pores to become sources for initiation of fatigue cracks, which reduces the fatigue strength of the joint. Emanuele Vincenzo Arcieri [22] et al. discussed the stress concentration at the notch of defective and defect-free TC4 titanium alloy specimens under axial fatigue loading using finite element analysis. The results revealed that the fatigue strength of the notched specimens was significantly lower than that of the smooth specimens. Md Mehide Hasan Tusher [23] et al. employed selective laser melting (SLM) to weld TC4 titanium alloy and systematically studied its fatigue behavior under high-cycle fatigue (HCF), and even very high-cycle fatigue (VHCF), conditions. They analyzed the effects of microstructure, defects, and process parameters on the fatigue performance.

Moreover, titanium alloy structural components may experience sudden overloads or random loads during service, which can reduce their fatigue life and accelerate structural failure. It is crucial to understand the impact of these sudden overloads or random loads on the fatigue behavior of titanium alloys. The highlight of this study is the use of different pre-deformation levels to simulate the sudden loading experienced by titanium alloy joints. Dissimilar welded joints of TC4 and Ti60 alloys are commonly used in components and structural parts of aero engines. Therefore, in this study, we conduct experiments on TC4/Ti60 dissimilar titanium alloy joints obtained through electron beam welding, examining their microstructural characteristics, tensile properties, and fatigue behavior. The results show that fatigue life decreases with increasing pre-deformation.

2. Materials and Methods

The experimental materials were forged TC4 and Ti60 alloys, with their chemical compositions shown in Table 1. First, the TC4 and Ti60 were machined into plates with dimensions of 140 mm length, 80 mm width, and 10 mm thickness. Before welding, the surface was cleaned to remove oil and impurities. Welding was carried out using a TECHMETA LARA 52 electron beam welding machine. The specific welding parameters are shown in Table 2. A total of 10 samples were welded in this study, with four pre-deformation conditions of 0 mm, 0.5 mm, 1 mm, and 1.5 mm, and two samples for each condition. The remaining two samples were used for tensile testing.

Table 1. TC4 and Ti60 chemical composition (w/%).

Alloy	Al	Sn	Zr	Mo	Cr	Fe	Nd	Ta	Si	O	C	N	Ti
TC4	5	2	2	4	4	0.3	-	-	-	0.15	0.05	0.04	Bal.
Ti60	5.13	3.97	1.98	0.78	-	-	1	0.3	0.3	-	0.09	-	Bal.

Table 2. Electron beam welding process parameters.

Accelerating Voltage (V)	Beam Current (I_b)	Focus Current (I_f)	Welding Speed (v)
60 kV	68 mA	2230 mA	500 mm/min

After electron beam welding, metallographic samples, tensile specimens, and fatigue specimens were cut using a wire-cutting device perpendicular to the welding direction. The weld seam was located in the center of these samples. The metallographic specimens were ground using sandpaper from 180 to 2000 grit, polished, and etched with Keller's reagent (12 mL HF, 36 mL HNO_3, and 42 mL H_2O). The microstructure observation is conducted using the VHX-600E optical microscope (OM) produced by KEYENCE Corporation, Japan, and the JSM 6380 LV scanning electron microscope (SEM) produced by JEOL Ltd., Japan. EBSD observation of the samples is performed using a TESCAN MIRA3 XMU field emission scanning electron microscope equipped with an HKL NordlysMax EBSD detector. Optical microscopy (OM) was used to observe the microstructure of materials, such as grain morphology, phase structure, and interfaces. The scanning electron microscope (SEM) allows for surface morphology and composition analysis, with a higher resolution than OM, typically reaching the nanoscale. Electron backscatter diffraction (EBSD) is a technique based on SEM that provides information on the crystal orientation, grain boundaries, and phase composition of materials. EBSD is particularly useful for grain orientation and texture analysis.

According to relevant standards, tensile specimens with a length of 140 mm and thickness of 1 mm, and fatigue specimens with a length of 110 mm and width of 3 mm were prepared. The final specimen dimensions are shown in Figures 1 and 2. To eliminate the influence of machining marks on the fatigue test, the specimens were ground and mechanically polished along the loading axis to achieve a smooth surface. Tensile tests were conducted at room temperature using a United tensile testing machine at a strain rate of 1×10^{-2} s^{-1}. Before the fatigue test, the sample was pre-stretched from 0 mm to 0.5 mm, 1 mm and 1.5 mm. Then, strained-controlled fatigue tests were performed at room temperature using an Instron 8801 fatigue testing system, with a strain amplitude of 0.8%. The waveform was triangular, with a strain ratio of R = −1. A constant strain rate of 1×10^{-2} s^{-1} was applied, where R is the ratio of minimum to maximum strain in strain-controlled fatigue tests. The surfaces of fatigue fractures were examined using SEM to investigate the crack initiation sites and crack propagation mechanisms.

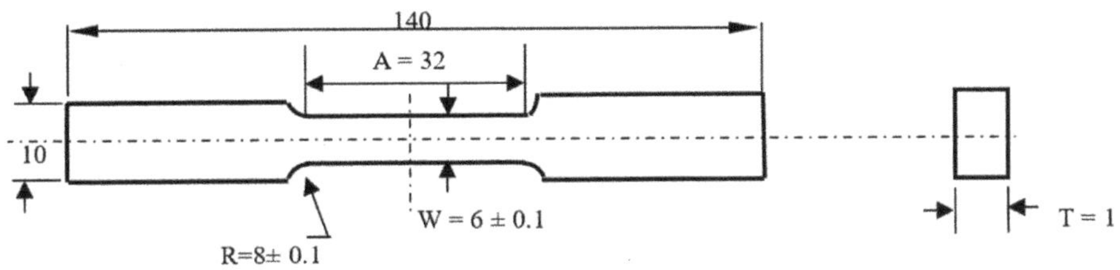

Figure 1. Specimen configurations of TC4/Ti60 for tensile test (mm).

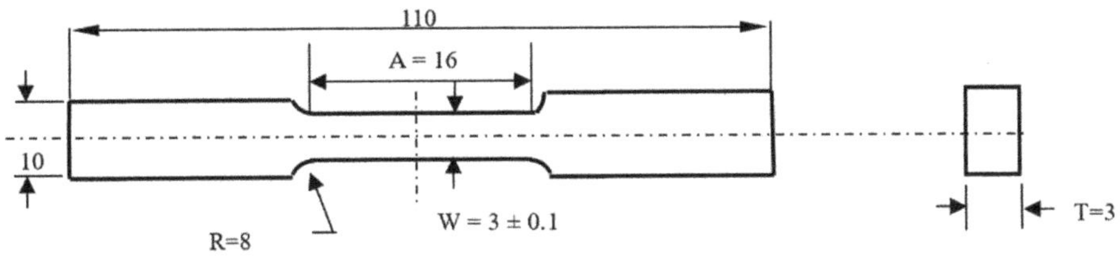

Figure 2. Specimen configurations of TC4/Ti60 for fatigue test (mm).

3. Results and Discussion

3.1. Microstructure

Figure 3 shows the microstructure of the TC4/Ti60 joint observed through scanning electron microscopy (SEM) in different regions. The base materials of TC4 and Ti60 alloys exhibit similar microstructures, primarily consisting of a large amount of α phase and a small amount of β phase, with fine equiaxed morphologies, as depicted in Figure 3a,c. Figure 3b,d present the microstructure derived from EBSD, clearly indicating that the grain size of TC4 is relatively uniform, while the grain size of Ti60 is notably non-uniform with significant size variation. Specifically, the average grain size of TC4 is approximately 4 μm, whereas that of Ti60 is around 13 μm. Furthermore, the equiaxed α phase in Ti60 is significantly larger than that in TC4.

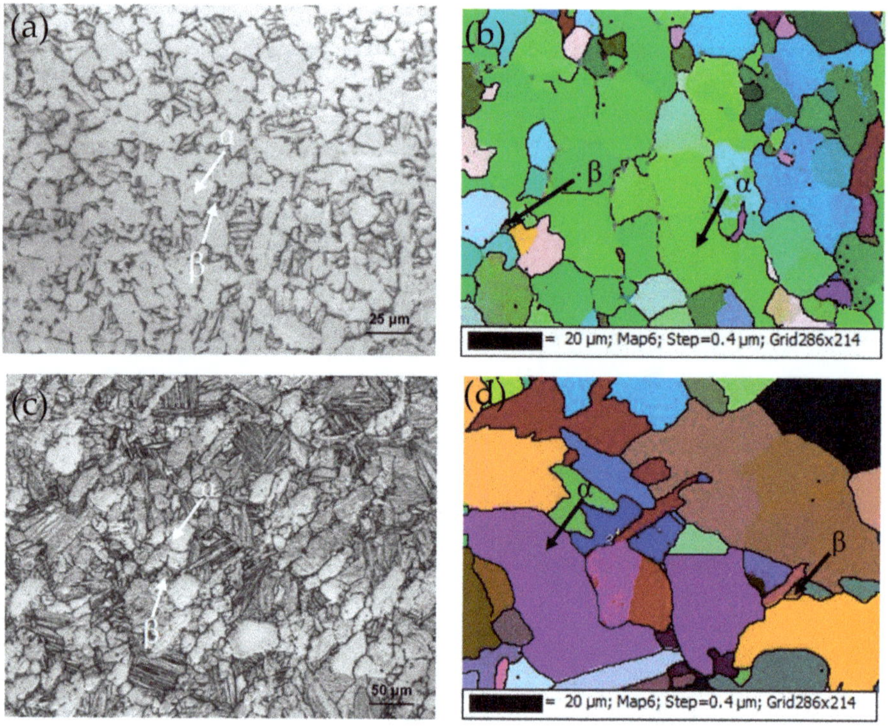

Figure 3. Microstructural characterization obtained by OM and EBSD. (**a**,**b**) TC4 titanium alloy, (**c**,**d**) Ti60 titanium alloy.

Figure 4a,b show the microstructures of the far heat-affected zone (closer to the Ti60 side) and the near heat-affected zone (closer to the weld zone), respectively. It can be observed that most of the α and β phase is transformed into needle-like α' phase at the near HAZ. While the inter-granular α and β lamellae are more readily transformed into α' phase at the far HAZ. Therefore, the far HAZ mainly consisted of α' phase and residual α phase. That is, as the distance from the weld seam increased, α' phase decreased.

Figure 4c shows the weld zone and the heat-affected zone of TC4. It can be clearly seen that there are obvious differences in microstructure between the weld zone and the heat-affected zone. The microstructure at the weld zone is relatively uniform, and composed of coarse α' phase, while α' phase in the heat-affected zone of TC4 side (Figure 4e) is slightly smaller than that in the weld zone (Figure 4d).

Figure 4e,f show the microstructures of the near heat-affected zone (closer to the weld zone) and the far heat-affected zone (closer to the TC4 side) respectively. It can be observed that small amounts of α' phase and residual equiaxed α phase appear on the far heat-affected zone of TC4 side. As the distance from the weld zone increases in the heat-affected zone, the amount of α' phase gradually decreases. The closer to the weld, the more α' phase appears.

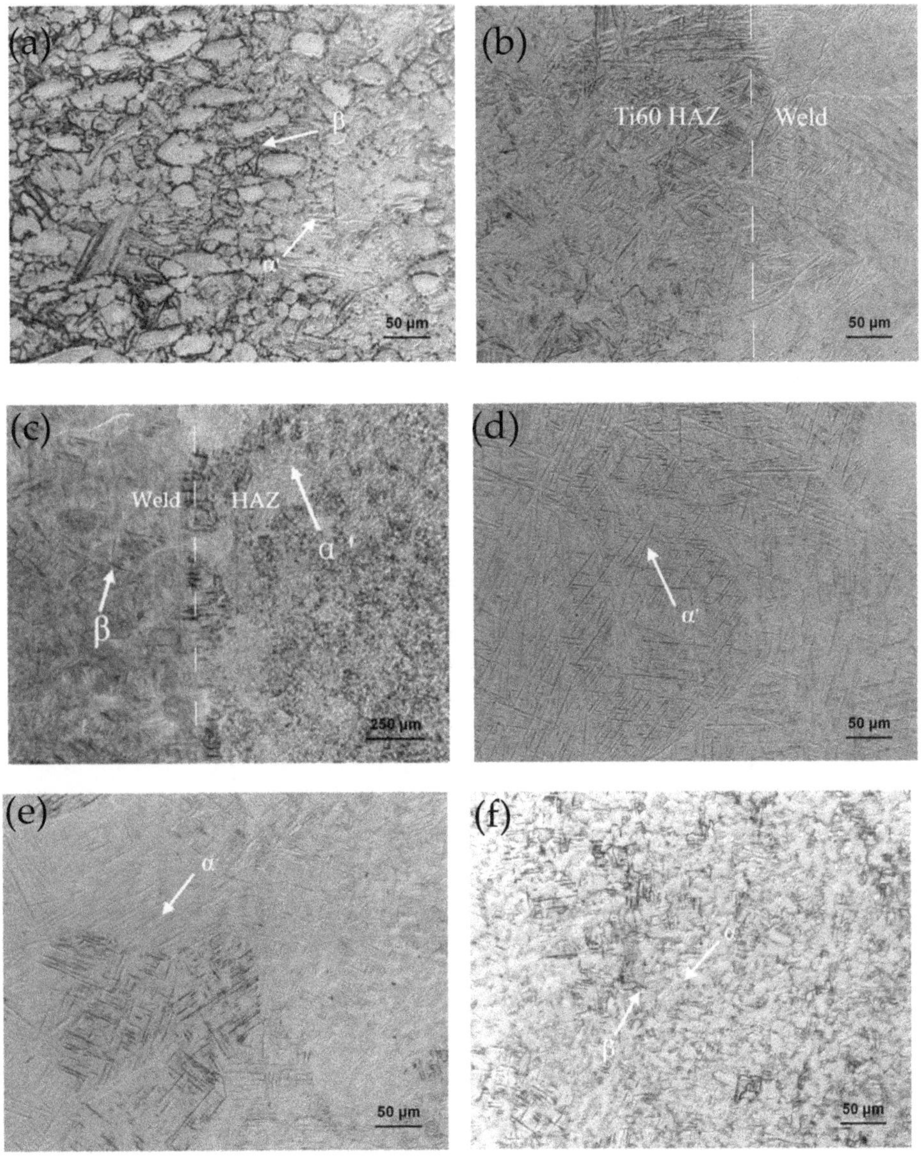

Figure 4. Microstructure of the joint between TC4 and Ti60 alloys. (**a**) TC4 far-HAZ, (**b**) TC4 near HAZ, (**c**) fusion line between WZ and TC4 HAZ, (**d**) WZ, (**e**) Ti60 near HAZ, (**f**) Ti60 far-HAZ.

The microstructure of the heat-affected zone on both sides is composed of mixed microstructure, which is caused by the welding thermal cycle. During the welding process, the heating temperature at the heat-affected zone is not enough to reach the β transition temperature of Ti60, especially the equiaxial α phase, so the equiaxial α phase is not completely transformed into the β phase, while the inter-granular α and β lamellae are small, and there are many nucleation sites; thus, it can change into α' phase easily at the HAZ.

3.2. Tensile Properties

Figure 5 shows the typical tensile stress–strain curve of the TC4/Ti60 joint at room temperature with a strain rate of 1×10^{-2} s^{-1}. As shown in the figure, the joint exhibits a continuous and smooth tensile stress–strain curve. As shown in Figure 5, the yield strength and tensile strength of the dissimilar joint are approximately 847 MPa and 895 MPa, respectively. Tensile test data indicate that the strength of the joint is slightly higher than that of the TC4 base metal. Based on the analysis of Figure 4, it can be observed that the primary microstructure in the weld is martensite, and some martensitic structures are also present in the heat-affected zones on both the TC4 and Ti60 sides. The hardness of the weld and the heat-affected zones on both sides is higher than that of the base material. Additionally, the Ti60 base material contains many hard particles, which increases its strength compared to the TC4 base material. Therefore, fractures occur on the TC4 side. A similar phenomenon was also observed by N. Farabi [24] and T.J. Ma [25] et al. in their study on the tensile properties of dissimilar joints. At an engineering strain of 0.8%, the joint is in the transition between elastic and plastic deformation, providing a reference for the subsequent fatigue tests.

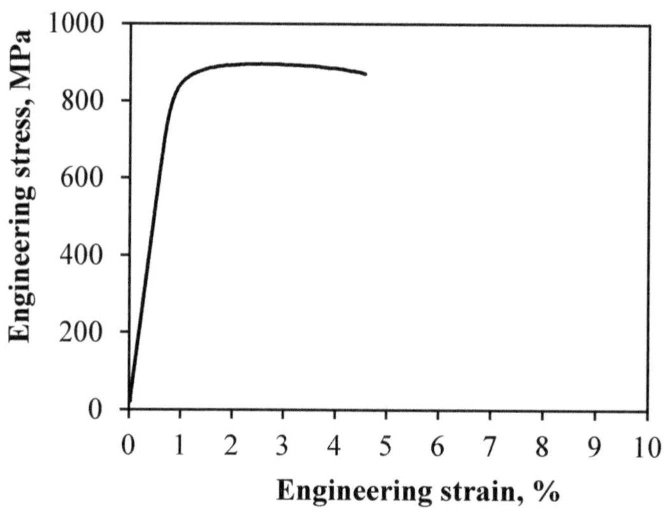

Figure 5. Tensile test results of the specimens.

3.3. Fatigue Properties

Figure 6 shows the hysteresis loops of the TC4/Ti60 joint under different pre-deformation. Figure 6a presents the hysteresis loops of the first cycle, which is asymmetric. In addition, with increasing the pre-deformation from 0 mm to 1.5 mm, the maximum/peak tensile stress of the first cycle increased from nearly zero or a few MPa to about 785, 818, 850, and 873 MPa, respectively, while the compressive stress of the first cycle decreased from 791 MPa to 772, 657, and 570 MPa, respectively. It can be seen that the maximum/peak tensile stress of the first cycle with the pre-deformation is slightly higher than that without pre-deformation. At the same time, with increases in pre-deformation, the maximum/peak tensile stress of the first cycle increases and the compressive stress of the first cycle decreases, which indicates that Bauschinger effect appears. The compressive stress (Figure 6b) is still obviously lower than the maximum/peak tensile stress at half-life ($N_f/2$). At the pre-deformation of 0.5 mm, the maximum/peak tensile stress of the first cycle is slightly lower than that of the second and third cycles, which demonstrates that cyclic hardening occurs at the initial stage of fatigue the test. However, during the fatigue test, the maximum/peak tensile stress of the first cycle was slightly higher than that of the mid-life, illustrating the appearance of cyclic softening. Throughout the entire cycling process, the joint exhibited cyclic hardening only during the early stages of the experiment, which accounts for a small proportion of the overall test. The majority of the test is characterized by cyclic softening. The same phenomenon is also observed at the pre-deformation levels of 1 mm and 1.5 mm. As the pre-deformation increases, the area enclosed by the hysteresis loop decreases, and the tensile and compressive stresses become nearly equal.

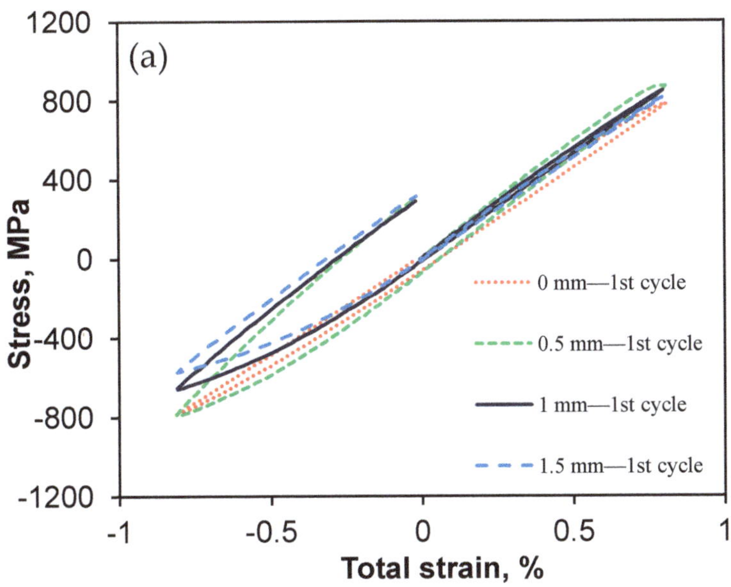

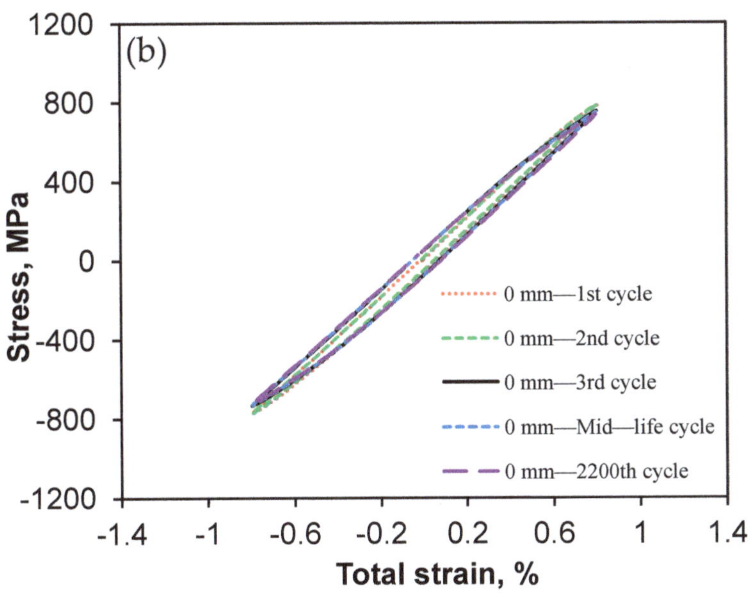

Figure 6. *Cont.*

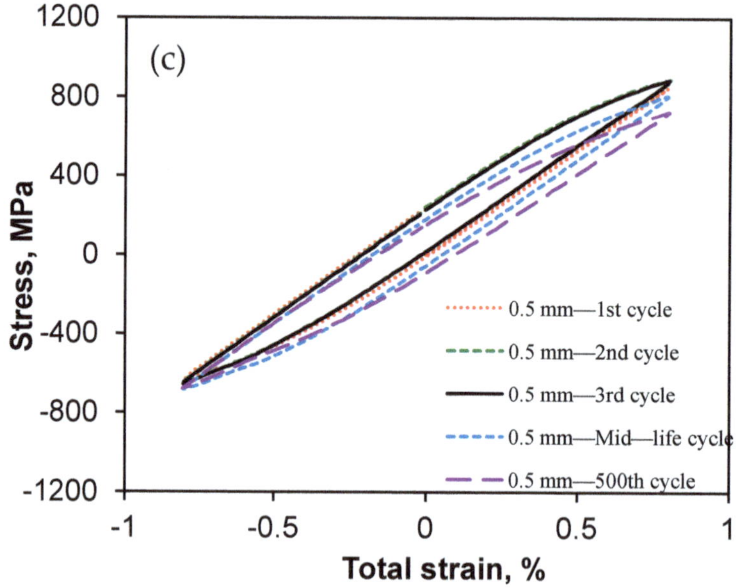

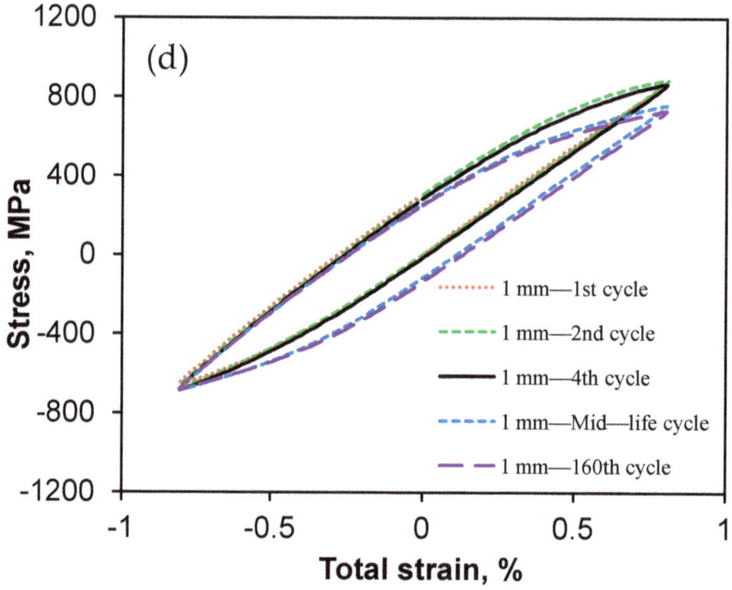

Figure 6. *Cont.*

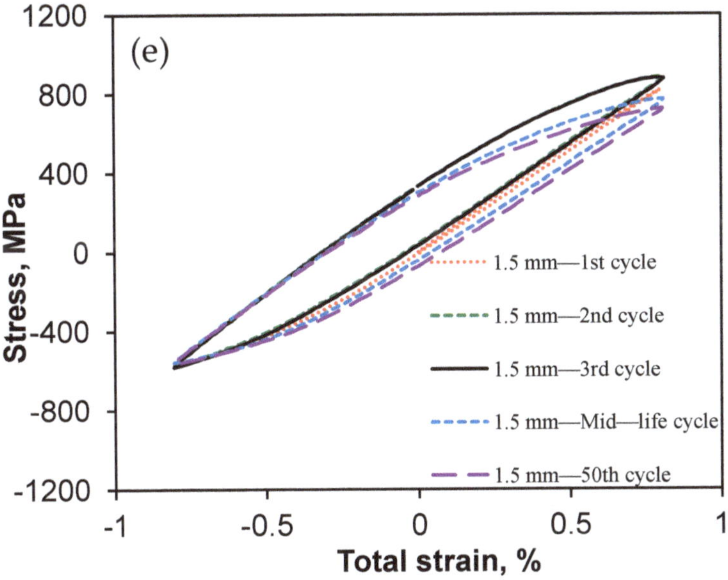

Figure 6. Typical stress–strain hysteresis loops of TC4/Ti60 joint. (**a**) The first cycle under different pre-deformations: (**b**) 0 mm cyclic hysteresis loop, (**c**) 0.5 mm cyclic hysteresis loop, (**d**) 1 mm cyclic hysteresis loop, (**e**) 1.5 mm cyclic hysteresis loop.

Figure 7 shows the relationship between stress and the number of cycles under different pre-deformation levels for the TC4/Ti60 joint. Under constant total strain, the change in cyclic stress with the number of cycles reflects the cyclic hardening or softening behavior of the TC4/Ti60 joint. The red line in Figure 7 and the data in Figure 6b both describe the experimental condition with pre-deformation of 0 mm. The difference in fatigue life is due to the fact that two samples were tested for each experimental condition, and the samples selected here are different. When the pre-deformation of the joint increased from 0 to 1.5 mm, the joint generally presented cyclic softening, which is also confirmed by Figure 6, where cyclic hardening accounted for a small proportion of the overall test. For example, in Figure 6d, under 1 mm pre-deformation, the joint exhibits cyclic hardening during the second and third cycles, with hardening only occurring in the initial few cycles. The remaining cycles predominantly exhibit cyclic softening, and the proportion of cyclic hardening is minimal. The majority of the test was characterized by cyclic softening. For example, the maximum/peak tensile stress of the first cycle (Figure 6e) is slightly higher than that of 50th and Mid-life cycles, which indicates that cyclic softening appears. The results of the cyclic stress in Figure 7 are consistent with those in Figure 6. Smith [26] et al. proposed an empirical formula to predict a material's tendency for cyclic hardening or softening based on its tensile properties. When the ratio of tensile strength to yield strength (σ_{UTS}/σ_{YS}) is greater than 1.4, the material exhibits cyclic hardening; when σ_{UTS}/σ_{YS} is less than 1.2, cyclic softening occurs; and when $1.2 < \sigma_{UTS}/\sigma_{YS} < 1.4$, the material may experience either hardening, softening, or both. According to Section 3.2, in this study, the yield strength and tensile strength of the dissimilar joint are approximately 847 MPa and 895 MPa, respectively, with a ratio of 1.05, which is less than 1.2, indicating that the joint undergoes cyclic softening. Although, TC4/Ti60 joint exhibits cyclic softening overall, it shows cyclic hardening at the start of the fatigue test. This is consistent with the results of the hysteresis loop, as shown in Figure 6. Cyclic hardening or softening is related to dislocations: cyclic hardening results from dislocation multiplication, while cyclic softening is caused by dislocation annihilation. When the rate of dislocation multiplication balances with the rate of dislocation annihilation, the joint reaches a cyclically stable state in which no significant cyclic hardening or softening occurs, achieving a dynamic balance between them. The initial stage of fatigue is characterized by rapid dislocation multiplication, which exceeds the rate of dislocation annihilation, resulting in cyclic hardening. In contrast, dislocation multiplication at the initial stage of the fatigue test without pre-deformation occurs slowly, with almost no dislocation annihilation, resulting in no significant cyclic-hardening process. In addition, it also can be seen that the fatigue life decreases as the pre-deformation increases.

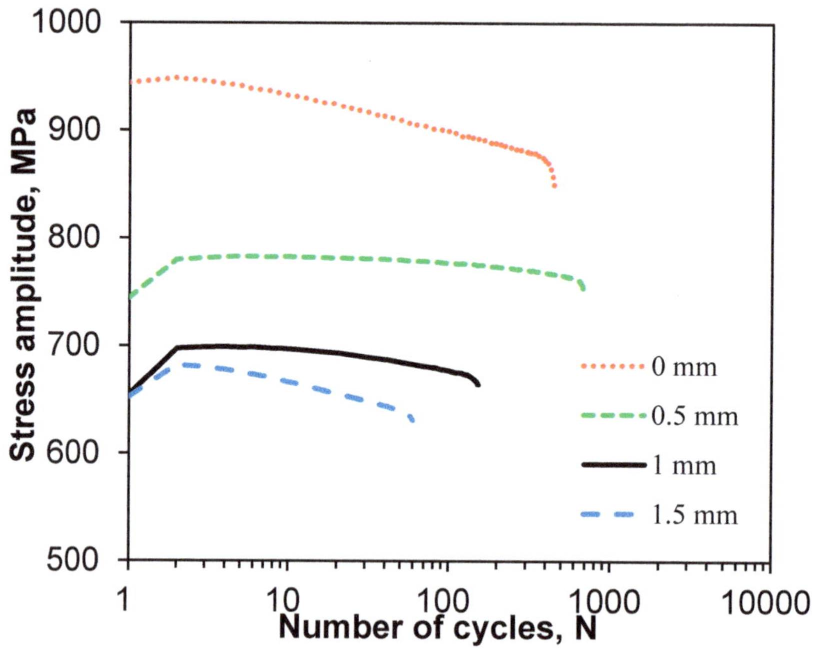

Figure 7. Stress amplitude vs. the number of cycles at different pre-deformation.

Figure 8 shows the relationship between plastic strain amplitude and the number of cycles. It can be observed from the figure that during the fatigue tests, all curves initially decrease and then increase, indicating that the joint undergoes cyclic hardening first, followed by signs of cyclic softening. This behavior is consistent with the cyclic hysteresis loops and the stress shown in Figures 6 and 7.

Figure 9 presents the relationship between the modulus and the number of cycles under different pre-deformations. From the figure, it can be seen that both the loading modulus and unloading modulus are stable during the whole fatigue. However, with increasing pre-deformation, both the loading and unloading modulus exhibit a decreasing trend, particularly showing a significant reduction from 0 mm to 0.5 mm. With the increase in pre-deformation, the Young's modulus exhibits slight variations. Therefore, it can be concluded that the elastic modulus is insensitive to microstructure.

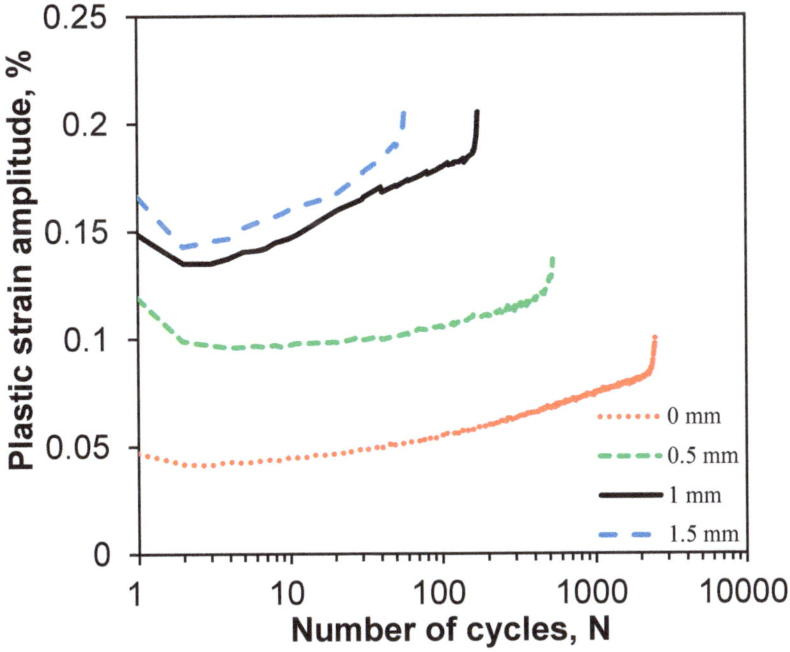

Figure 8. Plastic strain amplitude vs. the number of cycles at different pre-deformation levels.

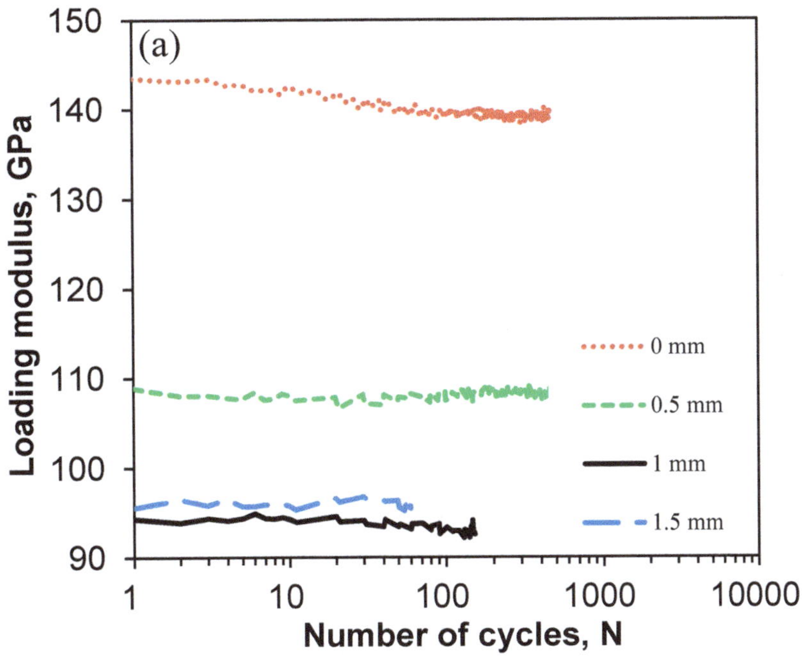

Figure 9. *Cont.*

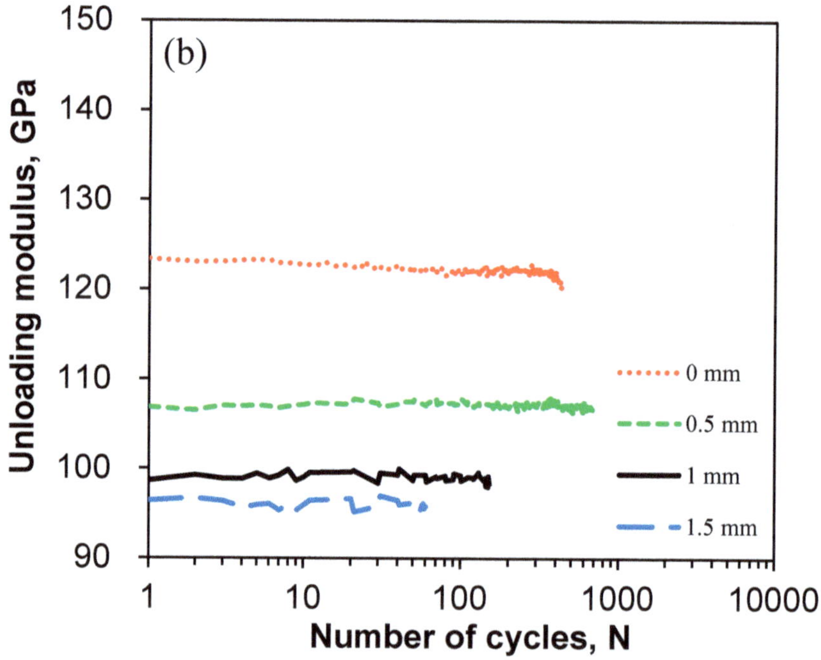

Figure 9. Modulus vs. the number of cycles at different pre-deformation levels. (**a**) Loading modulus, (**b**) unloading modulus.

Figure 10 shows the relationship between the number of failure cycles of the joint at different pre-deformations. Different pre-deformations are applied to simulate the sudden overload experienced by titanium alloy components during service. It can be observed that as the pre-deformation increases, the fatigue life decreases, and the joint becomes more susceptible to failure.When the pre-deformation is 0 mm and 1.5 mm, the fatigue life is 2624 and 64, which indicates that the sudden or accidental loads have a great influence on the fatigue life of the joint. In the two experiments with pre-deformation of 0 mm, the fatigue life shows significant differences. Microstructural observations reveal the presence of pores in the samples with short fatigue life (Figure 11). These pores are caused by the rapid cooling rate during the electron beam welding process. The crystallization rate of the liquid metal in the molten pool is faster than the gas expulsion rate, leading to the gas being trapped before it can escape. This pore defect results in a shorter fatigue life.

3.4. Fractography

Figure 12 presents the fracture morphology of the tensile specimen for the TC4/Ti60 joint at room-temperature. From Figure 12a, it can be seen that the fracture surface is relatively flat with smooth edges. The high-magnification tensile fracture microstructure images in Figure 12b,c reveal a large number of dimples, indicating significant plastic deformation before fracture. This suggests that the material absorbed more energy and exhibited good plasticity, which is consistent with the high tensile and yield strength observed in Figure 5. Based on the analysis of Figures 3 and 4, although the grain size of Ti60 is larger than that of TC4, the presence of numerous hard particles within the Ti60 matrix results in a higher strength compared to the TC4 matrix. Additionally, martensite is present in the heat-affected zones on both the TC4 and Ti60 sides, leading to tensile fracture occurring on the TC4 side.

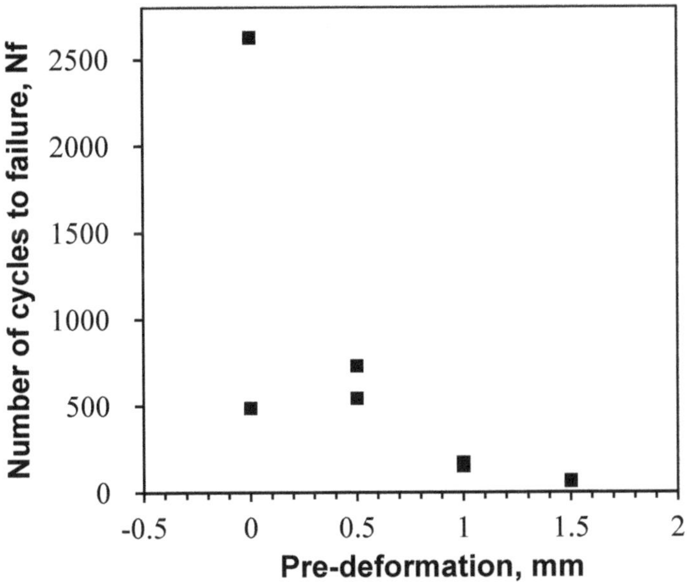

Figure 10. The number of cyclic to failure of the dissimilar joints under different pre-deformation levels.

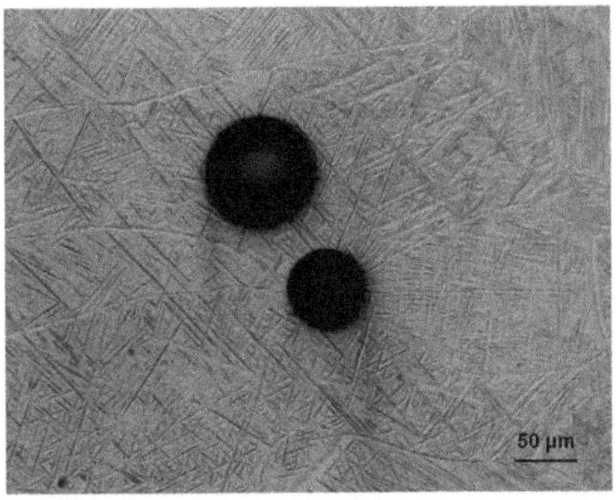

Figure 11. The pore defect.

Figure 13a,b display an overall view of the fracture surface of the fatigue samples for TC4/Ti60 dissimilar joints under pre-deformation of 0.5 mm and 1.5 mm, respectively. All of the samples of the joint failures occur in the TC4 base metal close to the HAZ, with no samples failing in the WZ. According to the literature [27], the presence of martensite increases the strength of the material. Combined with Figure 4, the large amount of martensite in the weld zone leads to a higher strength in the weld zone compared to the TC4 base material. Three regions can be observed from Figure 14: fatigue-crack-initiation region, propagation region, and final fast-fracture region. Usually, the fatigue crack initiates basically from the surface, or the near-surface defect, as shown in Figure 13. This is consistent with the findings of Liu [28] et al. regarding the tensile fatigue of TC17 at room temperature, where fatigue cracks initiate at the specimen surface and propagate, leading to failure. The fatigue propagation region is relatively flat, and many radial marks can be seen extending outward from the fatigue-crack-initiation region. In addition, fatigue striations are observed at the fatigue propagation region at the pre-deformation of 0.5 mm and 1.5 mm, as shown in Figure 14c,d. No secondary cracks are observed under the pre-deformation of 0.5 mm, whereas obvious secondary cracks are noted under the pre-deformation of 1.5 mm, which is related to the magnitude of pre-deformation.

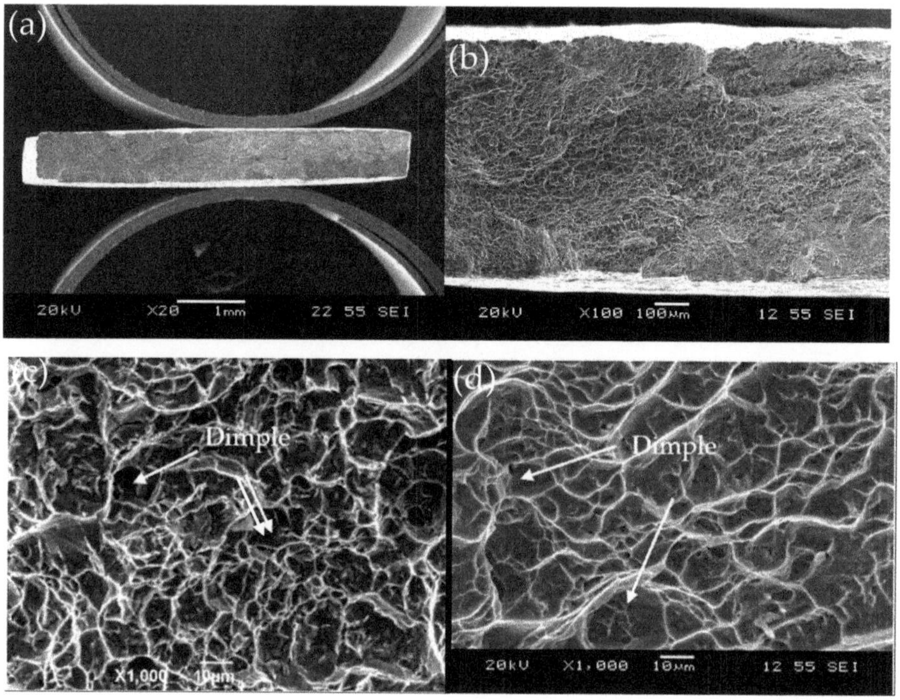

Figure 12. Surface morphologies of the tensile fracture surfaces of dissimilar joints under different magnifications at the TC4 side. (**a**,**b**) Fracture morphology under low magnification, (**c**,**d**) fracture morphology under high magnification.

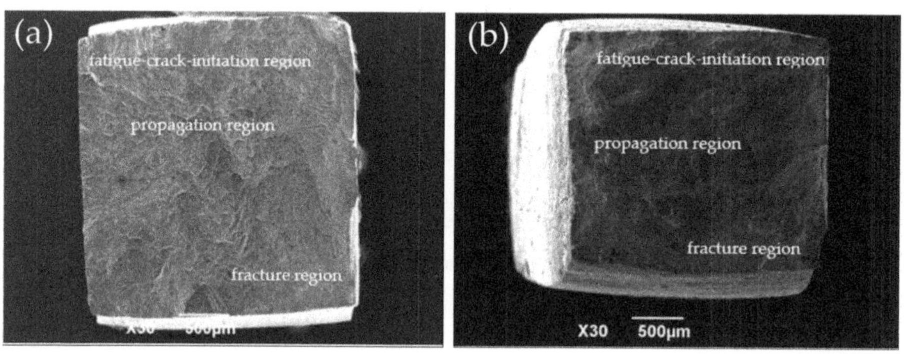

Figure 13. An overall view of fracture surfaces of the fatigued joint on the TC4 side under pre-deformations of (**a**) 0.5 mm and (**b**) 1.5 mm.

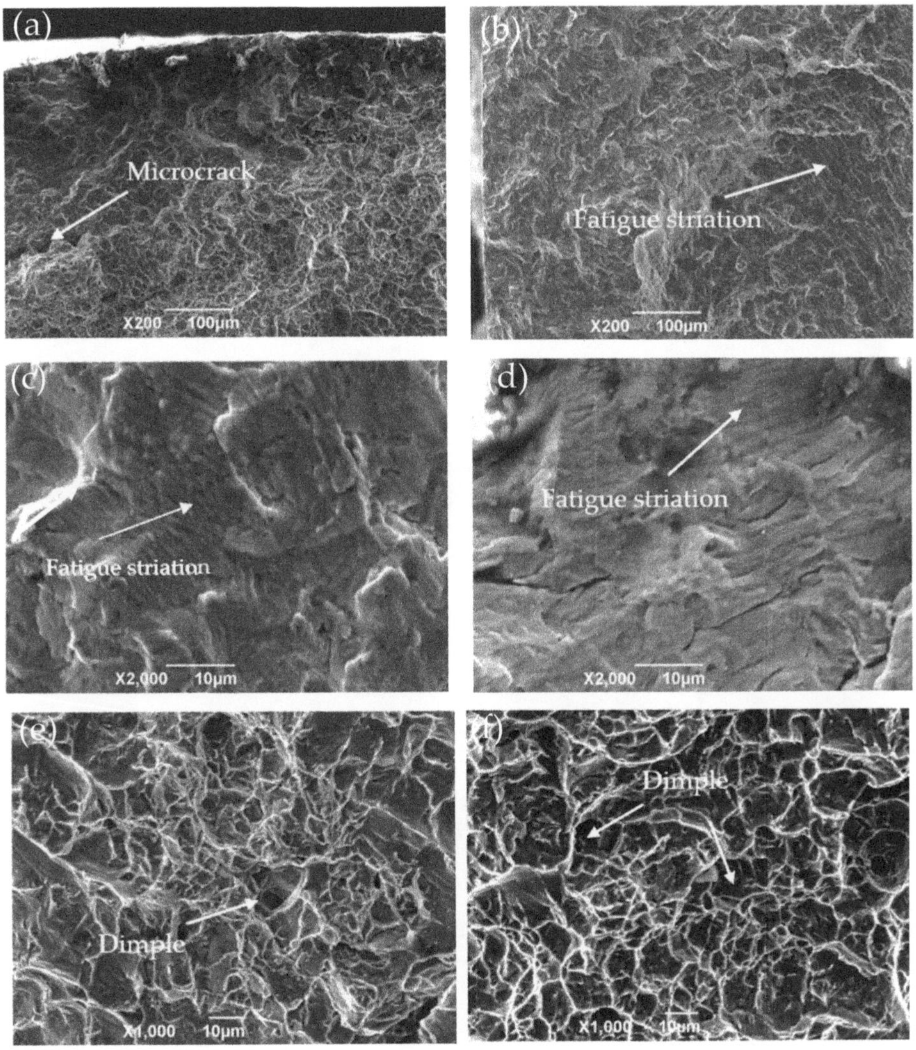

Figure 14. SEM images of fracture surfaces for the fatigued joint on the TC4 side at pre-deformations of (**a**,**c**,**e**) 0.5 mm, and (**b**,**d**,**f**) 1.5 mm. (**a**,**b**) Fatigue-crack-initiation region, (**c**,**d**) propagation region, (**e**,**f**) fast-fracture region.

Figure 14e,f show the images of the final fast-fracture region under the pre-deformations of 0.5 mm and 1.5 mm. Figure 14a,b present the microscopic morphology of the fatigue-crack-initiation zone, while Figure 14c,d show the crack-propagation zone of the joint. The continuous propagation of fatigue striations reduces the effective load-bearing area of the fatigue specimen, and it breaks when it can no longer withstand cyclic loading stresses. The rapid and unstable propagation of fatigue cracks in the final fast-fracture region results in a relatively small proportion compared to the other two regions, often exhibiting an uneven and fluctuating morphology. Therefore, the boundary between the fatigue propagation region and the final fast-fracture region is quite obvious and easy to identify.

4. Conclusions

This paper investigates the fatigue life and cyclic deformation characteristics of dissimilar titanium alloys joints welded by electron beam welding. Based on the research, the following conclusions can be drawn:

1. Throughout the entire cycling process, under constant conditions, as the pre-deformation increases, both the fatigue life and the number of cycles decrease. As the pre-deformation increases from 0 mm to 1.5 mm, the fatigue life decreases from 2626 cycles to 63 cycles. This indicates that during service, if titanium alloy components are subjected to sudden overloads or accidental loads, their fatigue life is likely to decrease.

2. Throughout the entire fatigue test, the joint predominantly exhibits cyclic softening, and the same result is observed in the cyclic hysteresis loops.

3. In all fatigue tests, failure occurs on the TC4 side, indicating that the strength of the joint is higher than that of the TC4 materials. The yield strength and tensile strength of the dissimilar joint are approximately 847 MPa and 895 MPa, respectively.

4. Fatigue cracks initiate at surface or near-surface defects, and secondary cracks are observed at a pre-deformation of 1.5 mm, which is related to the magnitude of the pre-deformation. In this study, different pre-deformations are applied to the joints to simulate the sudden overloads that titanium alloy structural components may experience during use, providing a reference for subsequent research on the fatigue life of the joints.

Author Contributions: Conceptualization, S.W. and W.Z.; methodology, S.W.; software, X.Z.; validation, S.W. and, X.Z.; formal analysis, S.W.; investigation, X.Z.; resources, W.Z.; data curation, Y.L. and Q.G.; writing—original draft preparation, X.Z.; writing—review and editing, S.W.; visualization, W.Z.; supervision, S.W. and Q.G.; project administration, Y.L.; funding acquisition, S.W. All authors have read and agreed to the published version of the manuscript.

Funding: This research was funded by the National Natural Science Foundation of China (No. 52475401, No. 51975469) and the Key Research and Development Program of Shaanxi Province (No. 2022GY-224).

Data Availability Statement: The original contributions involved in this study report can be obtained by directly contacting the corresponding author if needed.

Conflicts of Interest: The authors declare no conflict of interest.

References

1. Cheng, M.; Yu, B.B.; Guo, R.P.; Shi, X.H.; Xu, L.; Qiao, J.W.; Yang, R. Electron beam welding of a novel near α high temperature titanium alloy powder compact: Effect of post-welding heat treatment on tensile properties. *J. Mater. Res. Technol.* **2021**, *10*, 153–163. [CrossRef]
2. Zhao, X.L.; Lu, X.H.; Wang, K.; He, H. Investigation on the microstructure and mechanical properties of Ti6Al4V titanium alloy electron beam welding joint. *Proc. Inst. Mech. Eng. Part C J. Mech. Eng. Sci.* **2022**, *236*, 2957–2966. [CrossRef]
3. Wan, Y.; Song, L.M.; Zhang, X.Y.; Jiang, W.C.; Xie, X.F.; Qi, X.Y. Insight into the keyhole behaviour and its role on residual stress formation in vacuum electron beam welding of TC4 titanium alloy. *Vacuum* **2025**, *232*, 113835. [CrossRef]
4. Zhang, L.L.; Zhang, L.J.; Long, J.; Ning, J.; Zhang, J.X.; Na, S.J. Effects of titanium on grain boundary strength in molybdenum laser weld bead and formation and strengthening mechanisms of brazing layer. *Mater. Des.* **2019**, *169*, 107681. [CrossRef]
5. Rominiyi, A.L.; Peter, M.M. A critical review of microstructure and mechanical properties of laser welded similar and dissimilar titanium alloy joints. *J. Adv. Join. Process.* **2024**, *9*, 100191. [CrossRef]
6. Zhao, P.K.; Hu, Y.X.; Geng, P.H.; Shen, B.; Yuan, L. Effective toughening of dissimilar Ti alloy friction welded joint by combining electropulsing thermal and thermal effects. *Mater. Sci. Eng. A* **2024**, *909*, 146864. [CrossRef]
7. Zhao, P.K.; Chen, W.; Li, Y.; Tao, J.; Chen, C.; Xiao, X.; Zhang, M. Effect of heat treatment on the microstructure, microhardness and impact toughness of TC11 and TC17 linear friction welded joint. *Mater. Sci. Eng. A* **2021**, *803*, 140496. [CrossRef]
8. Singh, J.K.; Roy, G.G.; Majumdar, J.D. Studies of Wear and Tensile Behavior of Titanium-Based Alloy (Ti-6Al-4V) Joint Developed by Electron Beam Welding. *J. Mater. Eng. Perform.* **2024**, *280*, 1–16. [CrossRef]

9. Fu, P.F.; Mao, Z.Y.; Wang, Y.J.; Zuo, C.J.; Xu, H.Y. Mechanical properties of Ti6.5Al2Zr1Mo1V titanium alloy with EBW under different temperatures. *Mater. Sci. Eng. A* **2014**, *608*, 199–206. [CrossRef]
10. Lu, Y.; Turner, R.; Brooks, J.; Basoalto, H. Microstructural characteristics and computational investigation on electron beam welded Ti-6Al-4 V alloy. *J. Mater. Process. Technol.* **2021**, *288*, 116837. [CrossRef]
11. Ou, P.; Cao, Z.Q.; Hai, M.; Qiang, J.F.; Wang, Y.M. Microstructure and mechanical properties of K-TIG welded dissimilar joints between TC4 and TA17 titanium alloys. *Mater. Charact.* **2023**, *196*, 112644. [CrossRef]
12. Xu, W.; Jun, M.A.; Luo, Y. Microstructure and high-temperature mechanical properties of laser beam welded TC4/TA15 dissimilar titanium alloy joints. *Trans. Nonferrous Met. Soc. China* **2020**, *30*, 160–170. [CrossRef]
13. Li, L.Z.; Wang, S.G.; Huang, W.; Jin, Y. Microstructure and mechanical properties of electron beam welded TC4/TA7 dissimilar titanium alloy joint. *J. Manuf. Process.* **2020**, *50*, 295–304. [CrossRef]
14. Zhao, X.L.; Lu, X.H.; Wang, K.; He, F. Microstructure and mechanical properties of electron beam welded TC4 titanium alloy structure with backing plate. *Mater. Today Commun.* **2023**, *35*, 106160. [CrossRef]
15. Rajan, S.; Wanjara, P.; Gholipour, J.; Kabir, A.S. Fatigue Behavior of Linear Friction Welded Ti-6Al-4V and Ti-6Al-2Sn-4Zr-2Mo-0.1Si Dissimilar Welds. *Materials* **2021**, *14*, 3136. [CrossRef]
16. Long, J.; Zhang, L.J.; Zhang, L.L.; Ning, J.; Zhuang, M.X.; Zhang, J.X.; Na, S.-J. Analysis of heterogeneity of fatigue properties of double-sided electron beam welded 140-mm thick TC4 titanium alloy joints. *Int. J. Fatigue* **2021**, *142*, 105942. [CrossRef]
17. Sun, W.B.; Ma, Y.; Huang, W.; Zhang, W.H.; Qian, X.D. Effects of build direction on tensile and fatigue performance of selective laser melting Ti6Al4V titanium alloy. *Int. J. Fatigue* **2020**, *130*, 105260. [CrossRef]
18. Liu, H.Q.; Wang, H.M.; Zhang, Z.; Liu, Y.J.; Huang, Z.Y.; Wang, Q.Y.; Chen, Q. Tensile and fatigue behavior of electron beam welded TC17 titanium alloy joint. *Int. J. Fatigue* **2019**, *128*, 105210. [CrossRef]
19. Hu, Y.N.; Wu, S.C.; Xin, C.; Wu, W.W.; Zhang, J. Fatigue life evaluation of Ti-6Al-4V welded joints manufactured by electron beam melting. *Fatigue Fract. Eng. Mater. Struct.* **2021**, *44*, 2210–2221. [CrossRef]
20. Pasqualino, C.; Vincenzo, C. Fatigue analysis of TI6AL4V/INCONEL 625 dissimilar welded joints. *Ocean. Eng.* **2021**, *221*, 108582. [CrossRef]
21. Oh, J.; Kim, N.J.; Lee, S.; Lee, E.W. Correlation of fatigue properties and microstructure in investment cast Ti-6Al-4V welds. *Mater. Sci. Eng. A* **2003**, *340*, 232–242. [CrossRef]
22. Arcieri, E.V.; Baragetti, S.; Bozi, Z. Failure analysis of Ti6Al4V titanium alloy under fatigue loading: An experimental and numerical study. *Eng. Fail. Anal.* **2024**, *164*, 108715. [CrossRef]
23. Tusher, M.M.H.; Ince, A. High Cycle Fatigue and Very High Cycle Fatigue Performance of Selective Laser Melting Ti-6Al-4V Titanium Alloy—A Review. *Mater. Perform. Charact.* **2023**, *12*, 214–293. [CrossRef]
24. Farabi, N.; Chen, D.L.; Zhou, Y. Tensile properties and work hardening behavior of laser-welded dual-phase steel joints. *J. Mater. Eng. Perform.* **2012**, *21*, 222–230. [CrossRef]
25. Ma, T.J.; Zhong, B.; Li, W.Y.; Yang, S.Q.; Yang, C.L. On microstructure and mechanical properties of linear friction welded dissimilar Ti-6Al-4V and Ti-6.5Al-3.5Mo-1.5Zr 0.3Si joint. *Sci. Technol. Weld. Join.* **2012**, *17*, 9–12. [CrossRef]
26. Hirschberg, M.H.; Manson, S.S.; Smith, R.W. *Fatigue Behavior of Materials Under Strain Cycling in Low and Intermediate Life Range*; NASA: Washington, DC, USA, 1963.
27. Guo, Z.G.; Ma, T.J.; Yang, X.W.; Chen, X.; Tao, J.; Li, J.; Li, W.; Vairis, A. Linear friction welding of Ti60 near-α titanium alloy: Investigating phase transformations and dynamic recrystallization mechanisms. *Mater. Charact.* **2022**, *194*, 112424. [CrossRef]
28. Lin, F.L.; Chen, Y.; He, C.; Li, L.; Wang, C.; Li, H.Z.; Zhang, H.; Wang, Q.; Liu, Y. Tensile and very high cycle fatigue behaviors of a compressor blade titanium alloy at room and high temperatures. *Mater. Sci. Eng. A* **2021**, *811*, 141049. [CrossRef]

crystals

MDPI

Article

The Effect of Gradient Cooling Behavior on the Microstructure and Mechanical Properties of Al-2at.% Nd Alloy in a Vacuum Environment

Xiangjie Wang [1,2,*], Xinyu Zhang [1,2], Wenjie Wu [3] and Shuchen Sun [3]

1 Key Lab of Electromagnetic Processing of Materials, Ministry of Education, Northeastern University, Shenyang 110819, China; zxy_zhangxinyu_000@163.com
2 Advanced Manufacturing Technology and Engineering Research Centers, Northeastern University, Shenyang 110819, China
3 School of Metallurgy, Northeastern University, Shenyang 110819, China; 2271808@stu.neu.edu.cn (W.W.); sunsc@smm.neu.edu.cn (S.S.)
* Correspondence: wangxj@epm.neu.edu.cn

Abstract

Al-2at.% Nd alloy with a gradient cooling rate was prepared using a wedge-shaped mold in a vacuum environment. The relationship between gradient cooling behavior and the microstructure and properties of the Al-2at.% Nd alloy was investigated. The stability of the $Al_{11}Nd_3$ phase and the mechanical properties were confirmed through first-principles calculations. The results indicated that as the cooling rate decreased, the transformation of grain morphology in Al-2at.% Nd occurred as follows: a mixture of columnar grains and equiaxed grains→equiaxed grains. The grain size of the alloy increased. Discontinuous skeletal eutectic phases, α-Al and $Al_{11}Nd_3$, formed within the alloy, resulting in a reduction in the number of phase boundaries and grain boundaries. The hardness of the alloy decreased by 25.53%, and this pattern of change closely aligned with the calculated results.

Keywords: Al-2at.% Nd; wedge-shaped copper mold casting; microstructure; properties; first principles

1. Introduction

Aluminum alloys have garnered significant attention from researchers due to their low density, commendable corrosion resistance, and exceptional formability. These materials are extensively utilized in various sectors, including aerospace, rail transportation, automotive, maritime applications, and national defense technology [1–3]. However, with the rapid advancement of technology, the performance requirements for aluminum alloys are becoming increasingly stringent. Consequently, certain aluminum alloys can no longer meet specific application areas' demands. Therefore, there is a pressing need to develop new aluminum alloys that exhibit superior properties, such as high strength, enhanced corrosion resistance, and elevated temperature resistance, to satisfy the evolving requirements of both production and everyday life [4–7].

In recent years, the Al-Nd binary alloy system has gained significant attention among researchers as a target material in sputtering coating processes. However, the production and preparation of Al-Nd target materials present considerable challenges. To ensure the quality and performance of the target material, it is essential to use high-purity raw materials and employ precise processing techniques. Alloying represents a critical method for the development of new high-performance aluminum alloys. Research indicates that incorporating rare earth elements into aluminum and aluminum alloys can significantly modify the alloy composition and substantially enhance the microstructure of these material [8–12]. However, there is currently limited literature on Al-Nd target materials. One study indicates that the Al-2at.% Nd alloy is predominantly utilized in advanced thin-film transistor liquid crystal displays (TFT-LCDs) due to its exceptional resistance to hillock and whisker formation, alongside its having a low resistivity (<5 μΩcm) [13]. Therefore, conducting in-depth research on the preparation processes and performance characteristics of Al-Nd target materials is highly significant for advancing the development and progress of related industries. The challenge of homogenizing the microstructure of Al-Nd target materials lies in the formation and distribution of Al-Nd intermetallic compounds. In Al-Nd target materials, Nd primarily exists in the form of the eutectic compound $Al_{11}Nd_3$ [14]. Unlike the research on other Al-Nd intermetallic compounds, little is currently known about the $Al_{11}Nd_3$ phase. Xiao et al. [15] investigated the crystallization process of rapidly solidified Al-Nd-Ni alloys, which include $Al_{11}Nd_3$, Al_3Ni, and several unidentified phases. Liu et al. [16] examined the stability and mechanical properties of Al-RE intermetallic compounds (Al_2RE, Al_3RE, AlRE, $Al_{11}RE_3$, $AlRE_2$, and $AlRE_3$), demonstrating that these intermetallic compounds can act as strengthening phases, enhancing the properties of Al alloys. Therefore, it is essential to investigate the thermodynamic stability and mechanical properties of the $Al_{11}Nd_3$ phase to design and produce high-quality Al-Nd target materials and to advance the development of large-sized liquid crystal display panels.

First-principles calculations, commonly referred to as ab initio calculations, are methodologies grounded in the principles of quantum mechanics that address the Schrödinger equation utilizing only fundamental physical constants, thereby eschewing any reliance on empirical parameters [17–19]. Researchers have undertaken comprehensive investigations into the phase structures of binary aluminum alloy systems through these first-principles calculations. By evaluating physical quantities such as electronic structure, energy, and mechanical properties of materials, they can predict the macroscopic properties of these materials. In recent years, significant advancements in material design, synthesis, and simulation calculations have emerged from integrating first-principles calculations based on density functional theory (DFT) with molecular dynamics, establishing it as a crucial foundation and primary methodology in computational materials science. Michael C. Gao [20] conducted a systematic study on the formation and stability of Al-RE intermetallic compounds. Similarly, Michal Jahnátek [21] systematically investigated the plastic deformation behavior of Al_3(Sc, Ti, and V) intermetallic compounds in the L12 and D022 structures, providing a comprehensive analysis of various elastic constants.

Adding the element Nd can further improve the thermal stability and enhance the mechanical strength of aluminum alloys. Furthermore, the Al-2at.% Nd alloy exhibits favorable casting properties and is easily processed into various shapes and sizes of components. Therefore, due to considerations of processing efficiency and economic viability, we have chosen to study the Al-2at.% Nd alloy. However, there is still limited research on Al-Nd target materials and the $Al_{11}Nd_3$ phase. The stability, morphological characteristics, and inherent mechanical properties of $Al_{11}Nd_3$ in Al-Nd target alloys are not yet fully understood, and this will be essential in order to comprehend the material's properties. Therefore, to address this knowledge gap, this study employs wedge-shaped copper molds to cast an Al-2at.% Nd alloy. It examines the influence of cooling rate on the grain morphology, dendrite arm spacing, and eutectic phase morphology of the Al-2at.% Nd alloy. Furthermore, first-principles calculations are utilized to evaluate the alloy's phase stability and elastic properties, thereby providing a theoretical foundation for the fabrication of Al-Nd alloy targets.

2. Experimental Methods

2.1. Sample Preparation

Samples were prepared utilizing a commercially available 4N-grade bulk Al-2at.% Nd alloy. Initially, a high-purity graphite crucible was positioned within the coil of a vacuum induction melting furnace, and the interstices were filled with magnesia sand. Subsequently, the magnesia sand was combined with water glass to achieve a paste-like consistency, which was then applied around the crucible and allowed to dry. After thoroughly cleaning the furnace chamber and crucible, we placed the Al-Nd alloy raw material into the crucible. The furnace was subsequently evacuated to establish a vacuum conducive to melting. Upon reaching a temperature of 700 °C, the melts were poured into a wedge-shaped copper mold with an inside diameter of 65 × 50 mm and a height of 120 mm. After cooling for 30 min, we extracted the ingot.

Figure 1a illustrates the internal structure of the wedge-shaped copper mold. The cast ingot is segmented into four distinct regions, designated as A, B, C, and D, arranged from the bottom to the top, as depicted in the diagram.

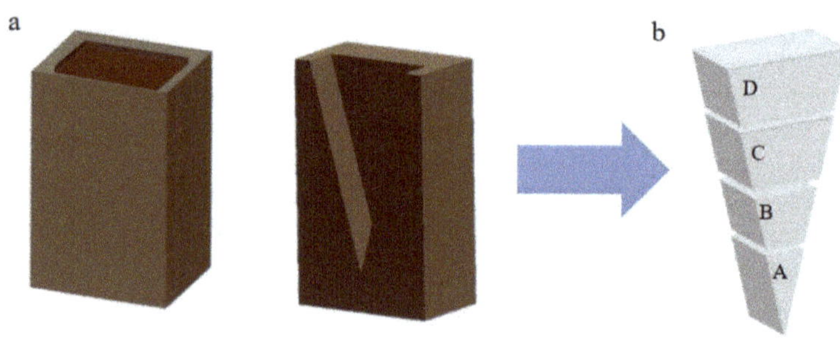

Figure 1. (**a**) Wedge-shaped copper mold and (**b**) schematic diagram of the sample cutting.

In order to observe the grain structure of the alloy, anodic oxidation was conducted for approximately 45 s utilizing a 10% HBF_4 solution at a voltage of 20 V and a current range of 0.3–0.5 mA. The microstructure of the anodic coating on the cast sample was examined using polarized light microscopy with a Leica DMI5000M optical microscope (Wetzlar, Germany). For deep etching, the 10% NaOH solution was employed at room temperature for a duration of 50 s. In order to study the phase composition of the alloy, rectangular samples measuring 15 mm × 15 mm × 10 mm were cut, polished, and analyzed using X-ray diffraction (XRD, Rigaku, DMAX-2500, Cu-Kα, 18 kW). The morphology and elemental composition of the phases within the cast alloy were characterized using a field emission scanning electron microscope (Tescan, Brno, Czechia, MIRA3) equipped with an X-ray spectrometer. The alloy's grain size and secondary dendrite arm spacing were statistically analyzed using Image-Pro Plus software (6.0, Media Cybernetics, Rockville, MD, USA). Using wire cutting technology, rectangular samples measuring 20 mm × 15 mm × 10 mm were cut out of areas of ingots A, B, C, and D. The hardness of the alloy at various positions was measured using a microhardness tester (FUTURE TECH, Kawasaki, Japan).

2.2. First-Principles Calculations

All first-principles calculations conducted in this study, based on Density Functional Theory (DFT), were executed using Vienna ab initio Simulation Package software (VASP 6.3). The Projector Augmented Wave (PAW) [22] method was utilized to represent the atomic pseudopotentials, while the Perdew–Burke–Ernzerhof (PBE) [23] functional was applied to characterize the exchange-correlation energy. Each unit cell was fully optimized until the total energy and force reached the convergence standards of 1.0×10^{-5} eV and 0.01 eV/Å, respectively [24]. Furthermore, all calculations presented in this paper underwent rigorous convergence testing.

The formation enthalpy of a compound serves as a valuable metric for evaluating the stability of alloy compounds. It is typically negative and represents the energy absorbed or released during the formation of a compound from its stable elemental constituents. A smaller formation enthalpy indicates greater stability of the compound. The formula for calculating the formation enthalpy of an alloy is as follows [25]:

$$\Delta H = 1/(x+y)(E_{tot} - xE^{A}_{bulk} - yE^{B}_{bulk}) \quad (1)$$

The Al_xNd_y alloy phase, E_{tot}, represents the total energy of the system. E^{A}_{bulk} and E^{B}_{bulk} are the energies of Al atoms and Cu atoms in the crystal structure, respectively. The variables x and y correspond to the quantities of Al and Nd atoms present in the crystal structure of the intermetallic compound, respectively.

Elastic constants are fundamental physical quantities that provide insight into the mechanical stability of materials. These constants characterize various mechanical properties, including strength, toughness, hardness, and brittleness. Within the linear deformation regime of materials, characterized by small strains, the relationship between stress and strain is linear, thereby conforming to Hooke's Law [26]:

$$\sigma_i = \sum_{j=1}^{6} C_{ij}\varepsilon_{ij} \quad (2)$$

In this context, σ and ε denote stress and strain, respectively, while C_{ij} represents the elastic stiffness constant. It is important to note that the indices $1 \leq i \leq 6$ indicate that both strain and stress possess six independent components.

3. Results and Discussion

3.1. The Influence of Gradient Cooling Behavior on the Microstructure of Al-2at.% Nd Alloy

Figure 2a–d illustrate the microstructural grain morphology of samples A, B, C, and D. It is evident that samples A, B, and C comprise a combination of columnar grains and equiaxed grains. In contrast, sample D is exclusively composed of equiaxed grains. As indicated by the arrows in direction 2 of Figure 2a–d, samples A, B, and C display columnar grains adjacent to the mold wall, with the growth direction of these columnar grains oriented perpendicularly to the mold wall. This is because the heat dissipation direction near the mold wall is perpendicular to the mold wall, and the grain growth rate in this direction is higher than that in other directions, resulting in preferential growth. The latent heat released by the preferentially growing dendrites inhibits the growth of dendrites in alternative directions, leading to the solidification of columnar grains in the preferred direction [27]. In contrast, sample D solidifies at the upper section of the mold, where heat dissipation is slower and lacks a specific direction for rapid heat dissipation. Consequently, the grains do not exhibit a preferred growth direction, resulting in the absence of columnar crystals along the mold wall. When the undercooling of the components at the tips of the columnar grain exceeds the critical undercooling necessary for nucleation, equiaxed crystal nuclei will be formed in the melt (as depicted in direction 1 of Figure 2a–d). By studying the grain size of the Al-2at.% Nd alloy under various cooling rates, we can gain insights into the mechanisms of grain growth during solidification. Additionally, by controlling the cooling rate, the grain size of the Al alloy can be adjusted, allowing for the regulation of the alloy's properties. The grain sizes of samples A, B, C, and D were statistically analyzed utilizing Image-Pro Plus software (6.0, Media Cybernetics, Rockville, MD, USA). For each sample, the grain sizes of 15 different regions were measured, and the average value was calculated. The results presented in Figure 2(a_1–d_1). Figure 2(a_1–d_1) illustrate the frequency distribution of grain sizes across the four sample regions: D, C, B, and A, respectively. As the cooling rate diminishes, the grain size progressively increases from 523.28 μm to 743.57 μm.

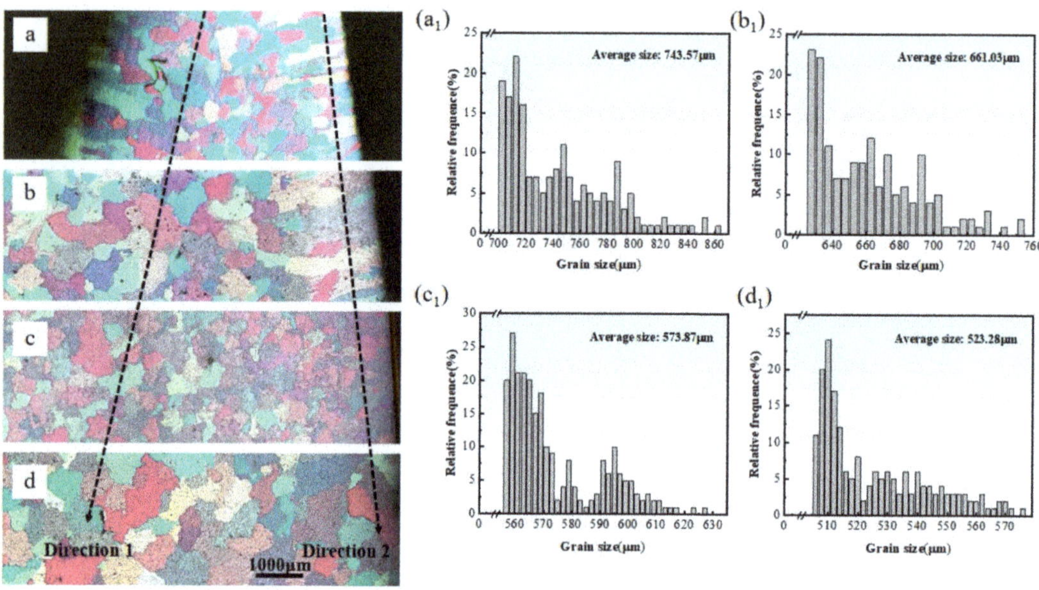

Figure 2. Grain morphology of Al-2at.% Nd at different cooling rates: (**a**) sample A; (**b**) sample B; (**c**) sample C; and (**d**) sample D. Relative frequency of grain size in Al-2at.% Nd alloy: (**a_1**) sample D; (**b_1**) sample C; (**c_1**) sample B; and (**d_1**) sample A.

Figure 3 shows the microstructure of the Al-2at.% Nd alloy at different cooling rates. It is evident that a gradual decrease in the cooling rate results in a notable coarsening of the alloy's structure. Furthermore, the secondary dendrite arm spacing of the Al-2at.% Nd alloy at different cooling rates was quantified, as depicted in Figure 4. The data indicate that, from region A to region D, the secondary dendrite arm spacing increases from 14.21 μm to 42.72 μm as the cooling rate decreases. This phenomenon occurs because when the alloy cools at a slower rate, the heat loss at the solidification front is slower, leading to a smaller degree of undercooling and a correspondingly reduced nucleation rate. As a result, fewer crystal nuclei are formed during the solidification process, and these nuclei have more time to grow, resulting in larger grains and increased secondary dendrite arm spacing. Conversely, when the cooling rate is rapid, the crystal nuclei do not have sufficient time to grow during solidification, leading to smaller grain sizes and decreased secondary dendrite arm spacing. The reduced secondary dendrite arm spacing results in a greater number of grain boundaries, which can more effectively impede the movement of dislocations and enhance the hardness of the alloy.

To thoroughly investigate the influence of cooling rate on the morphology of the eutectic phase in the Al-2at.% Nd alloy, scanning electron microscopy (SEM) was employed for observation. Based on the Al-Nd binary phase diagram (Figure 5), it is established that the predominant phase in the Al-2at.% Nd alloy is the eutectic $Al_{11}Nd_3$ phase. Using wire cutting technology, a square sample of 15 mm × 15 mm × 10 mm is extracted from the core of the solidified ingot. Point scanning and XRD detection were performed on various morphologies of the second phase to analyze its composition further. By integrating the results from point scanning and face scanning (Figure 6) with the XRD spectrum analysis (Figure 7), it is discernible that the black matrix (point B) corresponds to the α-Al phase. In contrast, the white eutectic phase (points A and C) is attributed to the $Al_{11}Nd_3$ phase.

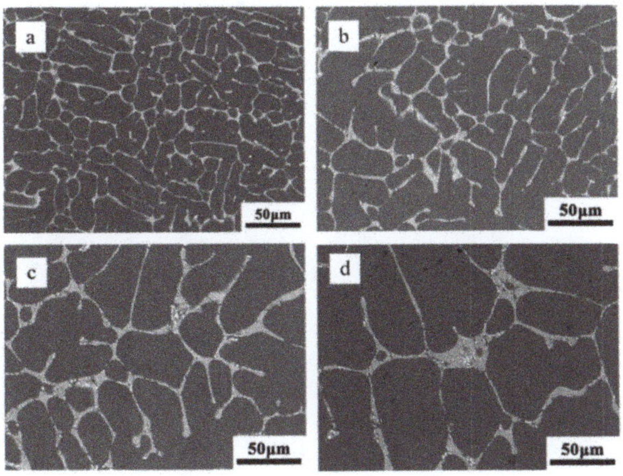

Figure 3. Microstructure of Al-2at.% Nd alloy at different cooling rates: (**a**) sample A; (**b**) sample B; (**c**) sample C; and (**d**) sample D.

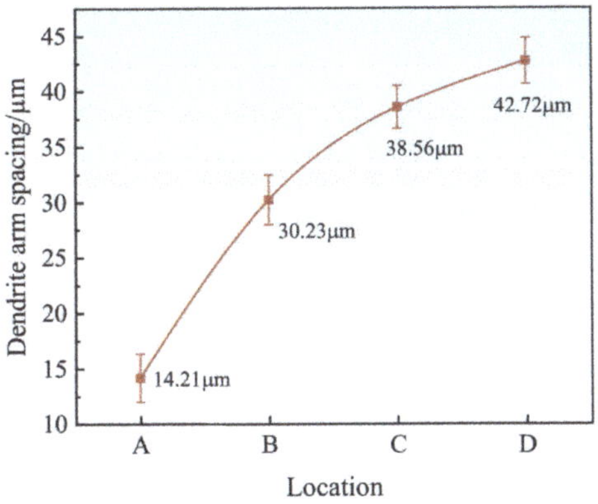

Figure 4. Variation in dendrite arm spacing in Al-2at.% Nd alloy at different cooling rates.

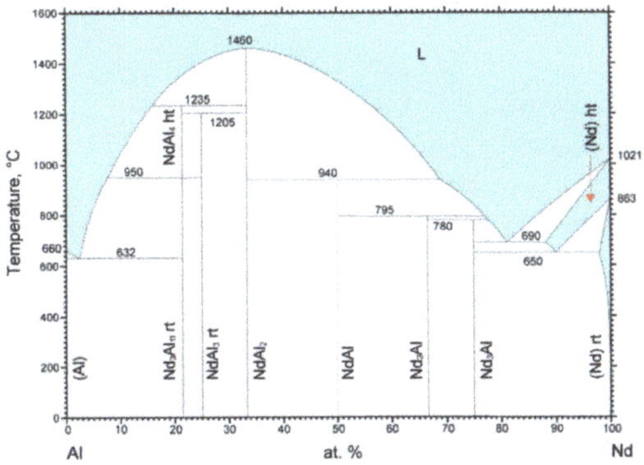

Figure 5. Phase diagram of Al-Nd system [28].

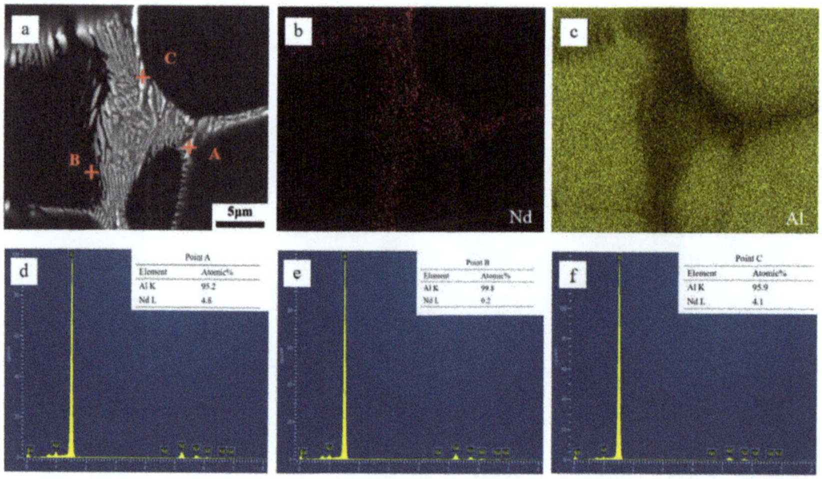

Figure 6. Microstructure of as-cast Al-2Nd alloy. (**a**) Backscattered electron (BSE) image. (**b**,**c**) Distributions of Al and Nd for the region indicated in the dashed box in a. EDS point scan results: (**d**) point A; (**e**) point B; and (**f**) point C.

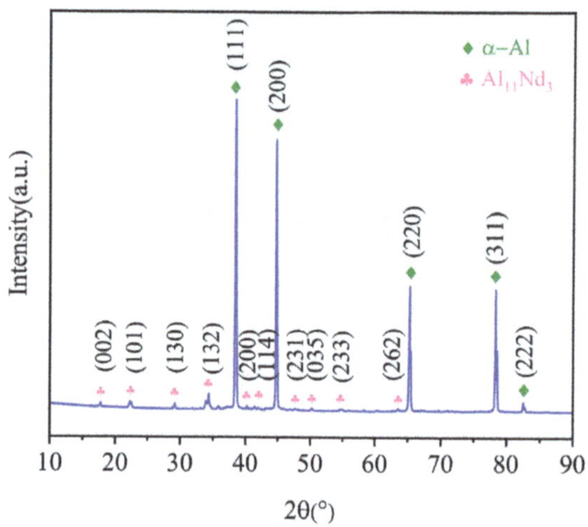

Figure 7. XRD spectrum of Al-2at.% Nd alloy.

Figure 8a–d illustrate the microstructure of the Al-2at.% Nd alloy following solidification at various cooling rates. Figure 8e–h correspond to the boxed portions in Figure 8a–d, respectively, illustrating the morphology of eutectic $Al_{11}Nd_3$ after deep etching with NaOH solution. The observations indicate that at higher cooling rates, the morphology of the eutectic $Al_{11}Nd_3$ exhibits a skeletal structure, which is embedded within the aluminum matrix. This phenomenon can be attributed to the high cooling rate and substantial undercooling, which result in an unstable growth interface of the lamellar eutectic, thereby producing a complex skeletal structure of $Al_{11}Nd_3$.

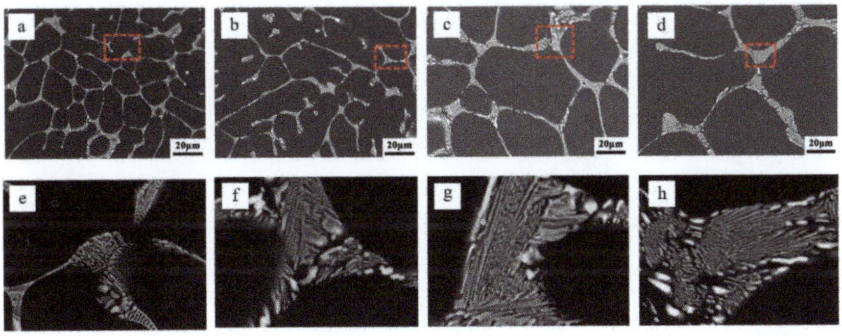

Figure 8. Morphology of eutectic $Al_{11}Nd_3$ at different cooling rates: (**a**) sample A; (**b**) sample B; (**c**) sample C; and (**d**) sample D. Morphology of eutectic $Al_{11}Nd_3$ after deep corrosion: (**e**) sample A; (**f**) sample B; (**g**) sample C; and (**h**) sample D. The red boxes in images (**a**–**d**) correspond to (**e**–**h**) respectively.

3.2. *The Effect of Gradient Cooling Behavior on the Properties of Al-2at.% Nd Alloy*

Figure 9 illustrates the evolution of hardness in Al-2at.% Nd at various cooling rates. Generally, the cooling rate has a positive influence on hardness [29]. The hardness of the Al-2at.% Nd alloy decreases with a decreasing cooling rate and increases by 25.53% with an increasing cooling rate. This is because, as the cooling rate increases, the number of alloy phases and grains also increases, leading to a higher number of phase boundaries and grain boundaries. The greater the resistance encountered during the dislocation process, the higher the hardness of the alloy. Additionally, the sub-eutectic grain refinement of Al-2at.% Nd, the finer eutectic $Al_{11}Nd_3$, and the reduced secondary dendrite arm spacing positively influence hardness.

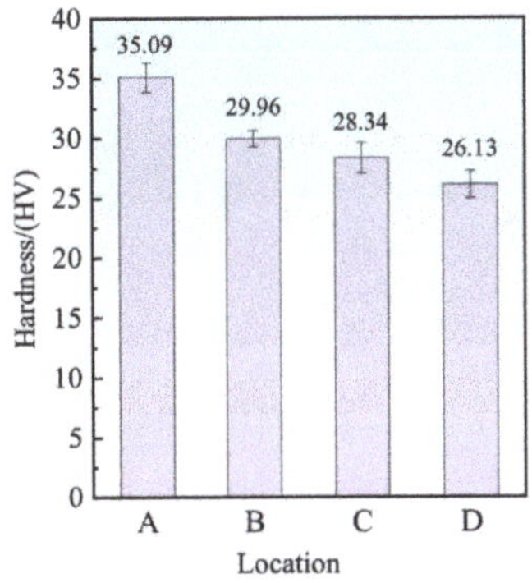

Figure 9. Hardness of Al-2at.% Nd alloy at different cooling rates.

3.3. *Crystal Structure and Phase Stability*

Figure 10a shows the crystal cell structure model of the Al matrix, which belongs to the Fm-3m space group and has a face-centered cubic structure, with a = b = c = 4.03893 Å and $\alpha = \beta = \gamma = 90°$. Figure 10b presents the unit cell structure model of the $Al_{11}Nd_3$ phase, which has an orthorhombic structure. The space group of $Al_{11}Nd_3$ is Immm, with lattice constants a = 4.3565 Å, b = 10.0679 Å, c = 12.9989 Å, and $\alpha = \beta = \gamma = 90°$. The lattice constants and formation enthalpy of Al and $Al_{11}Nd_3$ are listed in Table 1. The setting of the lattice constant is consistent with the results obtained from experimental testing. The Monkhorst–Pack method is used to divide the simple Brillouin zone K-point grid of the unit cell model, with the K-point for the Al matrix set to $5 \times 5 \times 5$ and the K-point for the $Al_{11}Nd_3$ phase set to 14 × 6 × 5. The plane wave cutoff energy (Ecut) is set to 400 eV for the structural optimization of the Al matrix and $Al_{11}Nd_3$. From Table 1, it is evident that the enthalpy of formation of $Al_{11}Nd_3$ is negative (−0.42 eV/atom), indicating that the $Al_{11}Nd_3$ phase can exist stably. This also confirms that the addition of the Nd element in the previous experiment can lead to the formation of a stable discontinuous eutectic phase at the grain boundary.

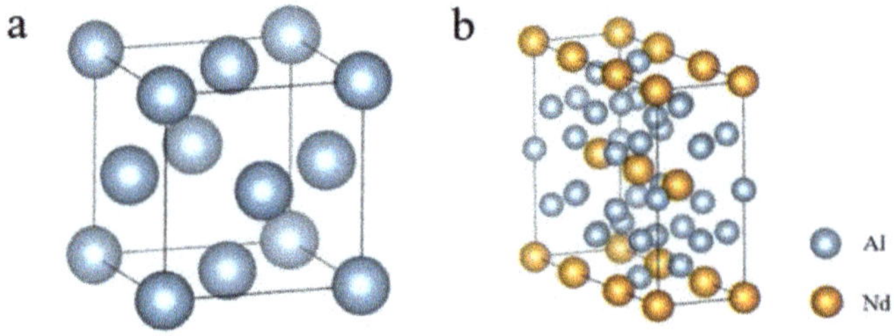

Figure 10. Schematic diagram of crystal structure: (**a**) Al and (**b**) $Al_{11}Nd_3$.

Table 1. Lattice constants (a, b, and c/Å), angles (α, β, and γ/°), and formation enthalpy ΔH (eV/atom) of Al and $Al_{11}Nd_3$.

Compounds	a	b	c	α	β	γ	ΔH
$Al_{11}Nd_3$	4.43	9.99	12.94	90	90	90	−0.42
Al	4.04	4.04	4.04	90	90	90	

3.4. Mechanical Properties

The elastic constant is a significant physical parameter for investigating the mechanical properties of compounds and comprehending the microscopic mechanical properties of crystal structures. Additionally, to ensure the accuracy of the calculations, we set the ion movement step size to POTIM = 0.015. The face-centered cubic structure of the Al matrix is characterized by three independent elastic constants (C_{11}, C_{22}, and C_{44}). In contrast, the body-centered orthorhombic structure of the $Al_{11}Nd_3$ compound is defined by nine independent elastic constants (C_{11}, C_{22}, C_{33}, C_{44}, C_{55}, C_{66}, C_{12}, C_{13}, and C_{23}). The conditions for mechanical stability are calculated by the following expressions [30]:

Cubic crystal system:

$$\left\{\begin{array}{l} C_{11} > 0 \\ C_{44} > 0 \\ C_{11} > |C_{12}| \\ (C_{11} + 2C_{12}) > 0 \end{array}\right\} \tag{3}$$

Orthorhombic crystal system:

$$\left\{\begin{array}{l} C_{ii} > 0 (i = 1,2,3,4,5,6) \\ (C_{11} + C_{22} - 2C_{12}) > 0 \\ (C_{11} + C_{33} - 2C_{13}) > 0 \\ (C_{22} + C_{33} - 2C_{23}) > 0 \\ [C_{11} + C_{22} + C_{33} + 2(C_{12} + C_{13} + C_{23})] > 0 \end{array}\right\} \tag{4}$$

This study calculates the elastic constants of the Al matrix and the $Al_{11}Nd_3$ phase to investigate the mechanical properties of the Al-Nd binary alloy. According to Table 2 and Formula (4), it can be established that $Al_{11}Nd_3$ satisfies the stability criterion for orthorhombic crystal systems, thereby confirming its stability. The compressibility of the crystal along the x, y, and z axes is directly indicated by the values of the elastic constants C_{11}, C_{22}, and C_{33}; a higher value signifies greater resistance to compression in that particular direction. The crystal's capacity to withstand pure shear deformation on the (100), (010), and (001) crystal planes is represented by the elastic constants C_{44}, C_{55}, and C_{66}. Table 2 presents the theoretically calculated elastic constants, revealing that $C_{11} > C_{22} > C_{33}$ for $Al_{11}Nd_3$ suggests that the material is more resistant to compression along the x-axis. Additionally, the relationship $C_{44} > C_{55} > C_{66}$ for $Al_{11}Nd_3$ indicates that it exhibits the highest resistance to shear deformation on the (100) crystal plane.

Table 2. Elastic constants Cij (GPa) of Al and $Al_{11}Nd_3$ in the ground state.

Compounds	C_{11}	C_{12}	C_{13}	C_{22}	C_{23}	C_{33}	C_{44}	C_{55}	C_{66}
Al	114	62					32		
$Al_{11}Nd_3$	130.8	43.2	43.3	126.7	51.8	125.9	55.4	50.3	47.9

This article also determines the bulk modulus (*B*), shear modulus (*G*), Young's modulus (*E*), and Poisson's ratio (σ) of the crystalline compound utilizing the Voigt–Reuss–Hill (VRH) approximation method. The values can be computed using the following formulas [31]:

$$B_H = \frac{1}{2}(B_V + B_R) \tag{5}$$

$$G_H = \frac{1}{2}(G_V + G_R) \tag{6}$$

$$E = \frac{9B_H G_H}{(3B_H + G_H)} \tag{7}$$

$$\sigma = \frac{(3B_H - 2G_H)}{[2(3B_H + G_H)]} \tag{8}$$

In the formula, B_H, B_V, B_R, G_H, G_V, and G_R represent the bulk and shear modulus calculated using the Voigt–Reuss–Hill, Voigt, and Reuss approximation methods [32].

The bulk modulus is a physical property that quantifies the compressive strength of an alloy when subjected to external forces, and it is intrinsically linked to the chemical bonds within the material. The shear modulus represents the capacity of a material to resist shear strain; a lower shear modulus indicates a diminished ability of the material to withstand shear strain. Young's modulus characterizes the stiffness or resistance to deformation of solid materials. Upon analyzing the data presented in Figure 11 and Table 3, it is observed that the *B* value of $Al_{11}Nd_3$ is marginally lower than that of the Al matrix, suggesting that $Al_{11}Nd_3$ possesses inferior compressive strength relative to the Al matrix. Conversely, the *G* and *E* values of $Al_{11}Nd_3$ exceed those of the Al matrix, indicating that $Al_{11}Nd_3$ can resist deformation and shear strain. Generally, the plasticity and brittleness of materials within a crystalline structure can be inferred from the ratio of the bulk modulus to the shear modulus and Poisson's ratio [33]. According to the Pugh criterion, the material is considered brittle if the Poisson's ratio is less than 0.26 or the *B/G* ratio is less than 1.75. As illustrated in Figure 11b, the *B/G* ratio of the Al matrix exceeds 1.75, indicating that the Al matrix exhibits favorable ductility. In contrast, the *B/G* ratio of $Al_{11}Nd_3$ is below 1.75, signifying that the presence of $Al_{11}Nd_3$ contributes to the material's brittleness. This also demonstrates the formation of $Al_{11}Nd_3$, which makes it difficult for Al-Nd alloys to undergo severe plastic deformation. The hardness changes observed in the previous experiments have been verified.

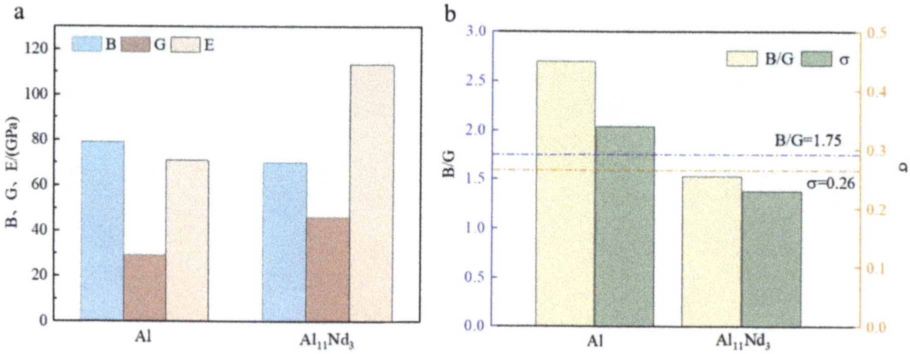

Figure 11. Elastic moduli of Al and $Al_{11}Nd_3$ (**a**) *B*, *G*, and *E* and (**b**) *B/G* and σ.

Table 3. Volume modulus, shear modulus, Young's modulus, Poisson's ratio, and B/G of Al and $Al_{11}Nd_3$.

Compounds	*B* (Gpa)	*G* (Gpa)	*E* (Gpa)	σ	*B*/*G*
Al	79.2	29.3	71.1	0.34	2.70
$Al_{11}Nd_3$	70.2	46.0	113.2	0.23	1.53

4. Conclusions

This article employs a wedge-shaped copper mold to fabricate Al-2at.% Nd alloy ingots and analyses the alloy's microstructure and phase structure in conjunction with first-principles calculations. The findings lead to the following conclusions:

1. Under varying cooling rates, the microstructure of the Al-2at.% Nd alloy displays distinct morphologies. The large grain morphology near the mold wall shows columnar crystals. Conversely, at the center of the ingot, a reduction in the cooling rate leads to a transition in grain morphology from a mixed structure of columnar and equiaxed grains to a predominantly equiaxed grain structure.
2. As the cooling rate diminishes, there is a corresponding increase in grain size, which ranges from 523.28 μm to 743.57 μm. Additionally, the spacing between secondary dendrite arms expands from 14.21 μm to 42.72 μm.
3. At higher cooling rates, the morphology of the $Al_{11}Nd_3$ phase is characterized as skeletal and embedded within the aluminum matrix. The eutectic phase of the alloy comprises α-Al and $Al_{11}Nd_3$. The hardness of the Al-2at.% Nd alloy decreases with a decreasing cooling rate and increases by 25.53% with an increasing cooling rate.
4. According to first-principles calculations, the Al-Nd binary intermetallic compound's formation enthalpy is negative, indicating that $Al_{11}Nd_3$ is thermodynamically stable. The Poisson's ratio of $Al_{11}Nd_3$ is measured at 0.23, and the *B/G* value is calculated to be 1.53, suggesting that the compound exhibits intrinsic brittleness in its ground state.

Author Contributions: Conceptualization, X.W. and S.S.; methodology, X.W.; validation, X.W.; formal analysis, X.W.; investigation, X.Z.; resources, X.W. and S.S.; data curation, X.Z.; writing—original draft preparation, X.Z.; writing—review and editing, W.W. and X.W.; visualization, X.W.; supervision, X.W.; project administration, X.W. and S.S.; and funding acquisition, X.W. and S.S. All authors have read and agreed to the published version of the manuscript.

Funding: This research was funded by the National Key R&D Program (2024YFB3714001, 2023YFB3710401, 2022YFB3504401), the National Natural Science Foundation (52271094), the Qingyuan City Science and Technology Plan Project (2023YFJH003), Central Basic Research Funds

(N2309003), the Bingtuan Science and Technology Program (2023AA003), and the Major Science and Technology Projects in Xinjiang Uygur Autonomous Region (2023A01001).

Institutional Review Board Statement: Not applicable.

Data Availability Statement: Some or all data that support the findings are available from the corresponding author upon reasonable request.

Conflicts of Interest: The authors declare that they have no known competing financial interests or personal relationships that could have appeared to influence the work reported in this paper.

References

1. Chaudhary, B.; Jain, N.K.; Murugesan, J. Development of friction stir powder deposition process for repairing of aerospace-grade aluminum alloys. *CIRP J. Manuf. Sci. Technol.* **2022**, *38*, 252–267. [CrossRef]
2. Xu, Z.; Liu, C.; Zhang, B.; Huang, H.; Cheng, W. Effects of base metal state on the microstructure and mechanical properties of Al–Mg–Si alloy friction stir-welded joints. *J. Manuf. Process.* **2020**, *56*, 248–257. [CrossRef]
3. Mo, Y.F.; Liu, C.Y.; Teng, G.B.; Jiang, H.J.; Chen, Y.; Yang, Z.X.; Han, S.C. Fabrication of 7075-0.25Sc-0.15Zr Alloy with Excellent Damping and Mechanical Properties by FSP and T6 Treatment. *J. Mater. Eng. Perform.* **2018**, *27*, 4162–4167. [CrossRef]
4. Berry, L.; Wheatley, G.; Ma, W.; Nejad, R.M.; Berto, F. The influence of milling induced residual stress on fatigue life of aluminum alloys. *Forces Mech.* **2022**, *7*, 100096. [CrossRef]
5. Chen, B.; Zhang, C.; Jin, Y. First-principles calculation of interface binding strength and fracture performance of β′/Al interface in Al–Mg–Si–Cu alloy. *J. Alloys Compd.* **2020**, *830*, 154515. [CrossRef]
6. Yin, Q.; Chen, G.; Shu, X.; Zhang, B.; Li, C.; Dong, Z.; Cao, J.; An, R.; Huang, Y. Analysis of interaction between dislocation and interface of aluminum matrix/second phase from electronic behavior. *J. Mater. Sci. Technol.* **2023**, *136*, 78–90. [CrossRef]
7. Huang, J.; Li, M.; Liu, Y.; Chen, J.; Lai, Z.; Hu, J.; Zhou, F.; Zhu, J. A first-principles study on the doping stability and micromechanical properties of alloying atoms in aluminum matrix. *Vacuum* **2023**, *207*, 111596. [CrossRef]
8. Lu, Z.; Li, X.; Zhang, L. Thermodynamic Description of Al-Si-Mg-Ce Quaternary System in Al-Rich Corner and Its Experimental Validation. *J. Phase Equilibria Diffus.* **2018**, *39*, 57–67. [CrossRef]
9. He, Y.; Liu, J.; Qiu, S.; Deng, Z.; Zhang, J.; Shen, Y. Microstructure evolution and mechanical properties of Al-La alloys with varying La contents. *Mater. Sci. Eng. A* **2017**, *701*, 134–142. [CrossRef]
10. Huang, X.; Yan, H. Effect of trace La addition on the microstructure and mechanical property of as-cast ADC12 Al-Alloy. *J. Wuhan Univ. Technol. Mater.* **2013**, *28*, 202–205. [CrossRef]
11. Pan, F.; Yang, M.; Chen, X. A Review on Casting Magnesium Alloys: Modification of Commercial Alloys and Development of New Alloys. *J. Mater. Sci. Technol.* **2016**, *32*, 1211–1221. [CrossRef]
12. Ding, W.; Zhao, X.; Chen, T.; Zhang, H.; Liu, X.; Cheng, Y.; Lei, D. Effect of rare earth Y and Al–Ti–B master alloy on the microstructure and mechanical properties of 6063 aluminum alloy. *J. Alloys Compd.* **2020**, *830*, 154685. [CrossRef]
13. Sarkar, J.; Saimoto, S.; Mathew, B.; Gilman, P. Microstructure, texture and tensile properties of aluminum2at.% neodymium alloy as used in flat panel displays. *J. Alloys Compd.* **2009**, *479*, 719–725. [CrossRef]
14. Zhu, S.; Gibson, M.; Nie, J.; Easton, M.; Abbott, T. Microstructural analysis of the creep resistance of die-cast Mg-4Al-2RE alloy. *Scr. Mater.* **2008**, *58*, 477–480. [CrossRef]
15. Xiao, Y.D.; Li, W.X.; Ma, Z.Q. Crystallization process in rapidly solidified Al-Nd-Ni amorphous alloy prepared by melt spinning. *Trans. Nonferrous Met. Soc. China* **2004**, *14*, 665–669.
16. Liu, T.; Ma, T.; Li, Y.; Ren, Y.; Liu, W. Stabilities, mechanical and thermodynamic properties of Al–RE intermetallics: A first-principles study. *J. Rare Earths* **2022**, *40*, 345–352. [CrossRef]
17. Liu, S.; Esteban-Manzanares, G.; LLorca, J. First Principles Prediction of the Al-Li Phase Diagram. *Metall. Mater. Trans. A* **2021**, *52*, 4675–4690. [CrossRef]
18. Guo, Y.; Wang, W.; Huang, H.; Zhao, H.; Jing, Y.; Yi, G.; Luo, L.; Liu, Y. Effect of doping Zn atom on the structural stability, mechanical and thermodynamic properties of AlLi phase in Mg–Li alloys from first-principles calculations. *Philos. Mag.* **2020**, *100*, 1849–1867. [CrossRef]
19. Gao, M.; Deng, Y.; Wen, D.; Zhao, H. A first-principles study on the site occupancy behavior of transition metals in L12-Al_3Li phase. *Mater. Res. Express* **2020**, *7*, 046519. [CrossRef]
20. Gao, M.C.; Rollett, A.D.; Widom, M. Lattice stability of aluminum-rare earth binary systems: A first-principles approach. *Phys. Rev. B* **2007**, *75*, 174120. [CrossRef]
21. Jahnátek, M.; Krajcí, M.; Hafner, J. Response of trialuminides to [110] uniaxial loading: An ab initio study for Al_3(Sc, Ti, V). *Phys. Rev. B* **2007**, *76*, 014110. [CrossRef]
22. Joubert, D.; Kresse, G. From ultrasoft pseudopotentials to the projector augmented-wave method. *Phys. Rev. B* **1999**, *59*, 1758–1775.

23. Burke, K.; Ernzerhof, M.; Perdew, J.P. Generalized Gradient Approximation Made Simple. *Phys. Rev. Lett.* **1997**, *78*, 1396.
24. Xue, B.; Xiao, W.; Li, X.; Gao, G.; Li, X.; Zhang, Y.; Wang, L.; Xiong, B. Comprehensive investigation on the structural, electronic and mechanical properties of T-$Mg_{32}(Al, Zn)_{49}$ phases in Al-Mg-Zn alloys. *J. Mater. Sci. Technol.* **2024**, *173*, 237–246. [CrossRef]
25. Lin, L.; Wang, F.; Yang, L.; Chen, L.; Liu, Z.; Wang, Y. Microstructure investigation and first-principle analysis of die-cast AZ91 alloy with calcium addition. *Mater. Sci. Eng. A* **2011**, *528*, 5283–5288. [CrossRef]
26. Fung, Y.C.; Drucker, D.C. Foundation of Solid Mechanics. *J. Appl. Mech.* **1966**, *33*, 238. [CrossRef]
27. Yao, X.; Wang, H.; He, B.; Zhou, X. Modeling of Columnar-to-Equiaxed Transition in Solidified Al-Si Alloys. *Mater. Sci. Forum* **2005**, *475–479*, 3141–3144. [CrossRef]
28. Okamoto, H. Al-Nd (Aluminum-Neodymium). *J. Phase Equilib.* **2000**, *21*, 206. [CrossRef]
29. Yang, Q.; Shi, W.; Wang, M.; Jia, L.; Wang, W.; Zhang, H. Influence of cooling rate on the microstructure and mechanical properties of Al–Cu–Li–Mg–Zn alloy. *J. Mater. Res. Technol.* **2023**, *25*, 3151–3166. [CrossRef]
30. Ahmad, S.; Ahmad, R.; Jalali-Asadabadi, S.; Ali, Z.; Ahmad, I. First principle studies of structural, magnetic and elastic properties of orthorhombic rare-earth diaurides intermetallics RAu_2 (R=La, Ce, Pr and Eu). *Mater. Chem. Phys.* **2018**, *212*, 44–50. [CrossRef]
31. Boucetta, S. Structural and elastic properties of $Mg_3CuH_{0.6}$ ternary hydride by ab initio study. *J. Magnes. Alloys* **2018**, *6*, 90–94. [CrossRef]
32. Liu, Y.; Hu, W.-C.; Li, D.-J.; Li, K.; Jin, H.-L.; Xu, Y.-X.; Xu, C.-S.; Zeng, X.-Q. Mechanical, electronic and thermodynamic properties of C14-type AMg_2 (A=Ca, Sr and Ba) compounds from first principles calculations. *Comput. Mater. Sci.* **2015**, *97*, 75–85. [CrossRef]
33. Abraham, J.A.; Pagare, G.; Chouhan, S.S.; Sanyal, S.P. Theoretical calculations of structural, electronic, optical, elastic, and thermal properties of YX3 (X = In, Sn, Tl, and Pb)compounds based on density functional theory. *J. Mater. Sci.* **2015**, *50*, 542–554. [CrossRef]

crystals

MDPI

Article

Study of the Hibridation of Ablation Casting and Laser Wire Metal Deposition for Aluminum Alloy 5356

Ana Isabel Fernández-Calvo [1,*], Mikel Madarieta [2], Ane Solana [2], Ibon Lizarralde [1], Mikel Rouco [1] and Carlos Soriano [2]

[1] AZTERLAN, Basque Research and Technology Alliance (BRTA), Aliendalde Auzunea, 6, 48200 Durango, Bizkaia, Spain; ilizarralde@azterlan.es (I.L.); mrouco@azterlan.es (M.R.)
[2] Fundación Tekniker, Basque Research and Technology Alliance (BRTA), Iñaki Goenaga, 5, 20600 Eibar, Gipuzkoa, Spain; ane.solana@tekniker.es (A.S.); carlos.soriano@tekniker.es (C.S.)
* Correspondence: afernandez@azterlan.es

Abstract

The rapidly growing field of metal additive manufacturing (AM) has enabled the fabrication of near-net-shape components with complex 3D structures in a more reliable, productive, and sustainable way compared to any other manufacturing process. The productivity of AM could be significantly increased combining conventional and AM technologies. However, the application at an industrial level requires the validation of the AM process itself and the assurance of the soundness of the junction between the substrate and the deposited metal at a sufficiently rapid metal deposition rate. In this work, the validation of additively manufactured samples of Al-5356 alloy was performed. These were manufactured partially via an ablation casting process and partially via laser metal deposition using a metallic wire (LMwD). The deposited material showed low porosity levels, i.e., below 0.04%, and a small number of lack-of-union defects, which are detrimental to the mechanical properties. In the tensile samples centred at the junction between the ablated and deposited materials, it was found that when the AM part of the sample exhibited no lack-of-union defects, the region manufactured using LMwD showed higher strength than the ablation-cast part. These results suggest that the combination of ablation casting and LMwD is a competitive technique for the manufacturing of Al-5356 alloy parts with complex geometries.

Keywords: hybrid manufacturing; aluminium 5356 alloy; additive manufacturing; laser metal wire deposition; casting; ablation casting

1. Introduction

Hybrid manufacturing, which integrates additive manufacturing (AM) with conventional techniques, is a relatively new trend. This approach combines the flexibility and precision of AM with the strength, scalability, and material properties attainable with traditional methods, such as casting and forging, followed by machining. The combination of ablation casting with laser metal deposition (LMwD) allows for the production of near-net-shape components in a more reliable, productive, and sustainable manner compared to any single manufacturing process [1]. Sefene et al. [2] reviewed hybrid manufacturing approaches using AM. Generally, AM is performed on a wrought or cast precursor of the final product [3–7]. An increased material yield, a reduction in the machining time, and chip removal are further advantages when using hybrid components. Hybrid manufacturing using AM implies the use of laser-based directed energy deposition (DED-LB) or

 https://doi.org/10.3390/cryst15020134

wire arc additive manufacturing (WAAM) rather than powder bed fusion (PBF) as additive technologies [2]. The latter allows for better dimensional accuracy and reduced heat inputs, thus being preferable for repair and maintenance work rather than hybridisation.

Ablation casting is an innovative casting process that offers higher flexibility in design than the conventional near-net-shape casting process [8–10] It allows for the manufacture of large, thin-walled parts and offers greater control over the presence of hot spots; this reduces the risk of hot cracking thanks to the possibility of locally modifying the cooling of the piece during solidification [11]. Ablation comprises two main successive steps. First, a liquid metal, e.g., aluminium, is poured into a mould, composed mainly of sand and other refractory particles held together by a water-soluble binder. Then, the mould is dissolved locally via the action of a water-spraying system. After the local collapse of the mould, the water directly impacts the surface of the metal, which results in efficient heat removal in this area and thus the rapid solidification of the metal. It should be noted that at the time of the mould's crumbling, the cast must already have a solid outer layer to prevent the liquid metal from leaking. Solidification occurs under a high cooling rate, achieving a very fine microstructure [12–15].

The microstructure of the component is strongly affected by the duration of the ablation cooling applied to the casting [14,16–18]. The refinement of microconstituents is the main reason for the improved mechanical properties achieved with this casting process, rather than the refinement of the SDAS [9,12,14,15] or the grain size [12]. Moreover, the ablation process allows for the fine-tuning of the grain size, the refinement of the eutectic phases, and a significant decrease in porosity in the manufactured alloys [12,19]. Ablated samples present improved mechanical properties compared to conventional sand-cast samples in terms of the quality index, Q, which is related [20] to the tensile strength and elongation. Their high Q index values are similar to those of permanent mould samples [12]. Another interesting feature of ablation technology is the possibility of casting wrought alloy compositions [8], such as the 5XXX [19] series. These alloys are not castable with any other technology when the typical complex casting geometries are used; this is due to their strong susceptibility to hot cracking, which can be overcome with ablation technology. Combined with the high mechanical properties achievable with ablation technology, these are the main reasons that ablation casting is selected for hybrid manufacturing using AM.

As recently stated in the ASTM 52900 standard, laser metal wire deposition (LMwD), or the DED-LB/M wire process, is an additive manufacturing technique that falls within the category of directed energy deposition (DED) processes. It consists of the controlled deposition of a metallic wire that is melted locally via a concentrated laser beam [21]. Different metals can be used in wire format, such as Fe-, Co-, Ni-, Ti-, and Al-based alloys. However, historically, aluminium alloys have been considered difficult to process with laser technology, mainly due to their high thermal conductivity, high reflectivity, and low viscosity [22]. Aluminium's high thermal conductivity facilitates heat dissipation throughout the treated component, hindering the concentration of energy in the interaction zone. However, some alloying elements, such as Mg, Si, and Zn, decrease the thermal conductivity and melting temperature of aluminium, facilitating the laser deposition process, although their influence is different in each alloy [23–25].

The high reflectivity of aluminium at typical industrial infrared laser wavelengths is influenced by the presence of an oxide layer and other surface contaminants and the surface roughness. Different conventional surface treatments, such as sandblasting or the application of dark coatings, have been shown to decrease the reflectivity of aluminium alloys. More sophisticated treatments based on laser texturing have also been developed to increase the absorption of laser radiation [26]. Moreover, the low viscosity of liquid aluminium prevents the expansion of the molten bath before its solidification, hindering the laser treatment of these alloys [26]. In addition to the high reflectivity and low viscosity of aluminium in the solid and liquid phases, respectively, there are other problems that can lead to defects in manufactured structures. For example, the 10–20 nm thick layers of hard aluminium oxide (Al_2O_3, or alumina) that form on the surface of aluminium can lead to the presence of inclusions in the deposited material once it solidifies. The solubility of hydrogen in aluminium is also a drawback, as hydrogen, when dissolved in the molten metal, can precipitate during solidification; this leads to the formation of hydrogen pores in the welded areas [27].

In recent years, laser welding has evolved into additive manufacturing using the DED-LB/M powder technique for part repair and component manufacturing [28–30]. However, the addition of aluminium in wire format is considered as a more accessible, efficient, and economical alternative, especially as there are more commercial alloys available in such a format. Thus far, there have been few studies related to the deposition of aluminium alloys using this technique. Those focused on the magnesium aluminium alloy Al-5087 are of particular interest; here, with the appropriate parameters, it is possible to control the geometric shape of the supplied weld seams, the dilution, and the aspect ratio and to obtain defect-free layers with geometric characteristics that are adapted to the specific requirements of the process [31–33]. Studies have also been carried out in which the microstructure formed and the refinement of the grain size through the deposition of controlled energy have been discussed [34,35]. Recently, substantial work has been published on the optimisation of the deposition of coaxial filling wires in aluminium 5356 alloy to minimise the presence of defects, especially regarding the lack of union between consecutive weld seams [36–40]. Moreover, compared to other conventional DED additive manufacturing technologies, DED laser techniques result in minimal distortion in the deposition thanks to the highly localised heat input inherent in laser technology. Thus, together with ablation technology, this constitutes an interesting alternative for the manufacture of structures and components from the aluminium 5XXX series.

In this study, a hybrid component was analysed. Ablation casting and additive manufacturing, namely, the LMwD technique, were used to manufacture aluminium 5356 hybrid components. Thus, the most beneficial aspects of conventional manufacturing technologies were combined with the advantages of additive manufacturing, and complex geometries could be achieved. An analysis of the microstructure, defects, and mechanical properties of the hybrid components was performed, and the results were compared with the findings of previous works on the microhardness [6,7,41–43] and tensile properties [41–43] of miniaturised samples [3,42] and hybrid components. The transition line between the substrate and the additively deposited material was also examined.

Hybrid manufacturing using these two technologies is a novelty at an industrial level; it exploits the benefits of each of the techniques, facilitating the manufacture of near-net-shape aluminium components and structures with certain complexity. This hybrid manufacturing solution could also be beneficial in repair and maintenance work or in the manufacturing of complex and unique parts whose manufacturing would incur substantial costs if conducted using ablation casting or LMwD independently. For example, in repair applications, hybrid manufacturing can be used to reduce the costs involved in redesigning moulds and manufacturing entirely new parts. In this way, it would only be necessary to manufacture a specific part of the component through LMwD. Moreover, although the ablation process itself already reduces the costs involved in manufacturing and redesigning casting moulds, hybrid manufacturing using an ablated substrate makes it possible to address the geometric limitations of the manufacturing process, allowing for the manufacturing of components with increasing complexity without the need for later machining operations.

2. Materials and Methods

2.1. Experimental Materials

Figure 1a shows the hybrid aluminium test piece manufactured to analyse the compatibility between the ablation casting and additive LMwD techniques. The base of the structure was cast following the ablation technique explained in Section 2.2; on top of this, the remaining material was deposited via LMwD, as described in Section 2.3. Both parts of the structure were manufactured from aluminium 5356 alloy. Table 1 shows the chemical composition of the ablation-cast substrate.

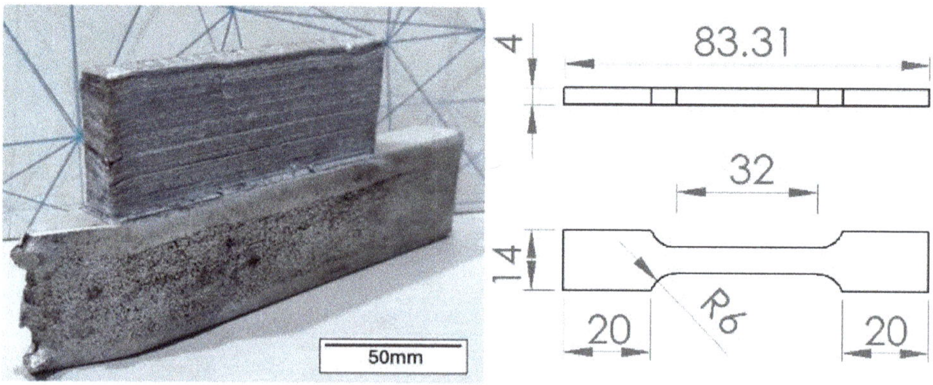

(a) (b)

Figure 1. (**a**) Test piece manufactured for microstructural analysis and for extraction of hybrid tensile specimens and (**b**) dimensions of tensile test specimens.

Table 1. Chemical composition of the aluminium 5356 alloy in the ablated substrate in weight percent (wt. %).

	Chemical Composition [wt. %]								
	Mg	Si	Fe	Mn	Cr	Cu	Zn	Ti	B
Ablated material	5.15	<0.10	0.12	0.070	0.076	<0.03	<0.04	0.11	0.013

2.2. Ablation Process

To manufacture the bases of the test samples, high-purity aluminium was melted in an electric furnace with a maximum capacity of 60 kg at temperatures between 740 °C and 750 °C. The composition of the 5356 alloy was achieved by adding pure Mg, along with Al-Mn (20 wt. % of Mn) and Al-Cr (80 wt. % of Cr) as master alloys. The melt was grain-refined with $AlTi_5B$. A deoxidising and covering flux was imposed on the melt's surface. Following this, the melt was subjected to rotary degassing with a rotor impeller with high-purity Ar (rotational speed: 50 rpm; argon flux: 8 L/min) for 30 min and then skimmed. A sample for chemical analysis was then cast; afterwards, the liquid metal was carefully transferred and poured into a filling basin attached to the sand mould using a ceramic saucepan. The mould was then tilted at 90° in less than 10 s so that it filled the specimens with the liquid metal.

The mould was produced using standard silica sand for foundry use, with an AFS Grain Fineness Number of 50–55, using a hydro-soluble binder. When the external surface of the casting was solidified, the mould was sprayed with water to accelerate the cooling rate during the solidification of the metal, hence obtaining the ablated test specimens.

Further details of the ablation procedure are given in Refs. [12,13]. The ablated specimens consisted of a parallelopipedon with a height of approximately 25 mm, a thickness of approximately 70 mm, and a length of approximately 200 mm.

2.3. LMwD Process

The ablated aluminium test pieces were machined to provide a flat surface for the additive manufacturing of the aluminium 5356 alloy via the LMwD process. After mechanisation, the flat surface was shot-blasted to improve the adsorption between the substrate and the filling material.

The additive manufacturing process was carried out on a 3-axis linear machine designed for the manufacture of metallic components with dimensions of up to 1900 mm × 900 mm × 750 mm using the LMwD technique. The machine incorporated a continuous-wave 4 kW fibre laser source and a coaxial wire laser head. A laser spot diameter of 3.5 mm was selected, resulting in defocusing of 6 mm from the focal plane. In this study, the deposition material was an aluminium wire with a diameter of 1.2 mm, composed of the same grade 5356 alloy that was used for the substrate.

To carry out this study, input parameters from a previous experimental work were used, seeking to obtain defect-free structures and deposition. In particular, it was necessary to ensure the absence of lack-of-union defects between the welding seams, as well as pores or cracks, while maximising the deposition rate [36]. Deposition was performed on the ablated specimens across a 25 mm × 200 mm surface, generating prismatic structures of approximately 150 mm × 50 mm × 20 mm (see Figure 1a). The optimised LMwD parameters are shown in Table 2.

Table 2. Process parameters used in the deposition of the Al-5356 wire on the ablated substrate.

Deposition rate	0.7 kg/h
Wire feeding rate	4 m/min
Linear speed	1.5 m/min
Laser power	3.5 kW
Welding seem width	3.5 mm
Overlapping degree	0.25%
Layer height	1.25 mm

2.4. Microstructural Characterisation

For the metallographic analysis, cross-sections obtained from the test sample indicated in Figure 1a were prepared via grinding and polishing using standard procedures. Optical and scanning electron microscopy (SEM) by using field emission scanning electron microscope (FESEM), ZEISS Ultra Plus microscope (CARL ZEISS AG, Oberkochen, Germany) were applied for the microstructural analysis of the sample cross-sections. Energy Dispersive Spectroscopy (EDS) X-Max 20mm^2, from Oxford Instruments, Abingdon, UK was employed for spectra and mappings to identify the alloying elements present in both the intermetallic particles and the matrix. The tensile test fracture surfaces were also analysed via SEM.

A cross-section of one manufactured test sample that contained both the ablated substrate and the filled material was metallographically prepared for the microstructural analysis. The metallographic preparation consisted of gradually sanding the surface, starting with waterproof sandpaper with a grain size of 120 pp and transitioning to a grain size of 2500 pp. After sanding, the surface was polished with a cloth lubricated with a [illegible] of alumina particles with a diameter of 0.06 μm, followed by etching using

Keller's reagent. Regarding the microstructural analysis, the morphologies of the grains and intermetallic precipitates were characterised via optical and electron microscopy.

2.5. Mechanical Behaviour

After the metallographic preparation process, microhardness measurements were taken around the junction between the base and deposited material to assess its soundness. These measurements were performed on 16 mm long chains centred on the interface, with indentations separated by 0.5 mm. The mechanical behaviour of the hybrid samples was determined by means of tensile tests on specimens with geometries that complied with ASTM E8/E8M, as shown in Figure 1b. The tensile specimens were extracted along the direction perpendicular to the interface between the ablated substrate and the LMwD material, leaving the interface directly in the middle of the sample. Based on the prior microstructural analysis, the results of the tensile tests, and a subsequent fractographic analysis, the soundness of the junction between the substrate and the deposition was determined. The fractographic analysis was performed based on the analysis of the fracture surface using a stereoscopic microscope (macroscopic) and on the SEM (microscopic) analysis, as well as on the microscopic inspection of the section perpendicular to the fracture surface near the rupture using an optical microscope.

3. Results and Discussion

In this section, the results are presented and discussed. The results are divided into the following sub-sections: microstructural analysis, mechanical properties, and fracture analysis.

3.1. Microstructural Characterisation

Defects such as pores and lack-of-union defects were found in the structures generated using LMwD. These defects are typical in samples manufactured using DED techniques. In the case of the additive manufacturing of aluminium alloys, pores are primarily associated with the entrainment of gas generated during the deposition process—mainly the dissolved hydrogen in the molten material. During the melting process, hydrogen gas from the atmosphere and from the dissolved aluminium surface accumulates during solidification and causes pore formation [27]. Meanwhile, a lack-of-union defect is typically due to gaps generated between the weld seams due to heterogeneous overlap between the seams. A cross-section of one of the prismatic structures is shown in Figure 2. Both pores and lack-of-union defects were observed. The pores had a spherical shape, with the diameter of the largest one being 85 μm (see image N. 1 in Figure 2). Regarding the lack-of-union faults, the largest had a length of approximately 300 μm and an area of 9250 μm, as shown in image N. 2 in Figure 2. In total, the density of the defects obtained in the added part of the structure shown in Figure 2 was 0.037%.

In terms of the microstructure, the Al-5356 alloy is formed as a matrix of aluminium and intermetallic Al-Fe precipitates due the unavoidable presence of iron impurities in the chemical composition (around 0.1 wt. %). The microstructure can vary considerably depending on the manufacturing process. As shown in Figure 3A,B, the ablated substrate, shown in the lower part of the image, consisted of aluminium grains with dendrites, marked by Al-Fe intermetallic precipitates in the interdendritic spaces and at the grain boundaries. As the primary aluminium solidified, the remaining liquid was enriched in the other alloying elements, mainly Mg, Mn, Fe, and Cr. On the other hand, in the LMwD area, the precipitates were dispersed in the aluminium matrix.

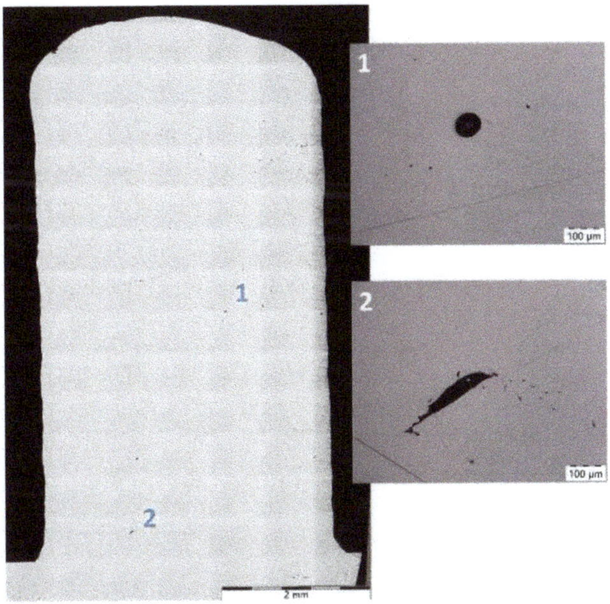

Figure 2. Cross-sectional analysis of LMwD-manufactured structure with details of the coarsest defects found in this section: (1) a hydrogen-type pore defect and (2) a lack-of-union defect generated between successive layers.

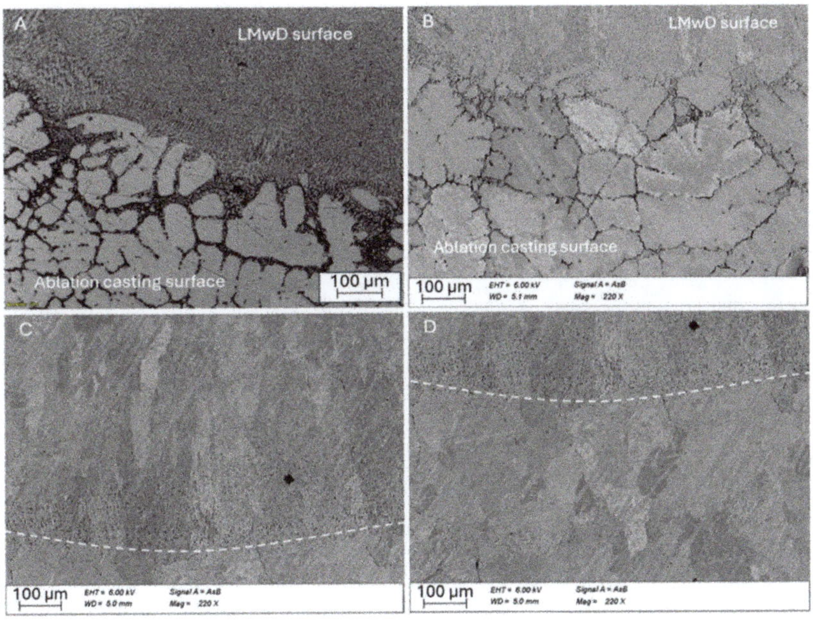

Figure 3. Microstructures in the different areas of the hybrid component: (**A**) optical microscopy image and (**B**) scanning electron microscopy of the junction area between ablation and LMwD deposition; (**C**,**D**) scanning electron microscopy images of the microstructural heterogeneity observed in the height of the LMwD-treated area at 15 mm from the ablated surface (**C**) and at the top of the bead (**D**). The dashed white lines indicate the observed microstructural differences between layers of subsequent deposition.

The grains of the ablated aluminium matrix were fully distinguishable, with an equiaxial shape and a size between 100 μm and 300 μm, as shown in Figure 3B. In contrast, the grains of the LMwD-treated zone were difficult to distinguish with Keller's reagent because the precipitates were not concentrated at the grain boundaries. Nevertheless, columnar grains were observed; these grew epitaxially from the joining area in the direction of cooling, i.e., perpendicular to the substrate base. The width of these columnar grains varied between 20 μm and 100 μm.

It should be noted that the microstructure did not remain uniform throughout the height of the deposition; this was mainly associated with the temperature changes that occurred during the manufacturing process. Figure 3C,D show the interlayered area of the LMwD region, separated by a dashed white line. In Figure 3C, in the region above the white line, corresponding to the lower part of a weld seam, columnar grains that formed in the cooling direction can be observed, as was the case in the first layer of the deposition. The width of these grains was between 30 μm and 90 μm. Meanwhile, Figure 3D shows the top of the bead that was reheated by the next layer. This reheating altered the microstructure of this area, generating larger equiaxed grains with sizes between 80 and 170 μm.

Regarding the Al-Fe intermetallic precipitates, large concentrations could be observed at the bottom of the weld seams, as shown in Figure 3C,D, where a dense area of dark particles is clearly distinguished. These particles are clustered above the dashed white lines. Moving along the weld seam, the density of these precipitates decreased until they were barely visible. As noted regarding the change in the shape of the grains in the matrix, this phenomenon was caused by the high reheating temperature achieved by the material present near the top layer, which was able to remelt the top surface of the previous layer. The Fe-rich particles were formed during the last stages of solidification of the remelt aluminium.

Figure 4a shows a general SEM image of the microstructure in the LMwD part of the sample, which consisted of a disperse distribution of small-sized precipitates. In Figure 4b, a detailed image of the area marked in Figure 4a is depicted. The comparison between the EDS spectra of the matrix (Figure 4c) and the precipitates (Figure 4d) indicates that the latter were intermetallic Al-Fe phases. This observation is supported by the spatial distribution measurements for Al, Mg, Fe, and Ti provided in the EDS mappings shown in Figure 4e–h. Figure 4f shows that the concentration of iron was higher in the precipitates, whereas titanium was homogeneously distributed throughout the material (Figure 4h). The heterogeneous distribution of Mg, as shown in Figure 4g, did not correlate with the position of the Fe-rich precipitates. Instead, it was likely associated with segregations that occurred during solidification.

3.2. Mechanical Properties

Vickers microhardness measurements were performed to analyse the junction between the areas created with the two manufacturing processes. Two samples were chosen for this test. The results of the two different microhardness tests are shown in Figure 5a. They ranged from 8 mm deep within the solidified substrate to a height of 8 mm in the LwMD area. The hardness of the ablated zone varied from 53 to 70 HV. The valleys in the microhardness were likely associated with the softer primary aluminium grains, while the peaks were associated with the interdendritic spaces and/or grain boundaries, which contained hard intermetallic precipitates. The hardness in the deposited zone was higher and more homogeneous than in the ablated zone, varying between 65.7 and 76.8 HV. This slight increase in hardness from the AM area to the ablated area was directly related to the observed microstructure. Al-Fe precipitates from the ablated substrate, which were harder than the matrix, were found in the casting, concentrated in the interdendritic spaces and grain boundaries, while the aluminium matrix was softer because it was free of precipitates. In the case of the deposited area, the precipitates were scattered throughout the matrix, so the indenter faced a harder material. The fact that the 5XXX series is non-thermically treatable means that the heat-affected zone (HAZ) of the ablated substrate did not present a clear decrease in hardness associated with the dissolution of age-hardening particles, as was observed, for example, in the AM of a 6082 forged substrate using WAAM [6]. A hardness increase in the first few layers of the AM part after the application of a heat treatment was also observed by Kučerová et al. [7]; this was associated with grain refinement during the printing of these layers on the cold substrate and with the precipitation of unique particles in the diffusion-affected area of these layers.

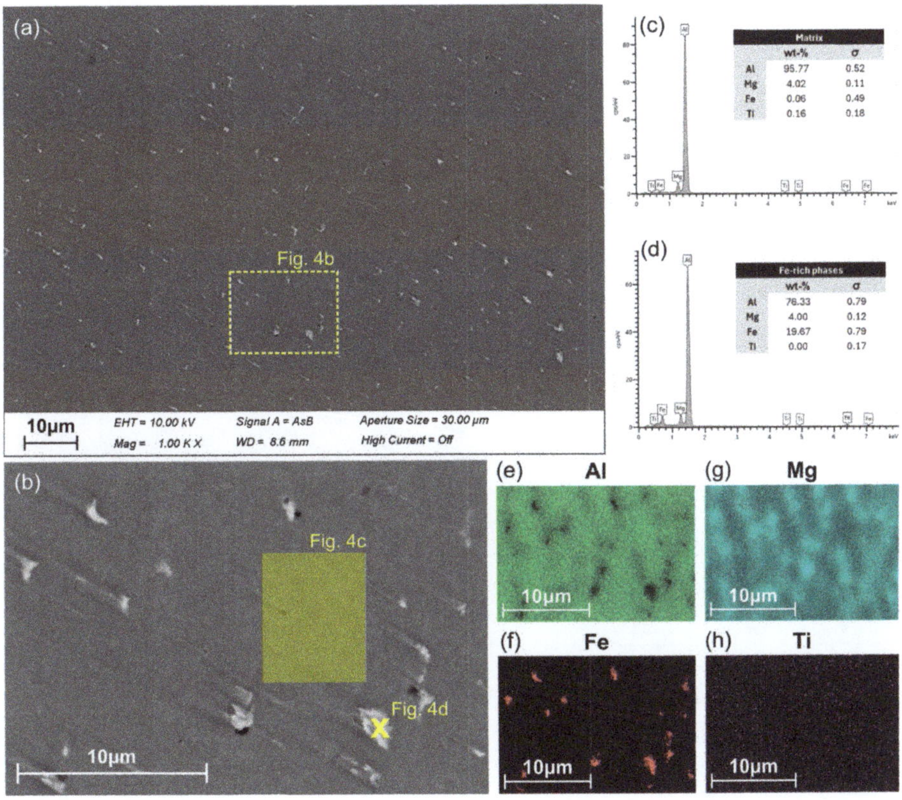

Figure 4. (**a**) Image of the intermetallic precipitates obtained via scanning electron microscopy; (**b**) a zoom of the intermetallic precipitates in the area framed in yellow in (**a**); (**c**) EDS spectra of the matrix area framed in yellow in Figure 4b; (**d**) EDS spectra of the intermetallic precipitate marked with an X in Figure 4b; and (**e–h**) spatial distribution of Al, Mg, Fe, and Ti obtained with EDS in the micrograph shown in Figure 4b (the relative intensity between different element mappings is meaningless).

The stress–strain curves obtained from the tensile tests are shown in Figure 5b, with the corresponding results shown in Table 3. The total elongation at maximum load, A_{gt}, is also indicated in Figure 5b. The inset in Figure 5b presents the fracture zone of each sample, taking the joint area between the ablation-cast and LMwD parts as a reference. These results show that three of the six analysed samples were fractured at the deposited area, while the remaining three samples were fractured in the ablation-cast part (two of them near the junction). The fracture zone did not have a major influence on the yield strength values, as can be observed in Table 3. This is because all specimens exhibited results close to the average (103 MPa), with one of the fractures in the junction yielding the highest value (112 MPa) and one fracture in the deposition zone yielding the lowest value (97 MPa). Samples 1, 4, and 6, corresponding to the ablated area, abruptly broke near the maximum load, with the shape of the three tensile/elongation curves being similar. Meanwhile, tensile samples 2, 3, and 5 broke in the additive area; they showed that after achieving the maximum load, the stress decreased continuously, while the elongation increased until complete fracture occurred.

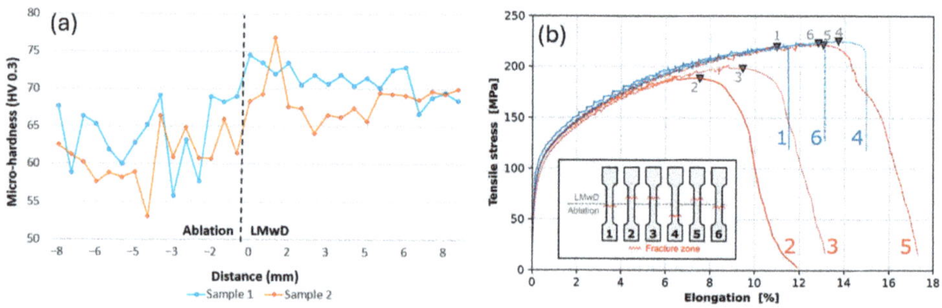

Figure 5. (**a**) Microhardness measurement chains across the ablation/LMwD interface and (**b**) stress–strain curves obtained from the tensile tests performed with the hybrid test specimens, with the points of maximum load indicated and (inset) the identification of the fracture zone. The stress–strain curves of the samples broken in the ablated and LMwD parts are represented in red and blue, respectively. The numbers 1–6 in Figure 5b indicate the sample numbers.

Table 3. Results of tensile strength, R_m, total elongation at maximum load, A_{gt} (according to ISO 6892-1), and yield strength, $R_{p0.2}$, of the tested hybrid samples.

	1	**2**	**3**	**4**	**5**	**6**	**Average Value**
Yield strength, $R_{p0.2}$ (MPa)	112	101	97	102	105	104	104 ± 5
Tensile strength, R_m (MPa)	222	190	202	225	223	224	214 ± 13
Tot. elong. at max. load, A_{gt} (%)	11.2	7.5	9.5	13.7	13.0	12.8	11 ± 2

Elongation greater than 10%, a yield strength greater than 97 MPa, and a tensile strength greater than 190 MPa were observed in all the hybrid test pieces analysed. The three test pieces fractured at the ablation zone showed similar yield strengths and maximum load values (see Figure 5b), as well as elongation between 11.2 and 13.7%. This exemplifies the mechanical behaviour of the 5356 alloy manufactured via the ablation casting method. However, the three test pieces that were fractured in the LMwD area showed greater dispersion in their mechanical properties. In particular, regarding the tensile strength, differences were found depending on the fracture zone.

3.3. Fracture Surface Analysis

Next, a fractographic analysis of the tensile test samples was performed on and around their fracture surfaces. This analysis was used to determine the fracture zone and the main fracture mode. These results can help to validate the soundness of the junction zone and determine which side of the hybrid structure is stronger.

Figure 6 shows a lateral view of both halves of each of the tensile test samples around the fracture surface, as observed under a stereoscopic microscope. Meanwhile, in Figure 7, the top view of the fracture surface in SEM is shown. The white numbers in these figures, ranging from 1 to 6, correspond to the sample numbers. At the macroscopic level, the rupturing of the six tensile test samples could be divided into two main categories. On the one hand, samples 1, 4, and 6 presented a rough fracture surface that formed a plane almost normal to the principal axis of the tensile sample (especially in samples 1 and 6), with no clear necking. On the other hand, samples 2, 3, and 5 showed clear necking near the fracture, with a surface with finer rugosity and a very irregular shape. Such an irregularly shaped fracture surface is an indication of the presence of defects in the material before the tensile test was performed.

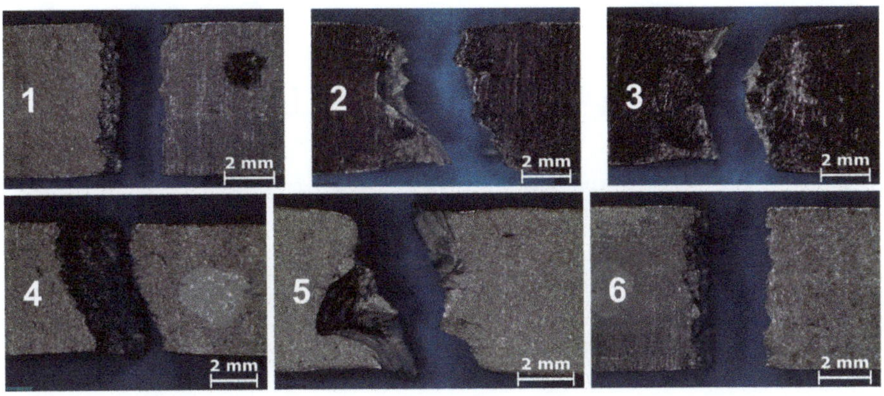

Figure 6. Lateral views of the six tensile test samples around their fracture surfaces, seen under a stereoscopic magnifier. The white numbers in the figures indicate the sample numbers.

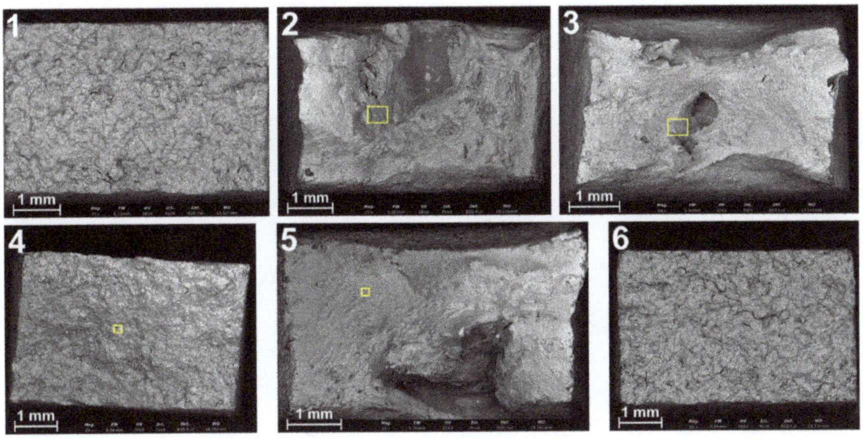

Figure 7. Top view of the fracture surface at one side of each of the tensile test samples. White numbers indicate the sample numbers. The yellow-framed boxes were selected for zoomed-in analysis in some of the following figures.

Figure 8 shows the reconstruction of the cross-section around the fracture surface of sample 1. Despite being near the junction zone, the fracture clearly advanced through the ablated substrate. The fracture in test sample 6 was located at a similar depth in the ablated substrate. From the combined analysis of Figures 6–8, it can be deduced that the samples with rough fracture surfaces (samples 1, 4, and 6) were those in which the fractures broke through the ablated aluminium substrate (see Figure 8), and those with irregularly shaped fracture surfaces (samples 2, 3, and 5) were those in which the fracture broke through the deposited material.

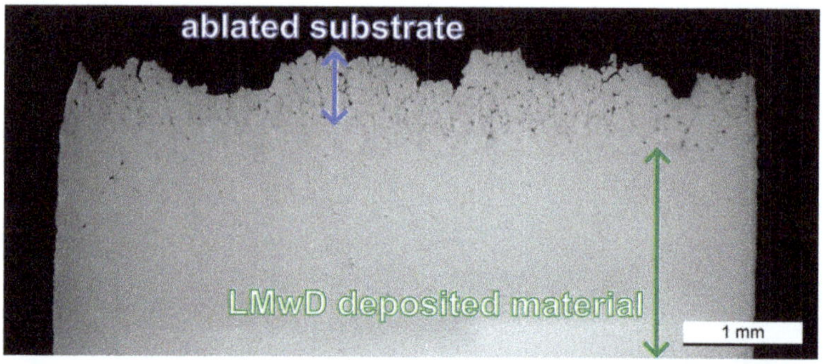

Figure 8. Reconstruction of the cross-section near the fracture zone of sample 1, with the ablated and LMwD-deposited parts of the test specimen indicated.

At the microscopic scale, the fracture surfaces of all samples in which the fracture broke through the ablated substrate (samples 1, 4, and 6) showed similar fractographic characteristics. The fracture surfaces in these samples showed some cracking, which advanced along the intergranular region (Figures 8 and 9), as well as a combination of two sets of dimples of different scales (Figure 9a). The set of large dimples was associated with the ductile fracture of the aluminium matrix, whereas the finer dimples were associated with the presence of Al-Mg intermetallic precipitates near the intergranular regions. Thus, the fractographic features indicate that samples 1, 4, and 6 exhibited a medium-energy ductile mode of fracture along the intergranular space [44,45]. The absence of defect indications in the fracture surface allows us to associate this fracture mode directly with the failure mode of ablated Al-5356 microstructures.

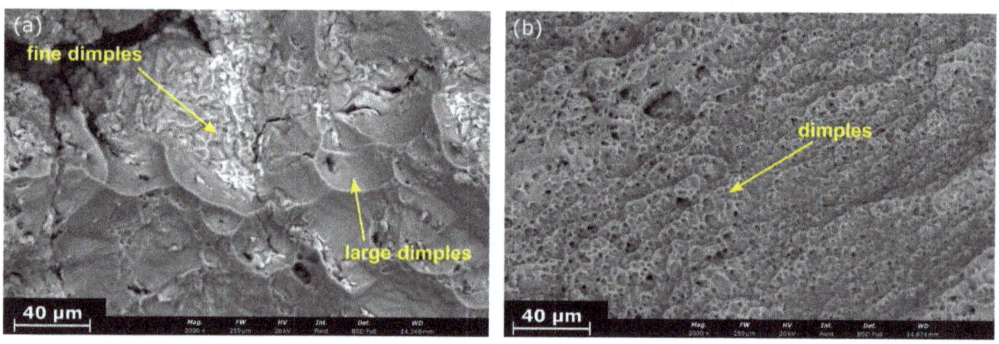

Figure 9. Micrographs of the fracture surfaces (zoom-in of the correponding yellow-framed boxes shown in Figure 7) of (**a**) sample 4, broken through the ablated substrate, and (**b**) a defect-free area in sample 5, broken through the filled material.

In contrast, samples 2, 3, and 5, where the fracture emerged across the deposited material, presented irregular fracture surfaces with large sinks and/or hills. Outside these areas of pronounced irregularity, the fracture surfaces of samples 2, 3, and 5 showed a pattern of dimples with very small sizes (see Figure 9b), which was consistent with a high-energy ductile fracture mode caused by microvoid coalescence [44,45]. In contrast, in Figure 10, the microscopic morphologies of the irregular areas of the fracture surfaces of samples 2 and 3 (indicated in the corresponding panels of Figure 7) are shown. Both features are associated with manufacturing defects generated by a lack of union between consecutive seams during the material's deposition. Moreover, as shown in Figure 10a, the surface was smooth due to the presence of an oxide film, while a dendritic morphology was clearly observed on the surface of the defect present in sample 3 (see Figure 10b).

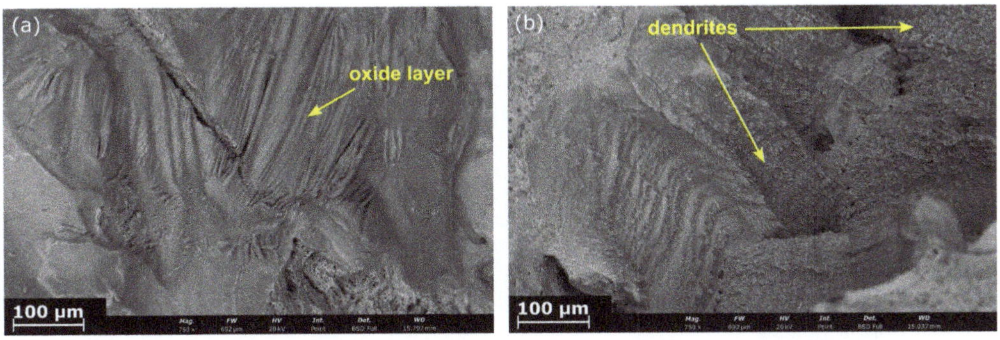

Figure 10. Micrographs (zoom-in of the correponding yellow-framed boxes shown in Figure 7) of pronounced irregularities on the fracture surfaces of (**a**) sample 2 and (**b**) sample 3, associated with manufacturing defects during the LMwD process.

On the other hand, the three test pieces that were fractured in the LMwD area showed large defects on the fracture surface and therefore a smaller resistant section. Thus, the mechanical properties of components manufactured via additive processes are entirely dependent on the defects (pores and lack-of-union defects) that form during their manufacture. As noted above, despite the manufacturing of walls with a high density, defects continued to appear, which can cause early fractures. The existence of a direct relationship between the superficial defect area on the fracture surface of a tensile sample and the elongation has been reported by Niklas et al. [46]. In addition, if a defect is present on the surface of a specimen, the consequences will be more severe [47]. It is concluded that specimens 2 and 3 contained more severe defects; consequently, the onset of breakage occurred at a lower level of stress, mainly affecting the breaking load and elongation.

The results of the fractographic analysis presented in this section allow us to conclude that the deposited material was generally stronger than the ablated aluminium substrate. Only those samples with relevant manufacturing defects in the deposited area broke in the part manufactured via LMwD. Therefore, after the further optimisation of the LMwD parameters to minimise the presence of such defects, the strength of parts manufactured via the hybrid ablation and LMwD technique would be governed by the strength of the ablated substrate.

It is worth noting that the machining treatment applied to prepare the casting substrate for LMwD deposition was sufficient to achieve good bonding, as was observed in the literature after tensile testing including the junction [41–43] or in other miniaturised test samples [3,42] at the junction of the hybrid components. It is not necessary to apply additional casting surface preparation, such as laser remelting, as required by Fan et al. [5], to facilitate selective laser melting (SLM) over a conventional die-cast substrate of AlSi7Mg, or laser texturing, as applied by Amaya et al. [26] to increase the absorption of laser radiation. The results regarding the mechanical properties confirm that the junction in a hybrid part does not need to be a critical region, provided that the high quality and high purity of the interface are maintained, as observed in previous works [3,7,41–43]. The fracture of the hybrid test parts typically occurred in the base material and not at the interface or in the heat-affected zone. When AM is performed over a forged substrate, it results in an area with poorer mechanical properties; thus, a fracture is expected in this area. However, in hybrid manufacturing over a conventional casting substrate, fractures are expected in the casting area. Nonetheless, the strong mechanical properties observed in ablated castings [12], associated with a fine microstructure and a minimal occurrence of defects, allow us to obtain mechanical properties that are comparable to those in the AM area.

4. Conclusions

In this study, the feasibility of a hybrid manufacturing approach was examined, where Al-5356 alloy was deposited using the laser wire metal deposition (LMwD) technique on an ablation-cast substrate of the same material. An analysis was performed based on the microstructure and mechanical properties. The main conclusions are summarised below.

- Regarding the microstructure, equiaxed grains were observed, with Al-Mg-rich particles deposited in the interdendritic spaces and at the grain boundaries. On the other hand, columnar grains and fine Mg-rich intermetallic precipitates (1 and 3 μm) were observed in the aluminium matrix, with a large quantity being vertically aligned at the bottom of the weld seam interlayer. The microstructure of the deposited area was also more heterogeneous than that of the ablated area.
- The junction between the areas created with the two manufacturing processes was not the weakest point.
- If the deposited area is free of large lack-of-union defects, the hybrid components fracture in the ablation region, with high mechanical properties. Here, the yield strength was higher than 100 MPa, the tensile strength was higher than 220 MPa, and the elongation was higher than 13%.
- Large lack-of-union defects were observed in the deposited parts of the hybrid tensile test samples that were broken across the deposited material. After the maximum load was achieved, the stress decreased continuously, while the elongation increased until complete fracture. Some necking surrounding the lack-of-union defect was observed on the fracture surface.

Author Contributions: Conceptualization, A.I.F.-C. and C.S.; methodology, I.L. and M.M.; validation, A.I.F.-C. and C.S.; investigation, A.S. and M.R.; writing—original draft preparation, M.M. and M.R.; writing—review and editing, A.I.F.-C. and C.S. All authors have read and agreed to the published version of the manuscript.

Funding: This research was funded by Basque Government's Department of Economic Development, Sustainability, and Environment, grant number KK-2023/00096 (ReIMAGIN project, ELKARTEK 2023 program).

Data Availability Statement: The original contributions presented in the study are included in the article, further inquiries can be directed to the corresponding author(s).

Conflicts of Interest: The authors declare no conflict of interest.

References

1. Ambrogio, G.; Gagliardi, F.; Muzzupappa, M.; Filice, L. Additive-incremental forming hybrid manufacturing technique to improve customised part performance. *J. Manuf. Process* **2019**, *37*, 386–391. [CrossRef]
2. Sefene, E.M.; Hailu, Y.M.; Tsegaw, A.A. Metal Hybrid Additive Manufacturing: State-of-the-art. *Prog. Addit. Manuf.* **2022**, *7*, 737–749. [CrossRef]
3. Miguel, I.; Artola, G.; Arrizubieta, J.I.; Fernández-Calvo, A.I.; Angulo, C. Inconel 718 Hybrid Laser-Based Directed Energy Deposition and Wrought Component Characterization through Small Punch. *Tests. Appl. Sci.* **2024**, *14*, 6420. [CrossRef]
4. Srinivasan, D.; Ananth, K. Recent Advances in Alloy Development of Metal Additive Manufacturing in Gas Turbine/Aerospace Applications: A Review. *J. Indian Inst. Sci.* **2022**, *102*, 311–349. [CrossRef]

5. Fan, H.; Witvrouw, A.; Wolf-Monheim, F.; Souschek, R.; Yang, S. Effects of substrate surface treatments on hybrid manufacturing of AlSi7Mg using die casting and selective laser melting. *J. Mater. Sci. Technol.* **2023**, *156*, 142–156. [CrossRef]
6. Hirtler, M.; Jedynaka, A.; Sydowa, B.; Sviridova, A.; Bambacha, M. A Study on the Mechanical Properties of Hybrid Parts Manufactured by Forging and Wire Arc Additive Manufacturing. *Procedia Manuf.* **2020**, *47*, 1141–1148. [CrossRef]
7. Kučerová, L.; Jeníček, Š.; Zetková, I.; Burdová, K. Heat treatment of bimetals produced by selective laser melting of MS1 maraging steel on conventionally produced 42SiCr martensitic steel. *Arch. Civ. Mech. Eng.* **2022**, *22*, 153. [CrossRef]
8. Grassi, J.R.; Campbell, J.; Kuhlmann, G.W. Mold-Removal Method and Apparatus. Patent US7216691B2, 7 July 2003.
9. Grassi, J.; Campbell, J.; Hartlieb, M.; Major, F. The ablation casting process. *Mater. Sci. Forum* **2009**, *618–619*, 591–594. [CrossRef]
10. Dudek, P.; Fajkiel, A.; Reguła, T.; Bochenek, J. Research on ablation casting technology for aluminium alloys. *Solid State Phenom.* **2014**, *223*, 70–77. [CrossRef]
11. Hosadyna-Kondracka, M.; Major-Gabryś, K.; Kamińska, J.; Grabarczyk, A.; Angrecki, M. Moulding Sand with Inorganic Cordis Binder for Ablation Casting. *Arch. Foundry Eng.* **2018**, *18*, 110–115. [CrossRef]
12. Barbarias, E.; Niklas, A.; Lizarralde, I.; Garat, M.; Bakedano, A.; Fernández-Calvo, A.I. Ablation technology applied to A356 alloys compared with conventional casting processes. *Mater. Sci. Eng. A* **2022**, *833*, 142528. [CrossRef]
13. Barbarias, E.; Bakedano, A.; Lizarralde, I.; Orden, S.; Fernández-Calvo, A.I. Chemical composition and melt treatment optimization of 5XXX and 6XXX alloys for ablation technology. *Int. Metalcast.* **2023**, *17*, 551–562. [CrossRef]
14. Han, Q. Ablation casting: Solidification characteristics, microstructure formation, and mechanical properties. *Int. Metalcast.* **2021**, *15*, 1213–1222. [CrossRef]
15. Boutorabi, S.M.A.; Torkaman, P.; Campbell, J.; Zolfaghari, A. Structure and properties of carbon steel cast by the ablation process. *Int. Metalcast.* **2021**, *15*, 306–318. [CrossRef]
16. Grassi, J.R.; Campbell, J.; Major, J.F. Solidification Microstructure of Aggregate Molded Shaped Castings. Patent PCT/US2007/018175, 14 August 2007.
17. Grassi, J.; Campbell, J.; Hartlieb, M.; Major, F. Ablation casting. *Foundry Trade J.* **2010**, 124–126.
18. Weiss, D.; Grassi, J.; Schultz, B.; Rohatgi, P. Testing the limits of ablation. *Mod. Cast.* **2011**, *101*, 26–29.
19. Salarvand, M.; Boutorabi, S.M.A.; Tamizifar, M.; Pourgharibshahi, M. Effect of cooling rate on the microstructure and mechanical properties of high zinc AA 5182 aluminium wrought ally cast by the ablation process. *Int. Metalcast.* **2021**, *15*, 1464–1475. [CrossRef]
20. Gruzleski, J.; Closset, B.M. *The Treatment of Liquid Aluminium-Silicon Alloys*; American Foundrymen's Society: Schaumburg, IL, USA, 1990.
21. Ding, D.; Pan, Z.; Cuiuri, D.; Li, H. Wire-feed additive manufacturing of metal components: Technologies, developments and future interests. *Int. J. Adv. Manuf. Technol.* **2015**, *81*, 465–481. [CrossRef]
22. Sánchez-Amaya, J.M.; Delgado, T.; de Damborenea, J.; López, V.; Botana, F.J. Laser welding of AA 5083 samples by high power diode laser. *Sci. Technol. Weld. Join.* **2009**, *14*, 78–86. [CrossRef]
23. Kuo, T.Y.; Lin, H.C. Effects of pulse level of Nd-YAG laser on tensile properties and formability of laser weldments in automotive aluminum alloys. *Mater. Sci. Eng. A* **2006**, *416*, 281–289. [CrossRef]
24. Sánchez-Amaya, J.M.; Delgado, T.; González-Rovira, L.; Botana, F.J. Laser welding of aluminum alloys 5083 and 6082 under conduction regime. *Appl. Surf. Sci.* **2009**, *255*, 9512–9521. [CrossRef]
25. Pierron, N.; Sallamand, P.; Matteï, S. Study of magnesium and aluminum alloys absorption coefficient during Nd: YAG laser interaction. *Appl. Surf. Sci.* **2007**, *253*, 3208–3214. [CrossRef]
26. Amaya, J.S.; Amaya-Vázquez, M.R.; Botana, F.J. Laser welding of light metal alloys: Aluminium and titanium alloys. In *Handbook of Laser Welding Technologies*; Woodhead Publishing: Cambridge, UK, 2013; pp. 215–254.
27. Forsman, T. Nd:YAG Laser Welding of Aluminium Alloys. Licentiate Thesis, Luleå University of Technology, Luleå, Sweeden, 1998.
28. Dinda, G.P.; Dasgupta, A.K.; Bhattacharya, S.; Natu, H.; Dutta, B.; Mazumder, J. Microstructural characterization of laser-deposited Al 4047 alloy. *Metall. Mater. Trans. A* **2013**, *44*, 2233–2242. [CrossRef]
29. Wang, X.; Li, L.; Qu, J.; Tao, W. Microstructure and mechanical properties of laser metal deposited AlSi10Mg alloys. *Mater. Sci. Technol.* **2019**, *35*, 2284–2293. [CrossRef]
30. Langebeck, A.; Bohlen, A.; Freisse, H.; Vollertsen, F. Additive manufacturing with the lightweight material aluminium alloy EN AW-7075. *Weld. World* **2020**, *64*, 429–436. [CrossRef]
31. Bock, F.E.; Froend, M.; Herrnring, J.; Enz, J.; Kashaev, N.; Kluseman, B. Thermal analysis of laser additive manufacturing of aluminium alloys: Experiment and simulation. *AIP Conf. Proc.* **2018**, *1960*, 140004. [CrossRef]
32. Froend, M.; Bock, F.E.; Riekehr, S.; Kashaev, N.; Klusemann, B.; Enz, J. Experimental investigation of temperature distribution during wire-based laser metal deposition of the Al-Mg alloy 5087. *Mater. Sci. Forum* **2018**, *941*, 988–994. [CrossRef]
33. Froend, M.; Riekehr, S.; Kashaev, N.; Klusemann, B.; Enz, J. Process development for wire-based laser metal deposition of 5087 aluminium alloy by using fibre laser. *J. Manuf. Process.* **2018**, *34*, 721–732. [CrossRef]

34. Froend, M.; Ventzke, V.; Riekher, S.; Kashaev, N.; Klusemann, B.; Enz, J. Microstructure and microhardness of wire-based laser metal deposited AA5087 using an Ytterbium fibre laser. *Mater. Charact.* **2018**, *143*, 59–67. [CrossRef]
35. Froend, M.; Ventzke, V.; Dorn, F.; Kashaev, N.; Klusemann, B.; Enz, J. Microstructure by design: An approach of grain refinement and isotropy improvement in multi-layer wirebased laser metal deposition. *Mater. Sci. Eng. A* **2020**, *772*, 138635. [CrossRef]
36. Madarieta, M.; Irazu, E.; Garmendia, I.; Flores, J.; Soriano, C. Geometrical study of 5356 aluminium alloy by means of a concentric wire laser metal deposition. *Procedia CIRP* **2022**, *113*, 361–366. [CrossRef]
37. Becker, D.; Boley, S.; Eisseler, R.; Stehle, T.; Möhring, H.C.; Onuseit, V.; Graf, T. Influence of a closed-loop controlled laser metal wire deposition process of S Al 5356 on the quality of manufactured parts before and after subsequent machining. *Prod. Eng.* **2021**, *15*, 489–507. [CrossRef]
38. Hagenlocher, C.; O'Toole, P.; Xu, W.; Brandt, M.; Easton, M.; Molotnikov, A. In process monitoring of the thermal profile during solidification in laser directed energy deposition of aluminium. *Addit. Manuf. Lett.* **2022**, *3*, 100084. [CrossRef]
39. Zapata, A.; Bernauer, C.; Stadter, C.; Kolb, C.G.; Zaeh, M.F. Investigation on the cause-effect relationships between the process parameters and the resulting geometric properties for wire-based coaxial laser metal deposition. *Metals* **2022**, *12*, 455. [CrossRef]
40. Hagenlocher, C.; O'Toole, P.; Xu, W.; Brandt, M.; Easton, M.; Molotnikov, A. The Effect of Heat Accumulation on the Local Grain Structure in Laser-Directed Energy Deposition of Aluminium. *Metals* **2022**, *12*, 1601. [CrossRef]
41. Bai, Y.; Zhao, C.; Zhang, Y.; Wang, H. Microstructure and mechanical properties of additively manufactured multi-material component with maraging steel on CrMn steel. *Mater. Sci. Eng. A* **2021**, *802*, 140630. [CrossRef]
42. Bhadrakali, A.S.; Sastry, D.V.A.R.; Chigilipalli, B.K.; Naik, K.S.; Kakaravada, T.I.; Archarya, A.; Kumar, K.L. Effect of heat input onmicrostructure and mechanical properties of bimetallic wire arc additive manufacturing of SS304L and ER308L prepared by hybridmanufacturing process. *Int. J. Interact. Des. Manuf.* **2023**, *19*, 263–274. [CrossRef]
43. Li, Z.; Wu, H.; Liu, B.; Wen, H.; Li, H.; Shi, J.; Tang, X. Interfacial microstructure and mechanical properties of forging and SLM hybrid manufacturing Ti-6Al-4V parts. *Mater. Lett.* **2023**, *339*, 134101. [CrossRef]
44. González-Velázquez, J.L. Fractography and Failure Analysis. In *Structural Integrity*; Springer International: New York, NY, USA, 2018; Volume 3.
45. Warmuzek, M. *Aluminum-Silicon Casting Alloys: Atlas of Microfractographs*; ASM International: Almere, The Netherlands, 2004.
46. Niklas, A.; Bakedano, A.; Orden, S.; da Silva, M.; Nogués, E.; Fernández-Calvo, A.I. Effect of Microstructure and Casting Defects on the Mechanical Properties of Secondary AlSi10MnMg(Fe) Test Parts Manufactured by Vacuum Assisted high Pressure Die Casting Technology. *Mater. Today Proc.* **2015**, *2*, 4931–4938. [CrossRef]
47. Du Plessis, A.; Yadroitsava, I.; Yadroitsev, I. Effects of defects on mechanical properties in metal additive manufacturing: A review focusing on X-ray tomography insights. *Mater. Des.* **2020**, *187*, 108385. [CrossRef]

crystals

MDPI

Article

Evaluation of Structural Transition Joints Cu-Al-AlMg3 Used in Galvanizer Hangers

Milan Marônek [1,*], Jozef Bárta [1], Katarína Bártová [1], Miroslav Sahul [2], Martin Sahul [1], Matej Pašák [1], Petr Nesvadba [3] and Petr Bezdička [4]

1 Faculty of Materials Science and Technology, Slovak University of Technology, J. Bottu 25, 917 24 Trnava, Slovakia; jozef.barta@stuba.sk (J.B.); katarina.bartova@stuba.sk (K.B.); martin.sahul@stuba.sk (M.S.); matej.pasak@stuba.sk (M.P.)
2 Faculty of Mechanical Engineering, Czech Technical University in Prague, Technická 4, 166 07 Dejvice, Czech Republic; miroslav.sahul@fs.cvut.cz
3 BBP Technologie, s.r.o., Na Vysluni 114, 533 52 Srch, Czech Republic; nesvadba@bbptech.cz
4 Institute of Inorganic Chemistry of the Czech Academy of Sciences, Řež 1001, 250 68 Husinec-Rez, Czech Republic; petrb@iic.cas.cz
* Correspondence: milan.maronek@stuba.sk

Abstract: The paper deals with the evaluation of the quality of Cu-Al-AlMg3 structural transition joints (STJ) made by explosion welding proposed for the renovation of galvanizer hangers. The three-layer joint consisted of electrolytic copper with a thickness of 25 mm, 2 mm of aluminium represented by the AW1050 alloy, and 25 mm of the EN AW 575 aluminium alloy. Light microscopy analysis confirmed the wavy pattern of both interfaces of the welded joint and significant plastic deformation in close proximity to the waves. Microhardness measurement revealed a partial strain hardening of the AW5754 copper-aluminium alloy near the interface and a significant increase in microhardness in the vortex zone of waves, reaching a value of up to 863 HV 0.025. Microcracks were also observed in these places. The intermetallic phase Al_2Cu was identified in the vortex zones by XRD analysis. As a continuous layer of intermetallic phase was not observed in the interface of the welded joint, it is possible to consider the used welding parameters as appropriate. A semi-quantitative EDX analysis revealed a diversity of chemical composition in the vortex zones, which does not correspond to the phase composition based on the equilibrium binary Al-Cu diagram due to non-equilibrium conditions in the formation of the welded joint interface. The bond strength of three-layer welded joint evaluated by the strength test ranged from 151 to 171 MPa, which represented approximately a two-fold increase in comparison to the ultimate tensile strength of alloy AW1050, while the failure occurred in all samples at the AW1050-AW5754 alloy interface.

Keywords: explosion welding; copper; aluminium; aluminium alloy; intermetallic phase; trimetal

1. Introduction

Structural transition joints (STJs) represent important structural nodes that allow for the connection of metals with different mechanical, physical, and chemical properties [1,2]. Direct joining of these materials by fusion welding methods is often not possible due to different melting temperatures or the formation of intermetallic phases causing embrittlement of the joint. Typical STJs are steel-aluminium bimetals [3,4], which find their application, e.g., in shipbuilding and allow joining the steel hull of the ship with an aluminium deck that better resists the corrosive environment of seawater [5]. In addition to improving corrosion resistance, transition joints allow financial savings to be made as it is possible to create combinations of cheaper and more expensive materials. Examples include Au-Cu, Ag-Cu, Pt-Cu, Al-Cu, and others.

Aluminium, copper, and their alloys are the most commonly used metals after steel. The joining of aluminium with copper has important role, because the combination of their properties provides benefits in numerous industrial applications. This material combination is utilized across a variety of applications, particularly those that benefit from the excellent thermal and electrical conductivity of these metals [6]. Aluminium's low density contributes to reduce the overall weight of components. The application of Al-Cu components can be found in various areas of industry, e.g., automobiles, aeronautics, electronics, refrigeration, electric power, household appliances, etc. [7–11].

The welding methods suitable for the production of STJs can be solid state welding methods [12]. These methods are characterized by a lower welding temperature than the melting temperature of the welded materials and the simultaneous application of pressure. The use of these methods significantly reduces the formation of intermetallic phases when joining a combination of materials prone to their formation, and the mutual mixing of the joined materials at the joint interface is reduced.

Among the solid state welding methods suitable for the production of STJ, beside roll bonded cladding [13], explosion welding has a dominant position [14,15]. This technology makes it possible to make large-scale semi-finished products, from which it is subsequently possible to produce a large number of STJs. The combinations of joined materials differ and usually depend on specific requirements of corrosion resistance, electrical and thermal conductivity, or saving precious metals [16–21]. In order to increase heat resistance, if necessary, it is also possible to create multi-layer joints (e.g., steel-Ti-Al) [22]. When creating large-area bimetals, the stability of the welding parameters along the entire length and width of the welded materials is important. There may be the formation of defects, or the formation of a very narrow molten areas at the interface of the welded joint and the formation of intermetallic phases. In addition, it is possible to observe microscopic or macroscopic cracks in the interface, which can decrease the mechanical and fatigue properties of the joint. The interface of the welded joint made by explosion welding has a typical wavy character, which is significantly strain-hardened to a distance of hundreds of micrometres. Another characteristic feature of welded joints made by explosion welding is their subsequent deformation after the welding, depending on the thickness of the materials used [23].

Equipment for galvanic coating uses STJ produced by explosion welding. Due to the high intensity of electric current, excellent electrical conductivity and minimal transition resistance of the welded joint are required. To save copper, the galvanizer hanger is made of aluminium alloy AW5754. This alloy has good mechanical properties and good weldability. The connection part of the hanger is equipped with a copper contact. The connection part (STJ) consists of three layers (Cu (25 mm)—AW1050 (2 mm)—AW5754 (25 mm)) produced by explosion welding. The STJ is welded on the AW5754 alloy side to the electrolyzer hanger by the MIG method, while the temperature is monitored to avoid overheating of the STJ and the formation of intermetallic phases (Figure 1). After the copper contact wears out, the transition piece can be cut off from the hinge and a new, unworn STJ can be welded. This results in a significant saving of material because the galvanizer hanger can be used repeatedly.

This paper deals with the quality evaluation of three-layer welded joints made by explosion welding and the assessment of the suitability of their use for the production of STJ for galvanizing equipment.

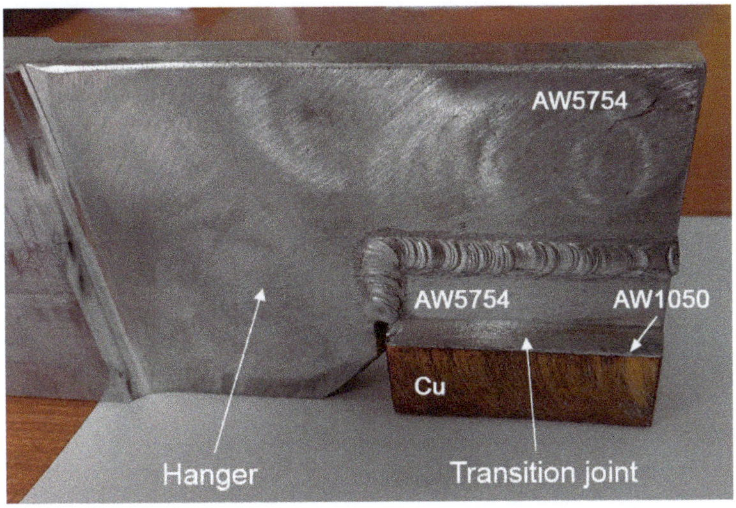

Figure 1. Detail of the electrolyser hanger with a welded structural transition joint.

2. Materials and Methods

The three-layer joint was made in two steps. First, a Cu joint with a copper of 25 mm thickness (stable material) and an EN AW1050 aluminium alloy with 2 mm thickness (accelerated material) was made. In the next step, a layer of aluminium alloy EN AW5754 with a thickness of 25 mm was plated on previously produced bimetal from the AW1050 side. All welded materials were in the form of rolled sheets. The stable material was placed on a woodblock board on a sand base, and the distance between the flyer plate with an explosive charge and the base material was realized with wooden spacers (Figure 2). The dimensions and designation of the materials used are listed in Table 1. The chemical composition and mechanical properties provided by manufacturers of the welded materials are listed in Tables 2 and 3, respectively. The welding parameters are listed in Table 4.

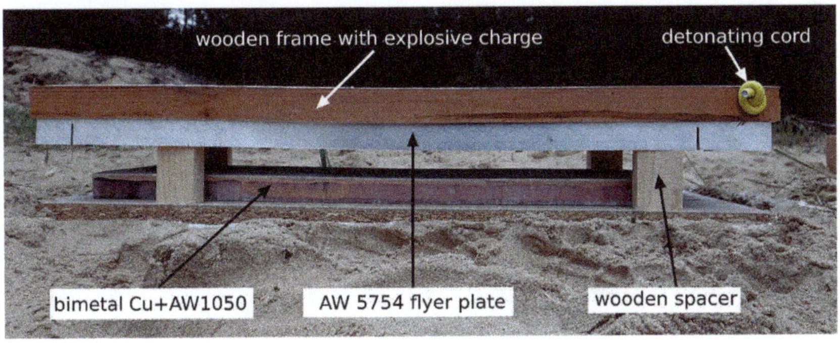

Figure 2. The assembly of materials to be welded and an explosive charge.

Table 1. Dimensions of welded materials.

Material	Designation	Dimensions (l × w × t) [mm]
Technically pure copper	Cu-ETP	604 × 150 × 25
Aluminium alloy	EN AW1050	700 × 220 × 2
Aluminium alloy	EN AW 5754	714 × 230 × 25

Table 2. Chemical composition of welded materials provided by manufacturers.

Element Content [wt. %]	Cu-ETP	AW 1050	AW 5754
Cu	99.990	0.050	0.039
Ag	0.0017	–	–
Bi	0.00004	–	–
Al	–	99.61	balanced
Mg	–	0.0170	2.80
Fe	–	0.26	0.29
Si	–	0.0850	0.22
Mn	–	0.0030	0.27
Cr	–	–	0.024
Ti	–	0.0070	0.036
Zn	–	0.0030	0.051
O	0.0034	–	–
P	0.00213	–	–

Table 3. Mechanical properties of welded materials provided by manufacturers.

Material	Rm [MPa]	Rp0.2 [MPa]	A5 [%]	A50 [%]	Hardness
Cu-ETP	215	55	56	–	40 HV10
AW 1050	71	45 *	–	51	66 HBW
AW 5754	197	101	–	23.8	68 HBW

* Rp.

Table 4. Welding parameters.

Parameter		Cu/AW 1050	AW 1050/AW 5754
Type of explosive used		Semtex S25	Semtex S25
Explosive charge density	[g·cm^{-3}]	1.202	1.201
Set-off distance	[mm]	2	25
Charge to flyer plate ratio	[-]	4.47	0.54
Time constant	[μs]	3.74	9.54
Deflection angle for selected stand-off distance	[deg]	28.65	8.09
Dynamic collision angle for selected stand-off distance	[deg]	28.65	8.09
Flyer plate velocity for selected stand-off distance	[m·s^{-1}]	938.54	305.77
Speed of movement of the collision point for selected stand-off distance	[m·s^{-1}]	1896.77	2167.34

After welding and cleaning the joint, an ultrasonic test was performed. No defects were identified at the welded joint interface.

A cutting plan, including samples for microstructural analysis and mechanical properties (tear test and microhardness test), is shown in Figure 3. To evaluate the consistency of the explosion welding process, samples representing the beginning, middle, and end part of the three-layer joint were selected for analysis in terms of its length and width.

The samples intended for optical microscopy were cleaned after cutting the three-layer joint with a band saw and subsequently a water-cooled grinding saw, and then embedded in epoxy resin. After hardening, the samples were prepared for metallographic examination on a PHOENIX 4000 device (Buehler, Lake Bluff, IL, USA). This consisted of wet grinding on metallographic papers with grain sizes of 240 μm, 600 μm, and 1200 μm, followed by polishing with diamond suspensions with grain sizes of 9 μm, 6 μm, 3 μm, and 1 μm. Due to the different properties of copper and aluminium alloys, two types of etchants were necessary. The copper microstructure was highlighted by electrolytic etching in a 10% ammonium persulfate solution ($(NH_4)_2(SO_4)_2$) at 6 V for 10 s. The structure of aluminium alloys AW1050 and AW5754 was highlighted by electrolytic etching in a Keller etchant (2.5 mL HNO_3, 1.5 mL HCl, 1 mL HF, 95 mL H_2O). Etching time was 30 s. The microstructures were observed on a Neophot 32 light microscope (Carl Zeiss, Jena, Germany) under both unpolarized and polarized light.

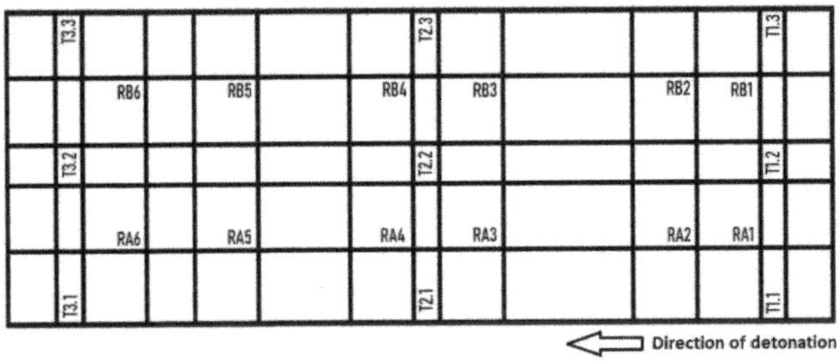

Figure 3. Cutting plan and distribution of samples for quality assessment of Cu-ETP-AW 1050-AW5754 three-layer joint. Designation legend: T—samples intended for microstructural analysis and microhardness test, RA, RB—samples intended for strength test).

The measurement of microhardness at the interface of the welded joint was carried out on a Buehler IndentaMet 1100 microhardness tester (Buehler, Lake Bluff, IL, USA) at a load of 0.025 N and a holding time of 10 s. Microhardness was measured in the direction perpendicular to the interface. The distance between individual indents was 100 μm. The second indentation line, 100 μm away from the first line, was realised in close proximity to the interface, where the plastic deformation is the most intense. The distance between indents was again 100 μm, but the indents were offset vertically by 50 μm, increasing the indentation density in the interface region.

A JEOL JSM 7600 F scanning electron microscope (JEOL, Nieuw-Vennep, The Netherlands) equipped with an Oxford Instruments EDX analyser with an X-max 50 mm^2 detector was used for a more detailed analysis of the chemical composition at the interface of the welded joint and fracture surfaces after the strength test.

The identification of the phases present in the region of the welded joint interface was carried out by X-ray diffraction analysis. Diffraction patterns were collected with a PANalytical X'Pert PRO diffractometer equipped with a conventional X-ray tube (Co $K\alpha$ radiation, 40 kV, 30 mA, point focus), an X-ray mono-capillary with the exit diameter of 0.1 mm, and a multichannel detector X'Celerator with an anti-scatter shield. A sample holder for single crystal XRD measurement was adapted by adding z-(vertical) axis adjustment (Huber 1005 goniometer head).

X-ray patterns were taken between 20 and 65° degrees 2theta with the step of 0.0334°. XRD patterns were not pre-treated before interpretation as no background correction was needed.

Qualitative phase analysis was performed with the HighScorePlus software package (Malvern PANalytical, The Netherlands, version 5.1.0) [24], with the Diffrac-Plus software package (Bruker AXS, Germany, version 8.0) and the PDF-4+ database, release 2022 [25].

The strength test was carried on cup-shaped samples (Figure 4) on the LabTest 5.250 SP1—VM device (Labortech Ltd., Prague, Czech Republic). The dimensions of the samples and the strength test scheme can be found in Figure 5. Two samples were tested from the beginning (RA1, RB1), the middle (RA4, RB4), and the end (RA6, RB6) of the three-layer welded joint with regard to the beginning of the explosive initiation. During the test, the samples were placed in the fixture and their destruction was realized by compressive force. The loading speed was 1 mm/min. The maximum force to failure and the elongation were recorded.

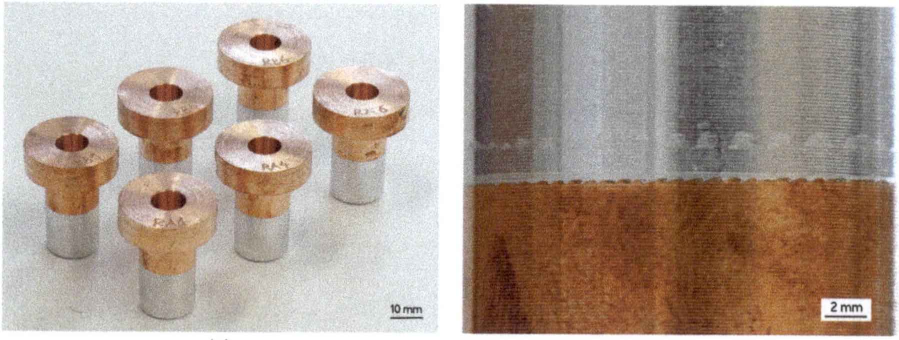

(**a**) (**b**)

Figure 4. Strength test samples. (**a**) Overall shape of the samples; (**b**) Detail of the sample surface with a visible wavy interface.

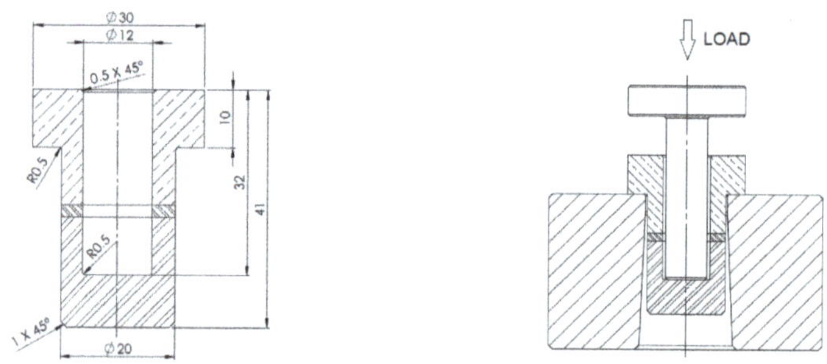

(**a**) (**b**)

Figure 5. Sample dimensions and strength test principle. (**a**) Dimensions of strength test samples; (**b**) Principle of strength test.

3. Results and Discussion

The results of light microscopy, microhardness measurement, semi-quantitative EDX analysis, micro-X-ray diffraction analysis, mechanical properties, and analysis of fracture surfaces by scanning electron microscope are presented in further subsections.

3.1. Light Microscopy

The copper microstructure consisted of polygonal grains of various sizes with no signs of grain deformation. Both aluminium alloys, AW1050 and AW 5754, were characterized by a fine grain structure.

The character of the interface of the samples from the beginning, middle, and end of the three-layer joint with respect to the place of initiation of the explosive was similar. Therefore, it can be concluded that there were no significant deviations in the welding parameters at the beginning and at the end of the joint during welding. In the direction parallel to the direction of propagation of the detonation wave, the interface had a periodic wavy pattern (Figure 6a); in the direction perpendicular to the direction of propagation of the detonation wave, the periodic wavy character of the interface disappeared (Figure 6b). This is typical for joints made by explosion welding due to high-velocity collision, plastic deformation, and jetting in the direction of propagation of the detonation wave.

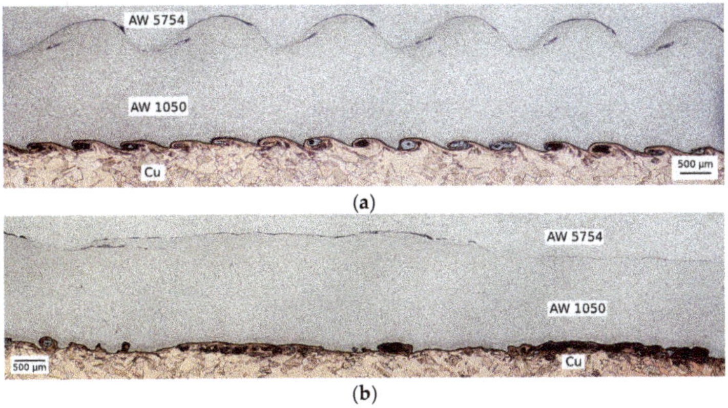

Figure 6. Character of the welded joint interface, sample T3.2. (**a**) Periodic wavy pattern, direction parallel to propagation of the detonation wave; (**b**) Non-periodic pattern, perpendicular direction to propagation of the detonation wave.

The amplitude of the wavy interface in the direction parallel to the direction of propagation of the detonation wave was approximately three times greater in the case of the interface between aluminium alloys compared to the interface copper—AW1050 (Figure 6). Similarly, the wavelength was about fourfold in the case of the interface between aluminium alloys compared to the interface of copper—AW1050 (Figure 6).

The character of the wave region neighbourhood corresponds to the typical welded joint made by explosion welding. The intense plastic deformation of all three layers near the waves was observed. Moreover, the deformation twins were found in the copper region (Figure 7). In addition to the intense plastic deformation, localized melt zones, mechanical mixing regions, and the presence of intermetallic compound regions were also discovered. The formation of vortexes, which is typical for explosion welding of dissimilar materials, was primarily observed at the copper-AW1050 interface (Figure 7). Micro cracks and cavities were detected in the vortexes, and their location was approximately perpendicular to the welded joint interface (Figure 7). These cracks could have been formed during the welding process, a phenomenon also reported by other authors [26], but they could also have developed during the metallographic preparation of the samples.

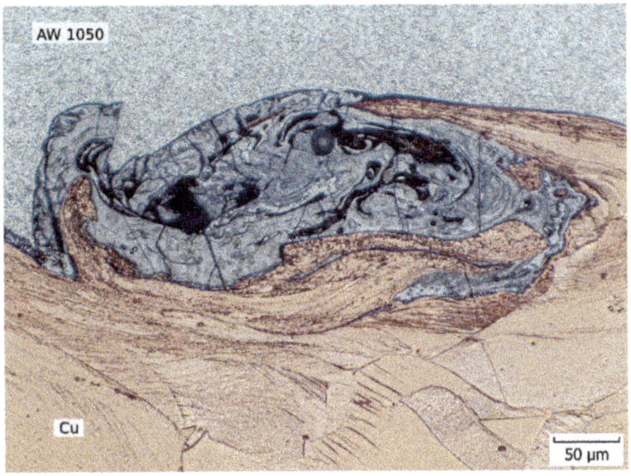

Figure 7. Vortex at Cu-AW1050 interface, parallel to the direction of propagation of the detonation wave, sample T3.2.

The lack of joining was revealed by light microscopy in the AW 1050-AW5754 interface of sample T1.2 at the beginning of the joint (Figure 8a), whereas its detail is showed in Figure 8b.

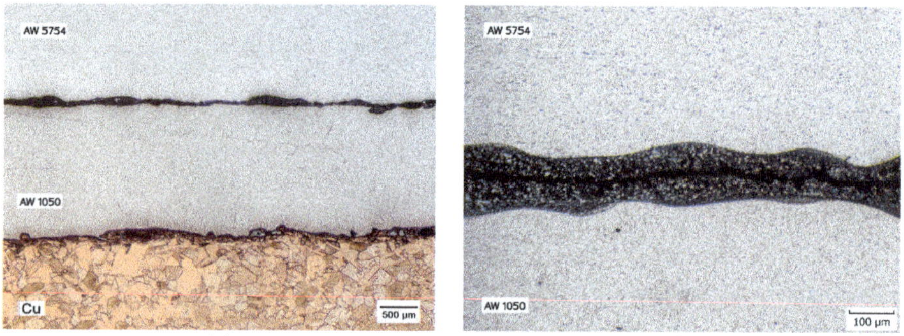

(**a**) (**b**)

Figure 8. (**a**) AW 1050-AW5754 welded joint interface, perpendicular to the direction of propagation of the detonation wave, sample T1.2. (**a**) The lack of joining; (**b**) Lack of joining close-up.

No discontinuities or lack of joining were observed in the other examined areas of the AW 1050-AW5754 interface, nor in the Cu-AW 1050 interface.

3.2. Microhardness Measurement

Explosion welding is characterized by intense plastic deformation of the welded materials at the welded joint interface and their strain hardening. Microhardness measurement indirectly allows the identification of the degree and localization of the strain hardening at the welded joint interface, as well as the detection of the presence of formed intermetallic phases. In Figure 9, the measurement of microhardness on sample T3.2 was realised in two different locations. The microhardness profiles are documented in Figure 10.

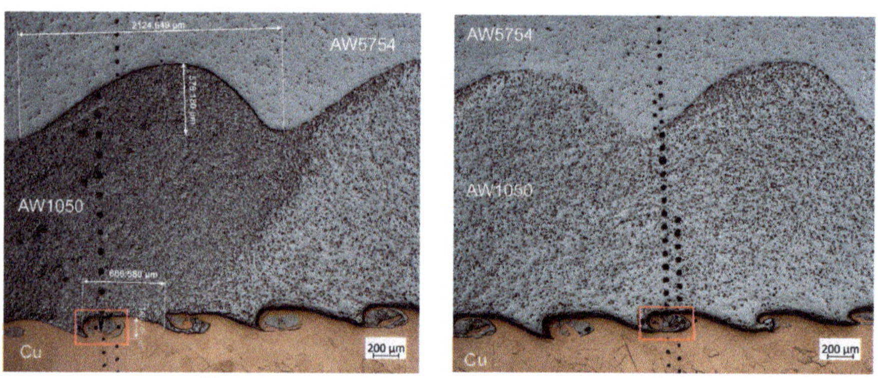

(**a**) (**b**)

Figure 9. Microhardness measurement line across the welded joint interface of sample T3.2. (**a**) Wave No. 2; (**b**) wave No. 7.

The course of microhardness was similar for both investigated waves (Figure 10a,b). In close proximity to the AW1050-AW5754 interface, a slight increase of the AW5754 aluminium alloy hardness can be seen due to its strain hardening during welding. In the area of the AW1050 alloy, an increase in microhardness was not observed. As a result of intense deformation of copper grains (Figure 9), a slight increase in hardness can be observed at the Cu-AW1050 interface up to a distance of about 400 µm. These results correspond to the results published by Huang et al. [27].

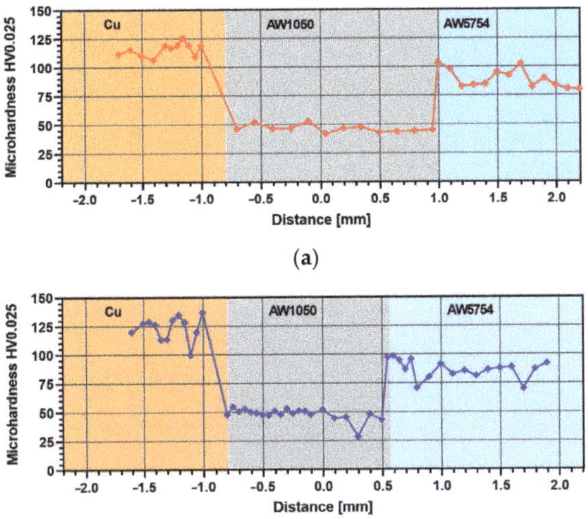

Figure 10. Course of microhardness across the interface of the welded joint, sample T3.2. (**a**) Wave No. 2; (**b**) Wave No. 7.

The measured microhardness HV 0.025 in the area of the vortex, marked in Figure 9a,b by a rectangle, is shown in Figure 11. Figure 11a,b shows that the interval of measured hardnesses is very wide. The reason may be a different degree of strain hardening, a different chemical composition in the measured areas, and the presence of formed intermetallic phases due to extreme pressures and temperatures and rapid solidification during the explosion welding process.

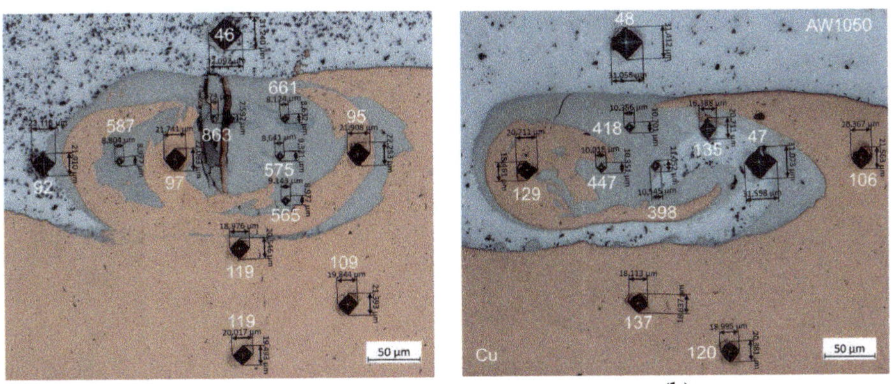

(**a**) (**b**)

Figure 11. Microhardness of the vortex area in the interface of the welded joint, sample T3.2, (**a**) wave No. 2; (**b**) wave No. 7.

The light grey areas (Figure 11a,b) correspond to the hardness of the AW1050 alloy, which has the lowest hardness among all three materials of the three-layer joint. Conversely, the dark grey areas show an enormous increase in hardness. In the case of wave No. 2

(Figure 11a), this value reaches 863 HV 0.025; for wave No. 7 (Figure 11b), the value is approximately half (447 HV 0.025). Both values significantly exceed the hardness of copper.

As the authors of [28] state, the hardness in the areas of intermetallic phase presence also depends on the thickness of the explosive charge. In the case of a larger explosive charge thickness, higher hardnesses were observed in the welded joint area due to the introduction of more kinetic energy into the welded joint interface. They also identified four types of Al-Cu intermetallic phases and different hardnesses in the area of the welded joint interface in the Al-Cu connection. Based on the results of these authors and our microhardness measurements, we can assume that the dark grey areas in Figure 11 are intermetallic phases rich in copper and aluminium.

3.3. *Semi-Quantitative EDX Analysis*

As some authors [29] state, the heating rate at the interfaces reaches approximately 10^9 K/s, while the cooling rate is around 10^7 K/s. That is why the solidification conditions at the interfaces of explosively bonded materials closely resemble those observed during rapid solidification, where metastable crystalline, quasicrystalline, or glassy phases could be observed.

The heterogeneity of the chemical composition in the region of the wave islands in the interface of the welded joint was analysed by semi-quantitative EDX analysis. The chemical composition at selected points of the Cu-AW1050 welded joint interface (Figure 12) is shown in Table 5.

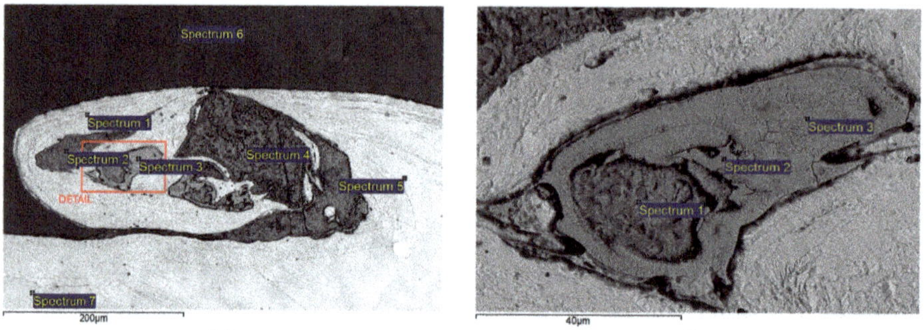

(**a**) (**b**)

Figure 12. EDX analysis of the chemical composition in the interface of the Cu-AW1050 welded joint. (**a**) Entire vortex; (**b**) Marked area.

Table 5. Chemical composition in selected points (Figure 12a) of welded joint interface Cu-AW1050 in atomic %.

Spectrum	Al	Cu	Total
Spectrum 1	0.51	99.49	100.00
Spectrum 2	72.51	27.49	100.00
Spectrum 3	49.16	50.84	100.00
Spectrum 4	81.22	18.78	100.00
Spectrum 5	0.00	100.00	100.00
Spectrum 6	100.00	0.00	100.00
Spectrum 7	0.22	99.78	100.00

The chemical composition (Table 5) in points 1, 5, and 7 corresponds to copper. The chemical composition of the area shown by the light microscope as light grey (Figure 11) is represented by Spectrum point 6 (Figure 12a, Table 5) and corresponds to alloy AW1050. The chemical composition in other points of analysis (Spectrum 2, 3 and 4) shows different proportions of aluminium and copper. The middle part of vortex, marked in Figure 12a as

"DETAIL", is magnified in Figure 12b and the chemical composition in selected points of this area is provided in Table 6. Even in this case, the chemical composition shows different proportions of aluminium and copper.

Table 6. Chemical composition in selected points of the welded joint interface Cu-AW1050 (in atomic %).

Spectrum	Al	Cu	Total
Spectrum 1	69.48	30.52	100.00
Spectrum 2	58.65	41.35	100.00
Spectrum 3	51.57	48.43	100.00

The microstructure of the Cu-AW1050 interface and the marking of individual points of the performed semi-quantitative EDX analysis is shown in Figure 13a. The chemical composition in these areas is shown in Table 7. The detailed chemical composition in the marked points of this area (Figure 13b) is shown in Table 8. The peripheral boundary of the island with the chemical composition Spectrum 2 was subjected to EDX line analysis in the area marked in Figure 13a as LINE PROFILE. The line profiles for Cu and Al are shown in Figure 13c.

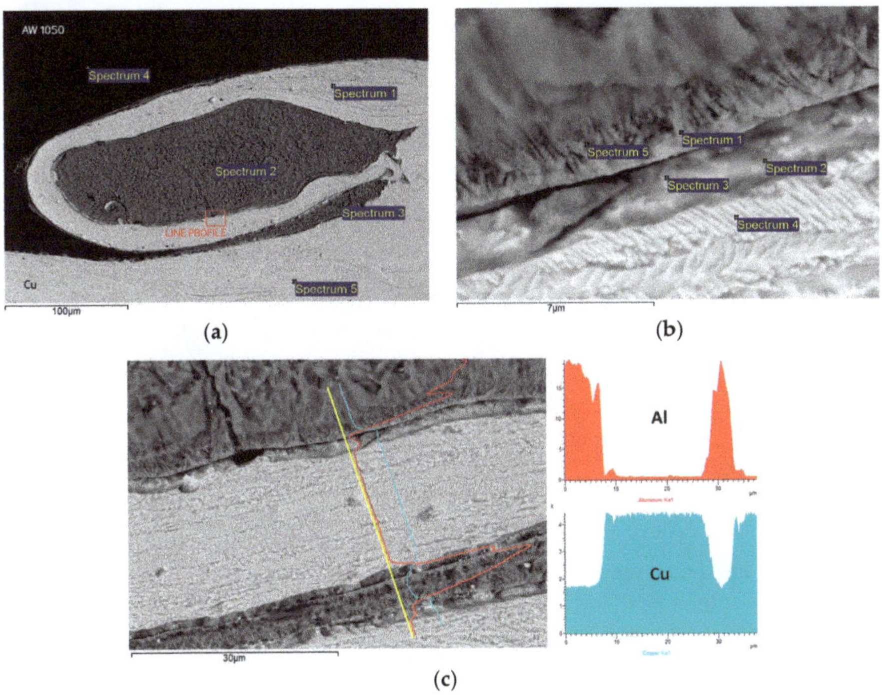

Figure 13. Detail of Cu-AW1050 welded joint interface microstructure, sample T1.2. (**a**) Marked points of EDX analysis; (**b**) Marked points for detailed analysis near boundary; (**c**) Line EDX analysis in marked area.

Table 7. Chemical composition at selected points of the Cu-AW1050 interface (Figure 13a) in atomic %.

Spectrum	Al	Cu	Total
Spectrum 1	0.81	99.19	100.00
Spectrum 2	75.89	24.11	100.00
Spectrum 3	75.30	24.70	100.00
Spectrum 4	99.88	0.12	100.00
Spectrum 5	0.21	99.79	100.00

Table 8. Chemical composition at selected points of the Cu-AW1050 interface (Figure 13b) in atomic %.

Spectrum	Al	Cu	Total
Spectrum 1	68.20	31.80	100.00
Spectrum 2	9.65	90.35	100.00
Spectrum 3	7.98	92.02	100.00
Spectrum 4	1.44	98.56	100.00
Spectrum 5	69.90	30.10	100.00

As stated by Mokhtari et al. [30], based on the Cu-Al binary diagram, intermetallic phases Al_2Cu, AlCu, Al_3Cu_4, Al_2Cu_3, and Al_4Cu_9 exist at temperatures below 300 °C. These phases were confirmed after annealing of Al-Cu joints by other authors [27,31]. Paul et al. [32] identified equilibrium phases CuAl, $CuAl_2$, and Cu_9Al_4 in the interface of Cu-Al welds. Moreover, they additionally identified the metastable intermetallic phases and Cu_mAl_n-type phases that are not present in the Al-Cu phase diagram under equilibrium conditions. Therefore, employing annealing heat treatment to forecast the exact phases created during welding is not feasible since certain compounds are formed under non-equilibrium conditions. The results of the performed EDX analysis in the interface of the Al-Cu welded joint correspond to the findings of the authors mentioned above. In our case, locations with chemical compositions corresponding to the AlCu phase (Table 5, Spectrum 3) and the Al_2Cu phase (Table 8, Spectrum 1) were observed. The chemical composition in most of the examined areas did not correspond to the stoichiometric ratio of phases present in the equilibrium binary Al-Cu diagram. That is in line with findings of other researchers [8]. As mentioned above, some areas of the Cu-Al1050 interface showed a higher copper content contrary to the minimum solubility of copper in aluminium, which is 5.65 wt.% [33]. Because the diffusion is thermally activated, diffusion coefficients $D_{Al\text{-}Cu}$ and $D_{Cu\text{-}Al}$ increase with increasing temperature. Despite the fact that diffusion coefficient of Cu in Al ($D_{Cu\text{-}Al}$) is about four orders of magnitude larger than that of Al in Cu ($D_{Al\text{-}Cu}$) [33], higher content of copper in Al1050 alloy can not be simply explained by solubility and diffusion capabilities of Cu and Al. There are undoubtedly several factors inherent to the explosion welding process. These factors include high temperatures and pressures, the formation of intermetallic phases, mechanical mixing, grain boundary diffusion, and non-equilibrium solidification. As a result, the microstructure becomes complex, exhibiting areas with elevated copper concentrations due to these non-equilibrium effects. Thermal exposure triggers the formation of new phases at the joint interface and promotes the growth of existing phases. Therefore, heat treatment of final trimetals is not recommended.

Details of the interface between AW1050 and AW5754 aluminium alloys are shown in Figure 14a, and the results of the EDX analysis are shown in Table 9. The areas marked as Spectrum 1 and Spectrum 2 correspond approximately to the chemical composition of AW5754 alloy and spectrum 3 to AW1050 alloy. The EDX line profiles for Mg and Al of AW5754 and AW1050 interface are shown in Figure 14b.

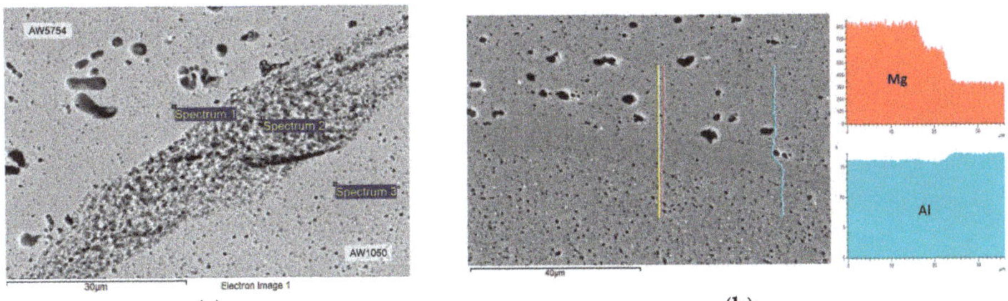

(a) (b)

Figure 14. Microstructure detail and EDX analysis of the AW1050-AW5754 interface, sample T1.2. (**a**) Marked points of EDX analysis; (**b**) Line EDX analysis Mg and Al elements in marked area.

Table 9. Chemical composition at selected points of the AW1050-AW5754 interface (Figure 14a) in atomic %.

Spectrum	Mg	Al	Si	Total
Spectrum 1	2.84	97.16	0.00	100.00
Spectrum 2	1.86	97.89	0.25	100.00
Spectrum 3	0.12	99.88	0.00	100.00

3.4. Micro X-Ray Diffraction Analysis

As already mentioned, the results of the EDX analysis revealed significant differences in the chemical composition in the welded joint interface. The chemical composition in some zones corresponded to chemical composition of AlCu and Al_2Cu intermetallic phases; however, it was possible to observe zones whose chemical composition did not correspond to the intermetallic phases identified by other authors. For this reason, an XRD analysis was performed. The analysis was performed in the regions of two waves (Figure 9) in the interface Cu-AW1050 and in one wave in the interface AW1050-AW5754. The results of XRD analyses for the Cu-AW1050 interface are shown in Figure 15a,b.

The diffraction lines of the Cu-AW1050 interface appear at 24.07°, 34.31°, 44.342°, 49.30°, 55.59°, and 56.18°, which correspond to the (110), (200), (211), (220), (112), (310), and (202) planes of the tetragonal structure of Al_2Cu, respectively. Nevertheless, the AlCu phase was not confirmed by XRD analysis. Similarly, in the AW1050-AW5754 interface, the presence of intermetallic phases was not confirmed.

Some studies [34,35] showed that IMF Al_2Cu first forms at the interface of the Al-Cu joint, which changes to IMF Cu_9Al_4 during heat treatment of the Al-Cu bimetal. It is therefore likely that the identified Al_2Cu phase would transform into the Cu_9Al_4 phase at the interface of the welded joint when the bimetal would be exposed to heat.

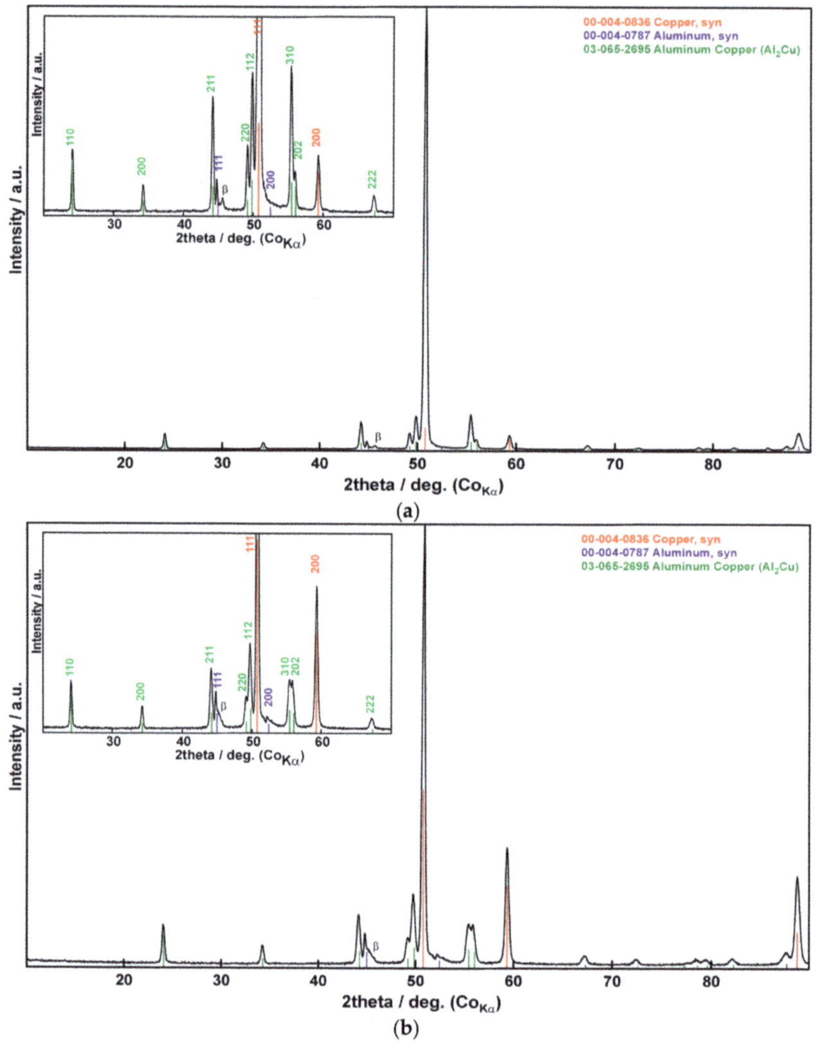

Figure 15. X-ray pattern of the vortex zone at the interface of the Cu-AW1050 welded joint. (**a**) Wave No. 2; (**b**) Wave No. 7.

3.5. Mechanical Properties

Improperly chosen explosion welding parameters can lead to an insufficient connection of the welded materials, as well as to the introduction of a large amount of kinetic energy, which turns into the heat during their collision. This subsequently enables the formation of intermetallic phases at the interface of the welded joint and its embrittlement. The thickness of the intermetallic phases thus affects the mechanical properties of the joints. A strength test was performed to assess the mechanical properties of the three-layer joints. The design of the samples (Figure 5) and the test procedure allowed for failure either in the welded materials or at the welded joint interfaces. The maximum force at failure, displacement, and the location of the failure were monitored. The samples used for the test represented the beginning, middle, and end of the three-layer joint with regard to the place of the explosive charge initiation. In all samples, failure occurred at the interface of

AW1050-AW5754. An example of failure of a sample with an exposed wavy interface of the welded joint after the strength test is shown in Figure 16. The results of the strength test are presented in Table 10.

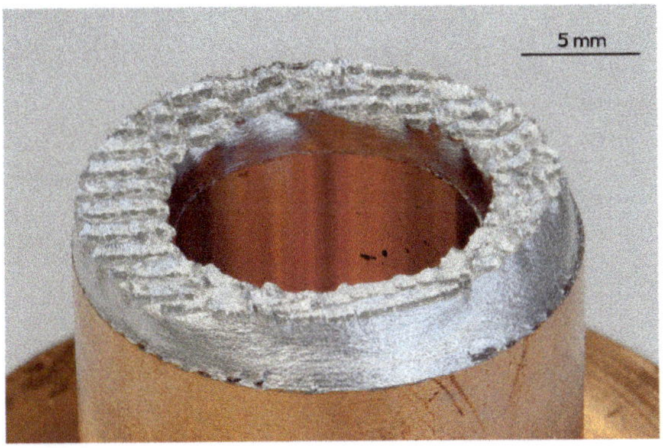

Figure 16. Detail of failure after strength test, sample RA1.

Table 10. Strength test results.

Sample	F_{max} [N]	D [mm]	d [mm]	S [mm^2]	A [%]	A_{avg} [%]	Rm [MPa]	Rm_{avg} [MPa]
RA1	28,519.5	20.1	11.95	205.15	8.83	9.16	139.02	151.12
RB1	34,039.8	20.09	11.75	208.56	9.48		163.21	
RA4	33,875.7	20.1	11.9	206.09	9.54	9.61	164.37	166.42
RB4	34,666.2	20.09	11.9	205.77	9.67		168.47	
RA6	35,388.9	20.08	11.96	204.33	9.72	9.71	173.19	171.33
RB6	35,291.7	20.08	11.75	208.24	9.69		169.47	

F_{max}—maximum force at failure, A—ductility, D—outer diameter of the sample, d—inner diameter of the sample, S—intercircle area, Rm—tensile strength, Rm_{avg}—average tensile strength of samples from the same zone, RA1, RB1—start of three-layer joint, RA4, RB4—middle of the three-layer joint, RA6, RB6—end of the three-layer joint (Figure 3).

Table 10 shows that the average strength related to the position of explosive charge initiation increased along the joint length. The increase in strength at the end of the three-layer joint compared to its beginning was 13.4%. The ultimate tensile strength of the AW5754 alloy (Table 3) was 197 MPa. Strength at the AW1050-AW5754 interface was thus approximately 77 to 87% of the ultimate tensile strength of the AW5754 alloy, depending on the locations within the three-layer joint. On the other hand, the obtained values from the strength test exceed the original ultimate tensile strength of the AW1050 alloy by 2.1 to 2.4 times. It is worth mentioning that the shape of the samples and the method of loading during the strength test and the static tensile test are different.

3.6. SEM Analysis of Fracture Surfaces

Analysis of the fracture surfaces after the strength test was performed with a scanning electron microscope. As was already mentioned, failure occurred at the interface of AW1050-AW5754 alloys. A detail of the fracture surface after failure by the strength test of sample RA6 is shown in Figure 17a. Morphology of the fracture of all samples had ductile morphology with typical dimples. However, the deposits at the bottom of the dimples with different morphology were observed as well. The EDX analysis of the areas marked as Spectrum 1 and Spectrum 2 (Figure 17b) revealed slightly increased content of magnesium in Spectrum 1 (Table 11). It can be concluded that the deposits represent the residues of AW5754 alloy, while the surrounding area belongs to the AW1050 alloy. This fact explains why the strength test exceeds the ultimate tensile strength of AW1050 alloy.

(a)

(b)

Figure 17. Fracture morphology, sample RA6. (**a**) After strength test; (**b**) EDX analysis of fracture area after strength test.

Table 11. Chemical composition at selected points of the AW1050-AW5754 interface fracture in wt. %.

Spectrum	Mg	Al	Fe	Total
Spectrum 1	0.37	99.32	0.31	100.00
Spectrum 2	0.00	99.73	0.27	100.00

4. Conclusions

This work describes the manufacturing procedure of Cu-Al-AlMg3 three-layer joints by explosion welding and evaluates their quality. The joints are used in the production of transition joints for electrolyser hangers. The obtained results can be summarized as follows:

1. Electrolytic copper with a thickness of 25 mm served as the base material. An AW1050 alloy, 2 mm thick, was initially cladded as a transition layer, followed by cladding of AW 5750 aluminium alloy with a thickness of 25 mm.
2. Visual and ultrasonic inspection of the produced three-layer joint revealed no defects.
3. The interface of the welded joint displayed a wavy structure aligned with the direction of the detonation wave propagation, a characteristic typical of explosion welding.
4. Intense plastic deformation resulted in observable grain deformation at both interfaces, indicated by increased microhardness extending approximately 400 μm into the copper side and 200–250 μm into the AW5754 alloy side.
5. Microhardness measurements in the vortex zones reached up to 863 HV 0.025.
6. Light and scanning electron microscopy revealed microcracks in the vortex areas exhibiting increased hardness; however, these microcracks remained isolated and did not propagate into the base or flyer plate.
7. Examination of the chemical composition through EDX analysis in wave zones and vortices indicated significant variability, which did not align with the phases shown in the Al-Cu equilibrium binary diagram, confirming the disequilibrium conditions during explosion welding.
8. XRD analysis identified the presence of the Al_2Cu intermetallic phase at the interfaces of the welded joint.
9. Mechanical property tests conducted on cup-shaped samples yielded strength values between 151 and 171 MPa, representing more than a twofold increase compared to the ultimate tensile strength of the AW1050 alloy. When compared to the ultimate tensile strength of the AW5754 alloy, the joint strength was found to be 77% to 87% of the original strength, depending on the sample location within the three-layer joint.

10. The welding parameters were low enough to prevent the formation of a continuous and brittle intermetallic layer while remaining high enough to achieve a continuous welded joint without defects.
11. The analyses confirmed the suitability of the proposed welding parameters and procedures in producing three-layer welded joints, providing substantial copper savings and excellent properties for their intended application. Further research in this area will focus on the thermal stability of welded joints.

Author Contributions: Conceptualization, M.M. and P.N.; methodology, M.M.; software, J.B. and P.B.; validation, J.B. and M.S. (Miroslav Sahul); formal analysis, J.B.; investigation, M.M., K.B., M.S. (Miroslav Sahul), M.S. (Martin Sahul), P.B. and M.P.; writing—original draft preparation, M.M. and J.B.; writing—review and editing, M.M., J.B. and K.B.; project administration, M.M.; funding acquisition, M.M. and J.B. All authors have read and agreed to the published version of the manuscript.

Funding: This work was financially supported by Cultural and Educational Grant Agency of Minsistry of Education, Science, Research and Sport of the Slovak Republic under Grant KEGA 019STU-4/2023. The authors acknowledge the assistance provided by the Research Infrastructure NanoEnviCz, supported by the Ministry of Education, Youth, and Sports of the Czech Republic under Project No. LM2023066.

Data Availability Statement: The research data collection is not shared on cloud, but is available upon request. In case of interest, please contact the corresponding author.

Acknowledgments: The authors would like to express their gratitude Martin Kusý, from Faculty of Materials Science and Technology for his assistance in XRD analysis and GAT-TI S.p.A., Italy for evaluating the technology in serial production.

Conflicts of Interest: Author Petr Nesvadba was employed by the company BBP Technologie. The remaining authors declare that the research was conducted in the absence of any commercial or financial relationships that could be construed as a potential conflict of interest.

References

1. Structural Transition Joint | Triclad—Triclad. Available online: https://triclad.com/structural-transition-joint/ (accessed on 5 November 2024).
2. Transition Joints, TriClad Supplier—Titanium International. Available online: https://www.cem-int.com.au/products/transition-joints/ (accessed on 5 November 2023).
3. Tricarico, L.; Spina, R. Mechanical strength of Fe/Al structural transition joints subject to thermal loading. *Arch. Mater. Sci. Eng.* **2009**, *37*, 85–93.
4. Tricarico, L.; Spina, R.; Sorgente, D.; Brandizzi, M. Effects of heat treatments on mechanical properties of Fe/Al explosion-welded structural transition joints. *Mater. Des.* **2009**, *30*, 2693–2700. [CrossRef]
5. Chao, R.M.; Yang, J.M.; Lay, S.R. Interfacial Toughness for the Shipboard Aluminum/Steel Structural Transition Joint. *Mar. Str.* **1997**, *10*, 353–362. [CrossRef]
6. Acarer, M. Electrical, Corrosion, and Mechanical Properties of Aluminum-Copper Joints Produced by Explosive Welding. *J. Mater. Eng. Perform.* **2012**, *21*, 2375–2379. [CrossRef]
7. Wei, Y.; Li, H.; Sun, F.; Zou, J. The Interfacial Characterization and Performance of Cu/Al-Conductive Heads Processed by Explosion Welding, Cold Pressure Welding, and Solid-Liquid Casting. *Metals* **2019**, *9*, 237. [CrossRef]
8. Kaya, Y. Investigation of Copper-Aluminium Composite Materials Produced by Explosive Welding. *Metals* **2018**, *8*, 780. [CrossRef]
9. Wei, Y.; Luo, Y.; Qu, H.; Zou, J.; Liang, S. Microstructure Evolution and Failure Analysis of an Aluminum–Copper Cathode Conductive Head Produced by Explosive Welding. *J. Mater. Eng. Perform.* **2017**, *26*, 6158–6166. [CrossRef]
10. Güleņç, B.; Kaya, Y.; Durgutlu, A.; Güleņç, I.T.; Yildirim, M.S.; Kahraman, N. Production of wire reinforced composite materials through explosive welding. *Arch. Civ. Mech. Eng.* **2016**, *16*, 1–8. [CrossRef]
11. Yang, M.; Cao, C.; Wang, J. Fabrication of high-quality platinum coating using explosive welding technology and its microstructure evolution mechanisms. *Mater. Des.* **2023**, *235*, 112372. [CrossRef]
12. Li, W.; Patel, V. Solid State Welding for Fabricating Metallic Parts and Structures. In *Encyclopedia of Materials: Metals and Allloys*; Elsevier: Amsterdam, The Netherlands, 2022; pp. 246–259.
13. Structural Transition Joints | Aluminum & Stainless Steel Clad Plates. Available online: https://www.powderclad.com/applications/structural-transition-joints (accessed on 14 November 2023).
14. Mori, A.; Bataev, I.; Wada, N.; Hokamoto, K. Explosive welding. *Explos. Shock. High-Strain-Rate Phenom. Adv. Mater.* **2021**, 35–70. [CrossRef]

15. Loureiro, A.; Carvalho, G.H.S.F.L.; Galvão, I.; Leal, R.M.; Mendes, R. Explosive welding. In *Advanced Joining Processes*; Elsevier: Amsterdam, The Netherlands, 2021; pp. 207–237.
16. Findik, F. Recent developments in explosive welding. *Mater. Des.* **2011**, *32*, 1081–1093. [CrossRef]
17. Bataev, I.A.; Ogneva, T.S.; Bataev, A.A.; Mali, V.I.; Esikov, M.A.; Lazurenko, D.V.; Guo, Y.; Jorge Junior, A.M. Explosively welded multilayer Ni-Al composites. *Mater. Des.* **2015**, *88*, 1082–1087. [CrossRef]
18. Liu, K.; Chen, P.; Ran, C.; Zhou, Q.; Feng, J.; Fan, X.; Zhu, L. Investigation on the interfacial microstructure and mechanical properties of the W-Cu joints fabricated by hot explosive welding. *J. Mater. Process. Technol.* **2022**, *300*, 117400. [CrossRef]
19. Parchuri, P.K.; Kotegawa, S.; Ito, K.; Yamamoto, H.; Mori, A.; Tanaka, S.; Hokamoto, K. Cladding of a crack-free W plate on Cu plates using explosive welding at higher collision velocity with lower collision angle. *Results Mater.* **2020**, *5*, 100023. [CrossRef]
20. Sun, C.; Wang, S.; Guo, W.; Shen, W.; Ge, C. Bonding Interface of W-CuCrZr Explosively Welded Composite Plates for Plasma Facing Components. *J. Mater. Sci. Technol.* **2014**, *30*, 1230–1234. [CrossRef]
21. Acarer, M.; Demir, B.; Dikici, B.; Salur, E. Microstructure, mechanical properties, and corrosion resistance of an explosively welded Mg–Al composite. *J. Magnes. Alloys* **2021**, *10*, 1086–1095. [CrossRef]
22. Aceves, S.M.; Espinosa-Loza, F.; Elmer, J.W.; Huber, R. Comparison of Cu, Ti and Ta interlayer explosively fabricated aluminum to stainless steel transition joints for cryogenic pressurized hydrogen storage. *Int. J. Hydrogen Energy* **2015**, *40*, 1490–1503. [CrossRef]
23. Urminský, J.; Jáňa, M.; Marônek, M.; Morovič, L. Analysis of weld joint deformations by optical 3D scanning. *Acta Polytech.* **2016**, *56*, 76. [CrossRef]
24. Degen, T.; Sadki, M.; Bron, E.; König, U.; Nénert, G. The HighScore suite. *Powder Diffr.* **2014**, *29*, S13–S18. [CrossRef]
25. Gates-Rector, S.; Blanton, T. The Powder Diffraction File: A quality materials characterization database. *Powder Diffr.* **2019**, *34*, 352–360. [CrossRef]
26. Huang, X.; Zang, Y.; Guan, B.; Wang, B.; Cai, H.; Cao, G. Tensile instability and limit strain behavior of explosion-welded steel/Al plate considering interface defect characteristics. *J. Mater. Res. Technol.* **2023**, *25*, 4842–4859. [CrossRef]
27. Honarpisheh, M.; Asemabadi, M.; Sedighi, M. Investigation of annealing treatment on the interfacial properties of explosive-welded Al/Cu/Al multilayer. *Mater. Des.* **2012**, *37*, 122–127. [CrossRef]
28. Shiran, M.K.G.; Khalaj, G.; Pouraliakbar, H.; Jandaghi, M.R.; Dehnavi, A.S.; Bakhtiari, H. Multilayer Cu/Al/Cu explosive welded joints: Characterizing heat treatment effect on the interface microstructure and mechanical properties. *J. Manuf. Process.* **2018**, *35*, 657–663. [CrossRef]
29. Bataev, I.A.; Lazurenko, D.V.; Tanaka, S.; Hokamoto, K.; Bataev, A.A.; Guo, Y.; Jorge, A.M. High cooling rates and metastable phases at the interfaces of explosively welded materials. *Acta Mater.* **2017**, *135*, 277–289. [CrossRef]
30. Mokhtari, O.; Kim, M.-S.; Nishikawa, H.; Kawashiro, F.; Itoh, S.; Maeda, T.; Hirose, T.; Eto, T. Investigation of Formation and Growth Behavior of Cu/Al Intermetallic Compounds during Isothermal Aging. *Trans. Jpn. Inst. Electron. Packag.* **2014**, *7*, 1–7. [CrossRef]
31. Wang, Y.; Song, R.; Yanagimoto, J.; Li, H. Effect of heat treatment on bonding mechanism and mechanical properties of high strength Cu/Al/Cu clad composite. *J. Alloys Compd.* **2019**, *801*, 573–580. [CrossRef]
32. Paul, H.; Morgiel, J.; Baudin, T.; Brisset, F.; Prazmowski, M.; Miszczyk, M. Characterization of explosive weld joints by TEM and SEM/EBSD. *Arch. Metall. Mater.* **2014**, *59*, 1129–1136. [CrossRef]
33. Bedjaoui, W.; Boumerzoug, Z.; Delaunois, F. Solid-State Diffusion Welding of Commercial Aluminum Alloy with Pure Copper. *Int. J. Automot. Mech. Eng.* **2022**, *19*, 9734–9746. [CrossRef]
34. Pelzer, R.; Nelhiebel, M.; Zink, R.; Wöhlert, S.; Lassnig, A.; Khatibi, G. High temperature storage reliability investigation of the Al–Cu wire bond interface. *Microelectron. Reliab.* **2012**, *52*, 1966–1970. [CrossRef]
35. Chen, J.; Lai, Y.S.; Wang, Y.W.; Kao, C.R. Investigation of growth behavior of Al–Cu intermetallic compounds in Cu wire bonding. *Microelectron. Reliab.* **2011**, *51*, 125–129. [CrossRef]

MDPI AG
Grosspeteranlage 5
4052 Basel
Switzerland
Tel.: +41 61 683 77 34

Crystals Editorial Office
E-mail: crystals@mdpi.com
www.mdpi.com/journal/crystals

www.ingramcontent.com/pod-product-compliance
Lightning Source LLC
Chambersburg PA
CBHW061437210926
51850CB00008B/1537